Atomic Spectroscopy

Atomic Spectroscopy

Uday Veer Singh

RANDOM

RANDOM PUBLICATIONS

NEW DELHI (INDIA)

Atomic Spectroscopy

ISBN 978-93-5111-250-1

Published in 2014 in India by

RANDOM PUBLICATIONS

4376-A/4B, Gali Murari Lal, Ansari Road
New Delhi-110 002
Phone : +91-11-43580356, +91-11-23289044
e-mail: randomexports@gmail.com, sales@randompublications.com,
info@randompublications.com

Reprinted 2026

Type Setting by : Keystoneprintads, Delhi-110051

Preface

Atomic Spectroscopy is closely related to other forms of Spectroscopy. It can be divided by atomization source or by the type of spectroscopy used. In the latter case, the main division is between optical and mass spectrometry. Atomic spectroscopy is the determination of elemental composition by using the Electromagnetic Spectrum or Mass Spectrum, respectively to optical or mass spectroscopy.

It can be divided by atomization source or by the type of spectroscopy used. In the latter case, the main division is between optical and mass spectrometry. Mass spectrometry generally gives significantly better analytical performance, but is also significantly more complex. This complexity translates into higher purchase costs, higher operational costs, more operator training, and a greater number of components that can potentially fail. Because optical spectroscopy is often less expensive and has performance adequate for many tasks, it is far more common Atomic absorption spectrometers are one of the most commonly sold and used analytical devices.

Atomic mass spectrometry is similar to other types of mass spectrometry in that it consists of an ion source, a mass analyzer, and a detector. Atoms' identities are determined by their mass-to-charge ratio (via the mass analyzer) and their concentrations are determined by the number of ions detected. Although considerable research has gone into customizing mass spectrometers for atomic ion sources, it is the ion source that differs most from other forms of mass spectrometry. These ion sources must also atomize samples, or an atomization step must take place before ionization. Atomic ion sources are generally modifications of atomic optical spectroscopy atom sources.

Raman spectroscopy provides information about molecular vibrations that can be used for sample identification and quantitation. The technique involves shining a monochromatic light source on a sample and detecting the scattered light. The majority of the scattered light is of the same frequency as the excitation source; this is known as Rayleigh or elastic scattering. This book would be of immense help to students and researchers.

I thank all members of my team who have helped in the preparation of the book. My special thanks go to "Random Publications" who have published the book.

– Uday Veer Singh

Contents

Preface ... *v-vi*

1. Introduction ... **1**

Absorption and Emission ... 1
Electron Configurations of Elements ... 12

2. Mass Spectrometry ... **21**

Mass Spectrometer Converts ... 21
Function of Spectroscopic Techniques ... 29

3. Models of Atomic Spectra ... **36**

Background of the Bohr Atom ... 37
Models of the Atom ... 41
Spectroscope ... 45
Atomic Orbitals ... 48
Atomic and Ionic Radius ... 61
The Atomic Hydrogen Emission Spectrum ... 66

4. Infrared Spectroscopy ... **78**

Near infrared ... 78
Quality control of Herbal Medicinal Products ... 83
NIR Spectrophotometers ... 87
Spectrometric Techniques Compared to Traditional Analytical Methods ... 91
Applications of NIR Spectroscopy in Pharmaceutical Technologies ... 96

5. Raman Spectroscopy ... **99**

Theoretical Basis ... 99
Application of Raman Spectroscopy ... 100
Particle of Light Energy ... 103
Electrons and Chemical Bonding ... 106
The Principles of Raman Spectroscopy ... 109
Effect of Raman ... 112
Attraction of Gravitation ... 118

6. Applications of Spectrum Analysis ... **126**

Current Research and Applications ... 126

Chemical Analysis 127
Astrophysical Applications 127
Origin of Elements 130
Abundance of Elements 131
Spectroscopic Techniques 133
The Quantum Atom 135
Development of Quantum Theory 139
Quantum Theory 140

7. Auger Electron Spectroscopy 150

Physical Basis 150
Applications of Vibrational Spectroscopy 152
Photoelectron Spectroscopy 159
The Auger Process and Auger Spectroscopy 161
X-ray Photoelectron Spectroscopy (XPS) 164
IR Spectroscopy 169

8. Scanning Tunnelin Spectroscopy 173

Introduction 173
SPM Environments and Ultra-high Vacuum 176
The Scanning Tunneling Microscope (STM) 178
Scanning Probe Microscopes 185
Convolution with Other Physics 199
Properties of Cantilevers 202

9. Theory of Electromagnetic Radiation 210

History of Electromagnetic radiation 211
Infrared Radiation 212
Ultraviolet Radiation 221

10. X-ray Emission Spectroscopy 230

Introduction 230
X-ray Emission Spectrum 231
X-ray Spectrograph 233
Total Reflection X-Ray Fluorescence 236
Electron Microprobe Analysis 240
Basic Theory of X-ray Fluorescence 243
Development of the Commercial X-Ray Spectrometers 245

11. Speed of Light Propagation 252

Introduction to Propagation of Light 252
Properties of Light 254
Law of Geometric Propagation 255
Phenomenon of Rainbows 259
Rays of Light Move 262
Newton's Laws and Geometrical Optics 269
Fundamental Principles of Dynamics 270
Two and Three Dimensions of Motion 276
Early Ideas about Light Propagation 279
The Nature of Light 284

***Index* 292**

1

Introduction

Atomic spectroscopy is the determination of elemental composition by using the Electromagnetic Spectrum or Mass Spectrum, respectively to optical or mass spectroscopy. It can be divided by atomization source or by the type of spectroscopy used. In the latter case, the main division is between optical and mass spectrometry. Mass spectrometry generally gives significantly better analytical performance, but is also significantly more complex. This complexity translates into higher purchase costs, higher operational costs, more operator training, and a greater number of components that can potentially fail. Because optical spectroscopy is often less expensive and has performance adequate for many tasks, it is far more common Atomic absorption spectrometers are one of the most commonly sold and used analytical devices.

Atomic spectroscopy exploits different energetic transitions experienced by atoms that are associated with either the absorption or emission of photons. When these transitions involve the excitation and relaxation of the valence (outer or bonding) shell electrons of metal atoms and ions, the corresonding photons have energies within the ultraviolet and visible regions of the spectrum. A good example of this is the dark absorption lines in the solar spectrum, which are caused by heavier elements present in the outer layers of the sun.

ABSORPTION AND EMISSION

In atomic absorption (AA) spectroscopy, absorption of a photon results in excitation of an electron from a lower to higher energy atomic orbital (AO). An instrument measures the absorbance, A, which is defined as the logarithm of the ratio of incident to transmitted radiant power of the photon beam, $A = \log (P0 \div P)$, at a wavelength specific to the element of interest. Samples are typically analysed using a flame atomic absorption spectrophotometer.

In atomic emission (AE) spectroscopy, thermal or electrical energy from an arc, flame, spark, or plasma is used to excite and electron from a lower to higher energy AO; when the excited electron returns to its original AO (*i.e.* the ground state), it may do so by emitting a photon. The instrument measures the intensity, I, of these emitted photons as a function of wavelength.

Because AO energies are well-defined, atomic absorption and emission spectra consist of discrete, narrow lines. This allows the concentration of metallic elements in different samples to be determined selectively, with lower limits at or below 1 mg/L (1 ppm). Techniques such as graphite furnace atomic absorption spectrophotometry (GFAAS) allow concentration to be measured down to µg/L (ppb) levels. Actual limits-of-detection vary with both element, technique, and sample matrix.

FREQUENCY, WAVENUMBER, WAVELENGTH

The photon energy due to an electron transition between an upper atomic level k (of energy E_k) and a lower level i is

$$\Delta E = E_k - E_i = h\nu = hc\sigma = hc/\lambda_{vac} ,$$

where í is the frequency, ó the wavenumber in vacuum, and λ_{vac} the wavelength in vacuum. The most accurate spectroscopic measurements are determinations of transition frequencies, the unit being the Hertz (1 Hz = cycle/ sec) or one of its multiples. A measurement of any one of the entities frequency, wavenumber, or wavelength (in vacuum) is an equally accurate determination of the others since the speed of light is exactly defined.

The most common wavelength units are the nanometer (nm), the Ångström ($1 = 10^{-1}$ nm) and the micrometer (µm). The SI wavenumber unit is the inverse meter, but in practice wavenumbers are usually expressed in inverse centimeters: $1\ cm^{-1} = 10^2\ m^{-1}$, equivalent to $2.997\ 924\ 58 \times 10^4$ MHz.

In addition to frequency and wavenumber units, atomic energies are often expressed in electron volts (eV). One eV is the energy associated with each of the following quantities:

$2.417\ 989\ 40(21) \times 10^{14}$ Hz
$8\ 065.544\ 45(69)\ cm^{-1}$
1 239.841 91(11) nm
11 604.505(20) K (kelvin)
$1.602\ 176\ 53(14) \times 10^{-19}$ J (joule)

We note that the basic unit of temperature, the kelvin, is equivalent to about $0.7\ cm^{-1}$, *i.e.*, the value of the Boltzmann constant k expressed in wavenumber units per kelvin is $0.695\ 035\ 6(12)\ cm^{-1}/K$. One reason for citing this particular equivalency involving the internationally accepted symbol for the kelvin (K) [1] is to suggest that use of the letter K as a symbol for $1\ cm^{-1}$ (1 kayser) should be discontinued.

The unit of energy in the system of atomic units (a.u.) often used for theoretical calculations is the hartree, which is equal to 2 rydbergs. The rydberg for an atom having nuclear mass M is

$$1\ Ry = R_M = M(M + m_e)^{-1} R_\infty ,$$

with

$$R_\infty = m_e c\alpha^2/(2h) = 10\ 973\ 731.568\ 525(73)\ m^{-1} ,$$

Which is equivalent to 13.605 692 3(12) eV. The Rydberg constant R_{∞} is thus the limit value for infinite nuclear mass.

HYDROGEN AND HYDROGEN-LIKE IONS

The quantum numbers n, l, and j are appropriate [4]. A particular *level* is denoted either by nl_j or by $nl\ ^2L_J$ with $L = l$ and $J = j$. The latter notation is somewhat redundant for one-electron spectra, but is useful for consistency with more complex structures. The L values are written with the same letter code used for l values, but with roman capital letters. The *multiplicity* of the *L term* is equal to $2S + 1 = 2s + 1 = 2$. Written as a superscript, this number expresses the *doublet* character of the structure: each term for $L \geq 1$ has two levels, with $J = L \pm {}^1/_2$, respectively.

The Coulomb interaction between the nucleus and the single electron is dominant, so that the largest energy sepa-rations are associated with levels having different n. The hyperfine splitting of the ^{1}H 1s ground level [1420.405 751 766 7(10) MHz] results from the interaction of the proton and electron magnetic moments and gives rise to the famous 21 cm line. The separations of the $2n - 1$ excited levels having the same n are largely determined by relativistic contributions, including the spin-orbit interaction, with the result that each of the $n - 1$ pairs of levels having the same j value is almost degenerate; the separation of the two levels in each pair is mainly due to relatively small Lamb shifts.

ALKALIS AND ALKALI-LIKE SPECTRA

In the central field approximation there exists no angular-momentum coupling between a closed subshell and an electron outside the subshell, since the net spin and orbital angular momenta of the subshell are both zero. The nlj quantum numbers are, then, again appropriate for a single electron outside closed subshells.

However, the electrostatic interactions of this electron with the core electrons and with the nucleus yield a strong l-dependence of the energy levels . The differing extent of "core penetration"" for ns and np electrons can in some cases, for example, give an energy difference comparable to or exceeding the difference between the np and $(n + 1)p$ levels. The spin-orbit fine-structure separation between the nl $(l > 0)$ levels having $j = l - {}^1/_2$ and $l + {}^1/_2$, respectively, is relatively small.

ATOMIC STATES, SHELLS, AND CONFIGURATIONS

A one-electron atomic *state* is defined by the quantum numbers nlm_lm_s or $nljm_j$, with n and l representing the principal quantum number and the orbital angular momentum quantum number, respectively. The allowed values of n are the positive integers, and $l = 0, 1, ..., n - 1$. The quantum number j represents the angular momentum obtained by coupling the orbital and spin angular momenta of an electron, *i.e.*, $j = l + s$, so that $j = l \pm {}^1/_2$. The magnetic quantum numbers m_l, m_s, and m_j represent the projections of the corresponding angular momenta along a particular direction; thus, for example, $m_l = -l, -l+1 \ldots l$ and $m_s = \pm {}^1/_2$. The central field approximation for a many-electron atom leads to wave functions expressed in terms of products of such one-electron states. Those electrons having the same principal quantum number n belong to the *shell* for that number. Electrons having both the same n value and l value belong to a *subshell*, all electrons in a particular subshell being *equivalent*. The notation for a *configuration* of N equivalent electrons is nl^N, the superscript usually being omitted for $N = 1$. A configuration of several subshells is written as $nl^N n2\, l2^M \ldots$. The numerical values of l are replaced by letters in writing a configuration, according to the code s, p, d for $l = 0, 1, 2$ and f, g, h ... for $l = 3, 4, 5$ ..., the letter j being omitted.

The Pauli exclusion principle prohibits atomic states having two electrons with all four quantum numbers the same. Thus the maximum number of equivalent electrons is $2(2l + 1)$. A subshell having this number of electrons is *full*, *complete*, or *closed*, and a subshell having a smaller number of electrons is *unfilled*, *incomplete*, or *open*. The $3p^6$ configuration thus represents a full subshell and $3s^2\, 3p^6\, 3d^{10}$ represents a full shell for $n = 3$. The *parity* of a configuration is *even* or *odd* according to whether $\Sigma_i l_i$ is even or odd, the sum being taken over all electrons (in practice only those in open subshells need be considered).

HELIUM AND HELIUM-LIKE IONS; *LS* COUPLING

The energy structure of the normal $1snl$ configurations is dominated by the electron-nucleus and electron-electron Coulomb contributions. In helium and in helium-like ions of the lighter elements, the separations of levels having the same n and having $l = s, p$, or d are mainly determined by direct and exchange electrostatic interactions between the electrons - the spin-orbit, spin-other orbit, and other relativistic contributions are much smaller. This is the condition for *LS* coupling, in which:

- The orbital angular momenta of the electrons are coupled to give a total orbital angular momentum $L = \Sigma_i l_i$.
- The spins of the electrons are coupled to give a total spin $S = \Sigma_i s_i$.

The combination of a particular S value with a particular L value comprises a spectroscopic *term*, the notation for which is ${}^{2S+1}L$. The quantum number $2S + 1$ is the *multiplicity* of the term. The S and L vectors are coupled to obtain the total angular momentum, $J = S + L$, for a *level* of the term; the level is denoted as ${}^{2S+1}L_J$. For $1snl$ configurations, $L = l$ and $S = 0$ or 1, *i.e.*, the terms are *singlets* ($S = 0$) or *triplets* ($S = 1$). As examples of the He I structure, the ionization energy (energy required to remove one of the 1s electrons in the $1s^2$ ground configuration) is 24.5874 eV, the $1s2s$ ${}^3S - {}^1S$ separation is 0.7962 eV, the $1s2p$ ${}^3P^\circ - {}^1P^\circ$ separation is 0.2539 eV, and the $1s2p$ ${}^3P^\circ_2 - {}^3P^\circ_0$ fine-structure spread is only 1.32×10^{-4} eV.

Hierarchy of Atomic Structure in LS Coupling: The centrality of *LS* coupling in the analysis and theoretical interpretation of atomic spectra has led to the acceptance of notations and nomenclature well adapted to discussions of particular structures and spectra.

The main elements of the nomenclature are shown in the table below, most of the structural entities having already been defined in the above discussions of simple spectra. The quantum numbers in the table represent a full description for complex configurations, and the accepted names for transitions between the structural elements are also given.

Table. Atomic Structural Hierarchy in LS Coupling and Names for the Groups of all Transitions Between Structural Entities.

Structural entity	Quantum numbers [a]	Group of all transitions
Configuration	$(n_i l_i)^{N_i}$	Transition array
Polyad	$(n_i l_i)^{N_i}\ \gamma S_1\ L_1\ nl\ S\ L,\ S\ L?...$	Supermultiplet
Term	$(n_i l_i)^{N_i}\ \gamma S\ L$	Multiplet
Level	$(n_i l_i)^{N_i}\ \gamma S\ L\ J$	Line
State	$(n_i l_i)^{N_i}\ \gamma S\ L\ J\ M$	Line component

[a] The configuration may include several open subshells, as indicated by the *i* subscripts. The letter γ represents any additional quantum numbers, such as ancestral terms, necessary to specify a particular term.

As an example, the Ca I $3d4p\ ^3D^\circ_2$ *level* belongs to the $^3D^\circ$ *term* which, in turn, belongs to the $3d4p\ ^3(P^\circ\ D^\circ\ F^\circ)$ triplet *triad*. The $3d4p$ *configuration* also has a $^1(P^\circ\ D^\circ\ F^\circ)$ singlet *triad*. The $3d4s$ configuration has only *monads*, one ^1D and one ^3D. The $3d4s\ ^3D_2 - 3d4p\ ^3D^\circ_3$ *line* belongs to the corresponding ^3D - $^3D^\circ$ triplet *multiplet*, and this multiplet belongs to the great Ca I $3d4s\ ^3D - 3d4p\ ^3(P^\circ\ D^\circ\ F^\circ)$ *supermultiplet* of three triplet multiplets discussed by Russell and Saunders in their classic paper on the alkaline-earth spectra. The $3d4s$ - $3d4p$ *transition array* includes both the singlet and triplet supermultiplets, as well as any (*LS*-forbidden) *intercombination* or *intersystem* lines arising from transitions between levels of the singlet *system* and those of the triplet *system*. The order of the two terms in the transitions as written above, with the lower-energy term on the left, is standard in atomic spectroscopy.

NOTATIONS FOR DIFFERENT COUPLING SCHEMES

The following examples make clear the meaning of the different coupling-scheme notations. Not all the configurations in the examples have been identified experimentally, and some of the examples of a particular coupling scheme given for heuristic purposes may be physically inappropriate. Cowan describes the physical conditions for the different coupling schemes and gives experimental examples.

LS Coupling (Russell-Saunders Coupling)®

Some of the examples given below indicate notations bearing on the order of coupling of the electrons.

1. $3d^7\ {}^4F_{7/2}$
2. $3d^7({}^4F)4s4p({}^3P^\circ)\ {}^6F^\circ_{9/2}$
3. $4f^7({}^8S^\circ)6s6p^2({}^4P)\ {}^{11}P^\circ_5$
4. $3p^5({}^2P^\circ)3d^2({}^1G)\ {}^2F^\circ_{7/2}$
5. $4f^{10}({}^3K2)6s6p({}^1P^\circ)\ {}^3L^\circ_6$
6. $4f^7({}^8S^\circ)5d\ ({}^7D^\circ)6p\ {}^8F^\circ_{13/2}$
7. $4f^7({}^8S^\circ)5d\ ({}^9D^\circ)6s\ ({}^8D^\circ)7s\ {}^9D^\circ_5$
8. $4f^7({}^8S^\circ)5d\ ({}^9D^\circ)6s6p({}^3P^\circ)\ {}^{11}F_8$
9. $4f^7({}^8S^\circ)5d^2({}^1G)\ ({}^8G^\circ)6p\ {}^7F_0$
10. $4f({}^2F^\circ)\ 5d^2({}^1G)6s\ ({}^2G)\ {}^1P^\circ_1$

In the second example, the seven $3d$ electrons couple to give a ^4F term, and the $4s$ and $4p$ electrons couple to form the $^3P^\circ$ term; the final $^6F^\circ$ term is one of nine possible terms obtained by coupling the ^4F grandparent and $^3P^\circ$ parent terms. The next three examples are similar to the second. The meaning of the index number 2 following the ^3K symbol in the fifth example is explained in the section *LS* Coupling.

The coupling in example 6 is appropriate if the interaction of the $5d$ and $4f$ electrons is sufficiently stronger than the $5d$-$6p$ interaction. The $^7D^\circ$ parent term results from coupling the $5d$ electron to the $^8S^\circ$ grandparent, and the $6p$ electron is then coupled to the $^7D^\circ$ parent to form the final ^8F term.

A space is inserted between the $5d$ electron and the $^7D^\circ$ parent to emphasize that the latter is formed by coupling a term ($^8S^\circ$) listed to the left of the space. Example 7 illustrates a similar coupling order carried to a further stage; the $^8D^\circ$ parent term results from the coupling of the $6s$ electron to the $^9D^\circ$ grandparent.

Example 8 is similar to examples 2 through 5, but in 8 the first of the two terms that couple to form the final ^{11}F term, *i.e.*, the $^9D^\circ$ term, is itself formed by the coupling of the $5d$ electron to the $^8S^\circ$ core term. Example 9 shows an $^8G^\circ$ parent term formed by coupling the $^8S^\circ$ and ^1G grandparent terms. A space is again used to emphasize that the following ($^8G^\circ$) term is formed by the coupling of terms listed before the space. A different order of coupling is indicated in the final example, the $5d^2\ {}^1G$ term being coupled first to the external $6s$ electron instead of directly to the $4f$ core electron. The $4f\ ({}^2F^\circ)$ core term is isolated by a space to denote that it is coupled (to the $5d^2({}^1G)6s\ {}^2G$ term) only after the other electrons have been coupled. The notation in this particular case (with a single $4f$ electron) could be simplified by writing the $4f$

electron after the ^{2}G term to which it is coupled. It appears more important, however, to retain the convention of giving the core portion of the configuration first. The notations in examples 1 through 5 are in the form recommended by Russell, Shenstone, and Turner, and used in both the *Atomic Energy States* and *Atomic Energy Levels* compilations.

The spacings used in the remaining examples allow different orders of coupling of the electrons to be indicated without the use of additional parentheses, brackets, etc. Some authors assign a short name to each (final) term, so that the configuration can be omitted in tables of classified lines, etc. The most common scheme distinguishes the low terms of a particular *SL* type by the prefixes *a, b, c, ...,* and the high terms by *z, y, x,* ..

jj Coupling of Equivalent Electrons

This scheme is used, for example, in relativistic calculations. The lower-case j indicates the angular momentum of one electron ($j = l \pm {}^1/_2$) or of each electron in an l_j^N group. Various ways of indicating which of the two possible j values applies to such a group without writing the j-value subscript have been used by different authors; we give the j values explicitly in the examples below.

We use the symbols J_i and j to represent total angular momenta.

11. $(6p\,^2{}_{1/2})_0$
12. $(6p\,^2{}_{1/2}\,6p_{3/2})^{\circ}{}_{3/2}$
13. $(6p\,^2{}_{1/2}\,6p\,^2{}_{3/2})_2$
14. $4d\,^3{}_{5/2}\,4d\,^2{}_{3/2}\;\;({}^9/_2,\,2)_{11/2}$

The relatively large spin-orbit interaction of the $6p$ electrons produces jj-coupling structures for the $6p\,^2$, $6p\,^3$, and $6p\,^4$ ground configurations of neutral Pb, Bi, and Po, respectively; the notations for the ground levels of these atoms are given as the first three examples above. The configuration in the first example shows the notation for equivalent electrons having the same j value l_j^N, in this case two $6p$ electrons each having $j = {}^1/_2$. A convenient notation for a particular level ($J = 0$) of such a group is also indicated. The second example extends this notation to the case of a $6p\,^3$ configuration divided into two groups according to the two possible j values. A similar notation is shown for the $6p\,^4$ level in the third example; this level might also be designated $(6p\,^{-2}{}_{3/2})_2$, the negative superscript indicating the two $6p$ holes. The $(J_1, J_2)_J$ term and level notation shown on the right in the fourth example is convenient because each of the two electron groups $4d\,^3{}_{5/2}$ and $4d\,^2{}_{3/2}$ has more than one allowed total J_i value. The assumed convention is that J_1 applies to the group on the left ($J_1 = {}^9/_2$ for the $4d\,^3{}_{5/2}$ group) and J_2 to that on the right.

$J_1\,j$ or $J_1\,J_2$ Coupling

15. $3d\,^9(^2\mathrm{D}_{5/2})4p_{3/2}\;\;({}^5/_2,\,{}^3/_2)^{\circ}{}_3$

16. $4f^{11}(^2H^\circ_{9/2}\,2)6s6p(^3P^\circ_1)$ $(^9/_2, 1)_{7/2}$
17. $4f^9(^6H^\circ)5d$ $(^7H^\circ_8)6s6p(^3P^\circ_0)$ $(8,0)_8$
18. $4f^{12}(^3H_6)$ $5d(^2D)6s6p(^3P^\circ)$ $(^4F^\circ_{3/2})$ $(6, {}^3/_2)^\circ_{13/2}$
19. $5f^4(^5I_4)6d_{3/2}$ $(4,{}^3/_2)_{11/2}7s7p(^1P^\circ_1)$ $(^{11}/_2, 1)^\circ_{9/2}$
20. $5f^4{}_{7/2}5f^5{}_{5/2}$ $(8,{}^5/_2)^\circ{}_{21/2}7p_{3/2}$ $(^{21}/_2, {}^3/_2)_{10}$
21. $5f^3{}_{7/2}$ $5f^3{}_{5/2}$ $(^9/_2, {}^9/_2)_9 7s7p(^3P^\circ_2)$ $(9,2)^\circ_7$

The first five examples all have core electrons in *LS* coupling, whereas *jj* coupling is indicated for the 5*f* core electrons in the last two examples.

Since the J_1 and J_2 values in the final (J_1, J_2) term have already been given as subscripts in the configuration, the (J_1, J_2) term notations are redundant in all these examples. Unless separation of the configuration and final term designations is desired, as in some data tables, one may obtain a more concise notation by simply enclosing the entire configuration in brackets and adding the final *J* value as a subscript. Thus, the level in the first example can be designated as $[3d^9(^2D_{5/2})\ 4p_{3/2}]^\circ_3$. If the configuration and coupling order are assumed to be known, still shorter designations may be used; for example, the fourth level above might then be given as $[(^3H_6)\ (^3P^\circ)\ (^4F^\circ_{3/2})]_{13/2}$ or $(^3H_6, {}^3P^\circ, {}^4F^\circ_{3/2})_{13/2}$. Similar economies of notation are of course possible, and often useful, in all coupling schemes.

J_1l or J_1L_2 Coupling (J_1K Coupling)

22. $3p^5(^2P^\circ_{1/2})5g$ $^2[^9/_2]^\circ_5$
23. $4f^2(^3H_4)5g$ $^2[3]_{5/2}$
24. $4f^{13}(^2F^\circ_{7/2})5d\ ^2(^1D)$ $^1[^7/_2]^\circ_{7/2}$
25. $4f^{13}(^2F^\circ_{5/2})5d6s(^3D)$ $^3[^9/_2]^\circ_{11/2}$

The final terms in the first two examples result from coupling a parent-level J_1 to the orbital angular momentum of a 5*g* electron to obtain a resultant *K*, the *K* value being enclosed in brackets.

The spin of the external electron is then coupled with the *K* angular momentum to obtain a pair of *J* values, $J = K \pm {}^1/_2$ (for $K \neq 0$). The multiplicity of such pair terms is usually omitted from the term symbol, but other multiplicities occur in the more general J_1L_2 coupling.

The last two examples are straightforward extensions of J_1l coupling, with the L_2 and S_2 momenta of the "external"' term (^1D and ^3D in examples 3 and 4, respectively) replacing the *l* and *s* momenta of a single external electron.

LS_1 Coupling (*LK* Coupling)

26. $3s\ ^23p(^2P^\circ)4f$ G $^2[^7/_2]_3$
27. $3d^7(^4P)4s4p(^3P^\circ)$ D° $^3[^5/_2]^\circ_{7/2}$

The orbital angular momentum of the core is coupled with the orbital angular momentum of the external electron(s) to give the total orbital angular momentum *L*.

The letter symbol for the final *L* value is listed with the configuration because this angular momentum is then coupled with the spin of the core (S_1) to obtain the resultant *K* angular momentum of the final term (in brackets). The multiplicity of the [*K*] term arises from the spin of the external electron(s).

Coupling Schemes and Term Symbols

The coupling schemes outlined above include those now most frequently used in calculations of atomic structure. Any term symbol gives the values of two angular momenta that may be coupled to give the total electronic angular momentum of a level (indicated by the *J* value).

For configurations of more than one unfilled subshell, the angular momenta involved in the final coupling derive from two groups of electrons (either group may consist of only one electron).

These are often an inner group of coupled electrons and an outer group of coupled electrons, respectively. In any case the quantum numbers for the two groups can be distinguished by subscripts 1 and 2, so that quantum numbers represented by capital letters without subscripts are total quantum numbers for both groups. Thus, the quantum numbers for the two vectors that couple to give the final *J* are related to the term symbol as follows:

Coupling Scheme	Quantum numbers for vectors that couple to give *J*	Term Symbol
LS	*L*,*S*	^{2S+1}L
$J_1 J_2$	J_1, J_2	(J_1, J_2)
$J_1 L_2$(? *K*)	*K*, S_2	$^{2S_2+1}[K]$
LS_1(? *K*)	*K*, S_2	$^{2S_2+1}[K]$

The parity is indicated by appended degree symbols on odd parity terms.

EIGENVECTOR COMPOSITION OF LEVELS

The wave functions of levels are often expressed as eigenvectors that are linear combinations of basis states in one of the standard coupling schemes.

Thus, the wave function $\psi(\alpha J)$ for a level labeled αJ might be expressed in terms of normalized *LS* coupling basis states $\Phi(\gamma SLJ)$:

$$\psi(\alpha \mathrm{J}) = \sum_{\gamma \mathrm{SL}} \mathrm{c}(\gamma \mathrm{SLJ}) \Phi(\gamma \mathrm{SLJ}).$$

The $c(\gamma SLJ)$ are expansion coefficients, and

$$\sum_{\gamma \mathrm{SL}} |\mathrm{c}(\gamma \mathrm{SLJ})|^2 = 2.$$

The squared expansion coefficients for the various γSL terms in the composition of the αJ level are conveniently expressed as percentages, whose sum is 100 per cent. Thus the percentage contributed by the pure Russell-Saunders state γSLJ is equal to $100 \times |c(\gamma SLJ)|^2$.

The notation for Russell-Saunders basis states has been used only for concreteness; the eigenvectors may be expressed in any coupling scheme, and the coupling schemes may be different for different configurations included in a single calculation (with configuration interaction). "Intermediate coupling" conditions for a configuration are such that calculations in both *LS* and *jj* coupling yield some eigenvectors representing significant mixtures of basis states.

The largest percentage in the composition of a level is called the *purity* of the level in that coupling scheme. The coupling scheme (or combination of coupling schemes if more than one configuration is involved) that results in the largest average purity for all the levels in a calculation is usually best for naming the levels. With regard to any particular calculation, one does well to remember that, as with other calculated quantities, the resulting eigenvectors depend on a specific theoretical model and are subject to the inaccuracies of whatever approximations the model involves. Theoretical calculations of experimental energy level structures have yielded many eigenvectors having significantly less than 50 per cent purity in any coupling scheme. Since many of the corresponding levels have nevertheless been assigned names by spectroscopists, some caution is advisable in the acceptance of level designations found in the literature.

ALLOWED TERMS OF LEVELS FOR EQUIVALENT ELECTRONS

LS Coupling

The allowed *LS* terms of a configuration consisting of two nonequivalent groups of electrons are obtained by coupling the *S* and *L* vectors of the groups in all possible ways, and the procedure may be extended to any number of such groups.

Thus the allowed terms for any configuration can be obtained from a table of the allowed terms for groups of equivalent electrons. The configuration l^N has more than one allowed term of certain *LS* types if $l > 1$ and $2 < N < 4l$ (d^3 - d^7, $f^3 - f^{11}$, etc.). The recurring terms of a particular *LS* term type from d^N and f^N configurations are assigned sequential index numbers in the tables of Nielson and Koster; the index numbers stand for additional numbers having group-theoretical significance that serve to differentiate the recurring terms, except for a few terms of f^5 and f^9, f^6 and f^8, and f^7.

These remaining terms, which occur only in pairs, are further labeled *A* or *B* to indicate Racah's separation of the two terms. The index numbers of Nielson and Koster are in practice the most frequently used labels for the recurring terms of f^N configurations. Use of their index numbers for the recurring terms of d^N configurations has perhaps the disadvantage of substituting an arbitrary number for a quantum number (the seniority) that itself distinguishes the recurring terms in all cases. The actual value of the seniority number is rarely needed, however, and a consistent notation for the d^N and f^N configurations is desirable. A table of the allowed *LS* terms of the l^N electrons for *l* d" 3 is given, with all recurring terms having the index numbers of Nielson and Koster as a following on-line integer. The theoretical group labels are also listed. Thus the d^3 ^{2}D term having seniority 3 is designated ^{2}D2, instead of D, in this scheme; and the level having $J = {}^3/_2$ is designated $^2D_{3/2}2$.

jj Coupling

The allowed *J* values for a group of *N* equivalent electrons having the same *j* value, l_j^N, are given in the table below for $j = {}^1/_2, {}^3/_2, {}^5/_2$, and ${}^7/_2$ (sufficient for *l* d" 3). The $l^4{}_{7/2}$ group has two allowed levels for each of the *J* values 2 and 4. The subscripts distinguishing the two levels in each case are the seniority numbers.

Table. Allowed *J* Values for l_j^N Equivalent Electrons (*jj* Coupling).

ljN	Allowed J values
11/2	½
121/2	0
13/2 and 133/2	3/2
123/2	0, 2
143/2	0
15/2 and 155/2	5/2
125/2 and 145/2	0, 2, 4
135/2	3/2, 5/2, 9/2
165/2	0
17/2 and 177/2	7/2
127/2 and 167/2	0, 2, 4, 6
137/2 and 157/2	3/2, 5/2, 7/2, 9/2, 11/2, 15/2
147/2	0, 22, 42, 24, 44, 5, 6, 8
187/2	0

The allowed levels of the configuration nl^N may be obtained by dividing the electrons into sets of two groups $nl^Q_{l+1/2^n} l^R_{l-1/2}$, $Q + R = N$. The possible sets run from $Q = N - 2l$ (or zero if $N > 2l$) up to $Q = N$ or $Q = 2l + 2$, whichever is smaller. The (degenerate) levels for a set with both *Q* and *R* nonzero have wave functions defined by the quantum numbers $(\alpha J_1, \beta J_2)$ *J*, with J_1 and J_2 deriving from the *Q* and *R* groups, respectively.

The symbols α and â represent any additional quantum numbers required to identify levels. The *J* values of the allowed levels for each $(\alpha J_1, \beta J_2)$ subset are obtained by combining J_1 and J_2 in the usual way.

ELECTRON CONFIGURATIONS OF ELEMENTS

GROUND LEVELS AND IONIZATION ENERGIES FOR THE NEUTRAL ATOMS

The ground-state electron configurations of elements heavier than neon are shortened in the table by using rare-gas element symbols in brackets to represent the corresponding electrons. The ground levels of all neutral atoms have reasonably meaningful *LS*-coupling names, the corresponding eigenvector percentages lying in the range from ~55 per cent to 100 per cent. These names are listed in the table, except for Pa, U, and Np; the lowest few ground-configuration levels of these atoms comprise better $5f^N(L_1S_1J_1)$, $6d_j7s^2$ $(J_1\,j)$ terms than *LS*-coupling terms. The relatively large spin-orbit interaction of the 6*p* electrons produces *jj*-coupling structures for the $(6p^2{}_{1/2})_0$, $(6p^2{}_{1/2}6p_{3/2})^o{}_{3/2}$, and $(6p^2{}_{1/2}6p^2{}_{3/2})_2$ ground levels of the $6p^2$, $6p^3$, and $6p^4$ configurations of neutral Pb, Bi, and Po, respectively.

As noted in the section *jj* Coupling of Equivalent Electrons, the *jj*-coupling names are more appropriate for these atoms than the alternative *LS*-coupling designations in the table. The ionization energies in the table are from recent compilations. The uncertainties are mainly in the range from less than one to several units in the last decimal place, but a few of the values may be in error by 20 or more units in the final place; *i.e.*, the error of some of the two place values could be greater than 0.2 eV. Although no more than four decimal places are given here, values for both the neutral and singly-ionized atoms are given to their full accuracies.

Zeeman Effect

The Zeeman effect for "weak" magnetic fields (the anomalous Zeeman effect) is of special interest because of the importance of Zeeman data in the analysis and theoretical interpretation of complex spectra. In a weak field, the *J* value remains a good quantum number although in general a level is split into magnetic sublevels. he *g* value of such a level may be defined by the expression for the energy shift of its magnetic sublevel having magnetic quantum number *M*, which has one of the $2J + 1$ values, $-J, -J + 1, \ldots, J$:

$$\Delta E = gM\ \mu BB\ .$$

The magnetic flux density is μB, and μB is the Bohr magneton (μB = $e\hbar/2me$). The wavenumber shift $\Delta\sigma$ corresponding to this energy shift is

$$\Delta s = gM\ (0.466\ 86\ B\ cm^{-1}),$$

with B representing the numerical value of the magnetic flux density in teslas. The quantity in parentheses, the Lorentz unit, is of the order of 1 or 2 cm^{-1} for typical flux densities used to obtain Zeeman-effect data with classical spectroscopic methods. Accurate data can be obtained with much smaller fields, of course, by using higher-resolution techniques such as laser spectroscopy. Most of the g values now available for atomic energy levels were derived by application of the above formula (for each of the two combining levels) to measurements of optical Zeeman patterns. A single transverse-Zeeman-effect pattern (two polarizations, resolved components, and sufficiently complete) can yield the J value and the g value for each of the two levels involved.

Neglecting a number of higher-order effects, we can evaluate the g value of a level ?J belonging to a pure LS-coupling term using the formula

$$g\beta SLJ = 1 + (ge - 1)\frac{J(J+1) - L(L+1) + S(S+1)}{2J(J+1)}$$

The independence of this expression from any other quantum numbers (represented by β) such as the configuration, etc., is important. The expression is derived from vector coupling formulas by assuming a g value of unity for a pure orbital angular momentum and writing the g value for a pure electron spin as ge.

A value of 2 for ge yields the Landé formula. If the anomalous magnetic moment of the electron is taken into account, the value of ge is 2.002 319 3. "Schwinger" g values obtained with this more accurate value for ge are given for levels of SL terms.

The usefulness of gSLJ values is enhanced by their relation to the g values in intermediate coupling. In the notation used in Equation for the wave function of a level β J in intermediate coupling, the corresponding g value is given by

$$g\beta J = \sum_{\gamma SL} gSLJ|c(\gamma SLJ)^2,$$

where the summation is over the same set of quantum numbers as for the wave function. The gβJ value is thus a weighted average of the Landé gSLJ values, the weighting factors being just the corresponding component percentages. Formulas for magnetic splitting factors in the J1 J2 and J1 L2 coupling schemes are given. Some higher-order effects that must be included in more accurate Zeeman-effect calculations are treated by Bethe and Salpeter and by Wybourne, for example.

Term Series, Quantum Defects, and Spectral-line Series

The Bohr energy levels for hydrogen or for a hydrogenic (one-electron) ion are given by

$$E_n = -\frac{Z^2}{n^2},$$

in units of RM, the Rydberg for the appropriate nuclear mass. For a multielectron atom, the deviations of a series of (core) nl levels from hydrogenic En values may be due mainly to core penetration by the nl electron (low l-value series), or core polarization by the nl electron (high l-value series), or a combination of the two effects. In either case it can be shown that these deviations can be approximately represented by a constant quantum defect δ_l in the Rydberg formula,

$$E_{nl} = -\frac{Z_c^2}{(n-\delta_1)^2}, = -\frac{z^2}{(n^*)^2},$$

where Zc is the charge of the core and n* = n - ? is the effective principal quantum number. If the core includes only closed subshells, the Enl values are with respect to a value of zero for the (core)1S0 level, *i.e.*, the 1S0 level is the limit of the (core)nl series.

If the quantities in Equation are taken as positive, they represent term values or ionization energies; the term value of the ground level of an atom or ion with respect to the ground level of the next higher ion is thus the principal ionization energy. If the core has one or more open subshells, the series limit may be the baricenter of the entire core configuration, or any appropriate sub-structure of the core, down to and including a single level. The Enl values refer to the series of corresponding (core)nl structures built on the particular limit structure. The value of the quantum defect depends to some extent on which (core)nl structures are represented by the series formula.

The quantum defect in general also has an energy dependence that must be taken into account if lower members of a series are to be accurately represented by Equation. For an unperturbed series, this dependence can be expressed by the extended Ritz formula

$$\begin{aligned}\delta &= n - n^* \\ &= \delta_0 + \frac{a}{(n-\delta_0)^2} + \frac{b}{(n-\delta_0)^4} + \cdots,\end{aligned}$$

with δ_0,a,b... constants for the series (δ_0 being the limit value of the quantum defect for high series members). The value of a is usually positive for core-penetration series and negative for core-polarization series. A discussion of the foundations of the Ritz expansion and application to high precision calculations in helium is given. A spectral-line series results from either emission or absorption transitions involving a common lower level and a series of successive (core)nl upper levels differing only in their n values. The principal series of Na I, 3s 2S1/2 z– np 2P°1/2,3/2 is an example. The regularity of successive upper term values with increasing n is of course observed in line series; the intervals between successive lines decrease in a regular manner towards higher wavenumbers, and the series of increasing wavenumbers converges towards the term value of the lower level as a limit.

Sequences

Several types of sequences of elements and/or ionization stages are useful because of regularities in the progressive values of parameters relating to structure and other properties along the sequences. All sequence names may refer either to the atoms and/or ions of the sequence or to their spectra.

Isoelectronic Sequence: A neutral atom and those ions of other elements having the same number of electrons as the atom comprise an isoelectronic sequence. (Note that a negative ion having this number of electrons is a member of the sequence.) An isoelectronic sequence is named according to its neutral member; for example, the Na I isolectronic sequence.

SPECTRAL WAVELENGTH RANGES, DISPERSION OF AIR

The ranges of most interest for optical atomic spectroscopy are:

~ 2 – 20 μm	mid-infrared (ir)
700 – 2000 nm	near ir
400 – 700 nm	Visible
200 – 400 nm	near ultraviolet (uv)
100 – 200 nm	vacuum uv or far uv
10 – 100 nm	extreme uv (euv or xuv)
< 10 nm	soft x-ray, x-ray

The above correspondence of names to ranges should not be taken as exact; the variation as to the extent of some of the named ranges found in the literature is considerable. Wavelengths in standard air are often tabulated for the region longer than 200 nm. These wavelengths can be related to energy-level differences by conversion to the corresponding (vacuum) wavenumbers or frequencies.

Wavelength (Frequency) Standards

In 2001 the Comité International des Poids et Mesures recommended values for optical frequency standards from stabilized lasers using various absorbing atoms, atomic ions, and molecules. These frequencies range from 29 054 057 446 579 Hz (10.318 436 884 460 μm, relative standard uncertainty 1.4×10^{-13} for a transition in OsO_4 to 1 267 402 452 889.92 kHz (236.540 853 549 75 nm; relative standard uncertainty 3.6×10^{-13}) for a transition in the $^{115}In^+$ ion.

Extensive tables of wavenumbers for molecular transitions in the mid-ir range 2.3 to 20.5 μm are included in a calibration atlas published in 1991. Wavenumbers of Ar I and Ar II emission lines having uncertainties as small as 0.0003 cm^{-1} are included in tables for these spectra covering a broad range from 222 nm to 5.865 μm. Measurements of U and Th lines (575 nm to 692 nm) suitable for wavenumber calibration at uncertainty levels of 0.0003 cm^{-1} or 0.0004$ cm^{-1} were reported. Comprehensive tables of lines for U, Th, and I_2 are useful for calibration at uncertainty levels of 0.002 cm^{-1} to 0.003 cm^{-1}, the atlas of the Th spectrum extending down to 278 nm.

A 1974 compilation gives reference wavelengths for some 5400 lines of 38 elements covering the range 1.5 nm to 2.5 μm, with most uncertainties between 10^{-5} and 2×10^{-4} nm. The wavelengths for some 1100 Fe lines selected from the Fe/Ne hollow-cathode spectrum have been recommended for reference standards over the range 183 nm to 4.2 μm, with wavenumber uncertainties 0.001 cm^{-1} to 0.002 cm^{-1}. Wavelengths for about 3000 vuv and uv lines (110 nm to 400 nm) from a Pt/Ne hollow-cathode lamp have been determined with uncertainties of 0.0002 nm or less .

More recent high-accuracy measurements of ultraviolet lines of Fe I, Ge I, Kr II, and Pt I, II, include some wavelengths with uncertainties smaller than 10^{-5} nm . The wavelengths tabulated for the Kr and Pt lines in extend from 171 nm to 315 nm, and the accuracies of earlier measurements of a number of spectra useful for wavelength calibration are discussed.

SPECTRAL LINES: SELECTION RULES, INTENSITIES, TRANSITION PROBABILITIES, VALUES, AND LINE STRENGTHS

Table. Selection Rules for Discrete Transitions.

		Electric dipole (E1) ("allowed")	Magnetic dipole (M1) ("forbidden")	Electric quadrupole (E2) ("forbidden")
Rigorous rules	1.	$\Delta J = 0, \pm 1$ (except $0 \nleftrightarrow 0$)	$\Delta J = 0, \pm 1$ (except $0 \nleftrightarrow 0$)	$\Delta J = 0, \pm 1, \pm 2$ (except $0 \nleftrightarrow 0$, $½ \nleftrightarrow 1/2$, $0 \nleftrightarrow 1$)
	2.	$\Delta M = 0, \pm 1$ (except $0 \nleftrightarrow 0$ when $\Delta J = 0$)	$\Delta M = 0, \pm 1$ (except $0 \nleftrightarrow 0$ when $\Delta J = 0$)	$\Delta M = 0, \pm 1, \pm 2$
	3.	Parity change	No parity change	No parity change
With negligible configuration interaction	4.	One electron jumping, with $\Delta l = \pm 1$, Δn arbitrary	No change in electron configuration; i.e., for all electrons, $\Delta l = 0$, $\Delta n = 0$	*No* change in electron configuration; *or* one electron jumping with $\Delta l = 0, \pm 2$, Δn arbitrary
For *LS* coupling only	5.	$\Delta S = 0$	$\Delta S = 0$	$\Delta S = 0$
	6.	$\Delta L = 0, \pm 1$ (except $0 \nleftrightarrow 0$)	$\Delta L = 0$ $\Delta J = \pm 1$	$\Delta L = 0, \pm 1, \pm 2$ (except $0 \nleftrightarrow 0$, $0 \nleftrightarrow 1$)

Emission Intensities (Transition Probabilities)

The total power ε radiated in a spectral line of frequency ν per unit source volume and per unit solid angle is λ

$$\epsilon_{line} = (4\pi)^{-1} h^{\nu} \, Aki \, Nk,$$

where Aki is the atomic transition probability and Nk the number per unit volume (number density) of excited atoms in the upper (initial) level k. For a homogeneous light source of length l and for the optically thin case, where all radiation escapes, the total emitted line intensity (SI quantity: radiance) is

$$I_{line} = \in_{line} l = int_{o}^{+\infty} I(\lambda) d\lambda$$
$$(4\pi)^{-1} (hc/\lambda_0) A_{ki} N_k l,$$

where I(λ) is the specific intensity at wavelength λ, and λ 0 the wavelength at line centre.

Absorption f Values

In absorption, the reduced absorption

$$W\lambda = [I(\lambda) - I'(\lambda)]I(\lambda),$$

is used, where I(λ) is the incident intensity at wavelength & ambda; *e.g.*, from a source providing a continuous background, and I′(λ) the intensity after passage through the absorbing medium. The reduced line intensity from a homogeneous and optically thin absorbing medium of length l follows as

$$W_{ik} = \int_{0}^{+\infty} W(\lambda) d\lambda = \frac{e^2}{4_{\in 0 mec}{}^{2}} \lambda_0^2 Ni dikl,$$

where f_{ik} is the atomic (absorption) oscillator strength (dimensionless).

Line Strengths

Aki and fik are the principal atomic quantities related to line intensities. In theoretical work, the line strength S is also widely used:

$$S = S(i,k) = S(k,i) = |R_{ik}|^2,$$
$$R_{ik} = \langle \psi k | P | \psi_i \rangle,$$

where ψ_i and ψ_k are the initial- and final-state wave functions and R_{ik} is the *transition matrix element* of the appropriate multipole operator *P* (R_{ik} involves an integration over spatial and spin coordinates of all *N* electrons of the atom or ion).

Relationships Between A, f, and S

The relationships between *A*, *f*, and *S* for electric dipole (E1, or allowed) transitions in SI units (*A* in s^{-1}, λ in m, *S* in m^2 C^2) are

$$Aki = \frac{2\pi e^2}{m_e c \in_0 \lambda^2} \frac{gi}{gk} fik = \frac{16\pi^3}{3h \in_0 \lambda^3 gk} S,$$

Numerically, in customary units (*A* in s^{-1}, λ in Å, *S* in atomic units),

$$Aki = \frac{6.6702 \times 10^{15}}{\lambda^2} \frac{gi}{gk} fik = \frac{2.0261 \times 10^{18}}{\lambda^3 gk} S,$$

and for S and ΔE in atomic units,

$$\text{fik} = \frac{2}{3}(\Delta E / gi)S.$$

g_i and g_k are the statistical weights, which are obtained from the appropriate angular momentum quantum numbers. Thus for the lower (upper) level of a spectral line, $g_{i(k)} = 2\,J_{i(k)} + 1$, and for the lower (upper) term of a multiplet,

$$\bar{g}i(k) = \sum_{i(k)} (2J_{i(k)} + 1)$$

$$= (2L_{i(k)} + 1)(2S_{i(k)} + 1)$$

The A_{ki} values for strong lines of selected elements are given. For comprehensive numerical tables of A, f, and S, including forbidden lines.

Table. Conversion Relations Between S and A_{ki} for the Most Common Forbidden Transitions.

	SI units [a]	Numerically, in Customary Units [b]
Electric quadrupole	$A_{ki} = \frac{16\pi^5}{15h\,\epsilon_0\,\lambda^5 g_k} S$	$A_{ki} = \frac{1.1199 \times 10^{18}}{g_k \lambda^5} S$
Magnetic dipole	$A_{ki} = \frac{16\pi^3 \mu_0}{3h\,\lambda^3 g_k} S$	$A_{ki} = \frac{2.697 \times 10^{13}}{g_k \lambda^3} S$

[a] A in s^{-1}, λ in m. Electric quadrupole: S in $m^4\,C^2$. Magnetic dipole: S in $J^2\,T^{-2}$. [b] A in s^{-1}, λ in Å. S in atomic units: $= 2.013 \times 10^{-79}\ m^4\,C^2$ (electric quadrupole), $= 8.601 \times 10^{-47}\ J^2\,T^{-2}$ (magnetic dipole). μ_B is the Bohr magneton.

Oscillator strengths f are not used for forbidden transitions, *i.e.*, magnetic dipole (M1), electric quadrupole (E2), etc.

[Numerical example: For the $1s2p\ ^1P - 1s3d\ ^1D_2$ (allowed) transition in He I at 6678.15 Å: $g_i = 3$; $g_k = 5$; $A_{ki} = 6.38 \times 10^7\ s^{-1}$; $f_{ik} = 0.711$; $S = 46.9\ a_0^2\,e^2$.]

Relationships Between Line and Multiplet Values

The relations between the total strength and f value of a multiplet (M) and the corresponding quantities for the lines of the multiplet (allowed transitions) are

$$S_M = \Sigma S_{line}$$

$$f_M = (\bar{\lambda}\bar{g}_i)^{-1} \sum_{J_k, J_i} g_i \lambda(J_i, J_k) f(J_i, J_k).$$

is the weighted ("multiplet") wavelength in vacuum:

$$\bar{\lambda} = n\bar{\lambda}_{air} = hc / \overline{\Delta E}$$

where

$$\overline{\Delta E} = \overline{Ek} - \overline{Ei} = (\bar{g}_k)^{-1} \sum_{J_k} g_k E_k - (\bar{g}_i)^{-1} \sum_{J_i} g_i E_i,$$

and n is the refractive index of standard air.

SPECTRAL LINE SHAPES, WIDTHS, AND SHIFTS

Observed spectral lines are always broadened, partly due to the finite resolution of the spectrometer and partly due to intrinsic physical causes. The principal physical causes of spectral line broadening are Doppler and pressure broadening.

Doppler Broadening

Doppler broadening is due to the thermal motion of the emitting atoms or ions. For a Maxwellian velocity distribution, the line shape is *Gaussian;* the full width at half maximum intensity (FWHM) is, in Å,

$$\Delta\lambda^{D}_{1/2} = \left(7.16\times10^{-7}\right)\lambda\left(T/M\right)^{1/2.}$$

T is the temperature of the emitters in K, and *M* the atomic weight in atomic mass units (amu).

Pressure Broadening

Pressure broadening is due to collisions of the emitters with neighboring particles. Shapes are often approximately Lorentzian, *i.e.*, $I(\lambda)\ \{1 + [(\lambda - \lambda_0)/\Delta\lambda_{1/2}]^2\}^{-1}$. In the following formulas, all FWHM's and wavelengths are expressed in Å, particle densities N in cm^{-3}, temperatures T in K, and energies E or I in cm^{-1}. Resonance broadening (self-broadening) occurs only between identical species and is confined to lines with the upper or lower level having an electric dipole transition (resonance line) to the ground state. The FWHM may be estimated as

$$\Delta\lambda^{D}_{1/2} = 8.6\times10^{-30}\left(gi/gk\right)^{1/2.}\ \lambda^2\lambda_r fr\, Ni$$

where λ is the wavelength of the observed line. f_r and λ_r are the oscillator strength and wavelength of the resonance line; g_k and g_i are the statistical weights of its upper and lower levels. N_i is the ground state number density.

For the $1s2p\,^{0}_{1}\ {}^1P - 1s3d\ {}^1D_2$ transition in He I [$\lambda = 6678.15$ Å; λ_r $(1s^2\ {}^1S_0 - 1s2p\ {}^1P^{0}_{1})$ = 584.334 Å; $g_i = 1$; $g_k = 3$; $f_r = 0.2762$] at $N_i = 1\times10^{18}\ cm^{-3}$: $\Delta\lambda^R_{1/2} = 0.036$ Å. *Van der Waals broadening* arises from the dipole interaction of an excited atom with the induced dipole of a ground state atom. (In the case of foreign gas broadening, both the perturber and the radiator may be in their respective ground states.) An approximate formula for the FWHM, strictly applicable to hydrogen and similar atomic structures only, is

$$\Delta\lambda^{W}_{1/2}\ 30\times10^{16}\,\lambda^2 C_6^{2/5}\left(T/\mu\right)^{3/10} N,$$

where μ is the atom-perturber reduced mass in units of u, *N* the perturber density, and C_6 the interaction constant. C_6 may be roughly estimated as follows: $C_6 = C_k - C_i$, with $C_{i\,(k)} = (9.8\times10^{10})$ ($\alpha_d\ R^2{}_i\ (k)\ \alpha_d$ in cm^3, R^2 in $a_0{}^2$). Mean atomic polarizability α_d H (6.7×10^{-25}) $(3I_H/4E^*;)^2\ cm^3$, where I_H is the ionization energy of hydrogen and E^*; the energy of the first excited level of the perturber atom. $R^2_{i\,(k)}$ H″ 2.5 $[I_H/(I-E_{i\,(k)})]^2$, where I is the ionization energy of the radiator. Van der Waals broadened lines are red shifted by about one-third the size of the FWHM. For the $1s2p\ ^1P^0_1 - 1s3d\ ^1D_2$ transition in He I, and with He as perturber: $\lambda = 6678.15$ Å; $I = 198\ 311\ \mathrm{cm}^{-1}$; $E^* = E_i = 171\ 135\ \mathrm{cm}^{-1}$; $E_k = 186\ 105\ \mathrm{cm}^{-1}$; $\mu = 2$. At $T = 15\ 000$ K and $N = 1 \times 10^{18}\ \mathrm{cm}^{-3}$: $\Delta\lambda^W_{1/2} = 0.044$ Å. *Stark broadening* due to charged perturbers, *i.e.*, ions and electrons, usually dominates resonance and van der Waals broadening in discharges and plasmas. The FWHM for hydrogen lines is

$$\Delta\lambda^{S,H}_{1/2}\left(2.50\times10^{-9}\right)\alpha_{1/2}N_e^{2/3},$$

where N_e is the electron density. The half-width parameter $\alpha_{1/2}$ for the H_β line at 4861 Å, widely used for plasma diagnostics, is tabulated in the table below for some typical temperatures and electron densities.

$$\epsilon_{cont} = \frac{\epsilon^6}{2\pi\,\epsilon_0^3\,(6\pi\,me)^{3/2}}Ne\,Nz\,Z^2\times\frac{1}{(kT)^{1/2}}\exp\left(-\frac{hc}{\lambda kT}\right)\frac{\Delta\lambda}{\lambda^2}$$

$$\times\left\{\frac{2Z^2I_H}{kT}\sum_{n\ge(z^2\ln\lambda/hc)^{1/2}}^{n'}\frac{\gamma fb}{n^3}\left[\exp\left(\frac{Z^2I_H}{(n'+1)2kT}\right)-1\right]+\gamma ff\right\},$$

This reference also contains $\alpha_{1/2}$ parameters for other hydrogen lines, as well as Stark width and shift data for numerous lines of other elements, *i.e.*, neutral atoms and singly charged ions (in the latter, Stark widths and shifts depend linearly on N_e). Other tabulations of complete hydrogen Stark profiles exist.

Table. Values of Stark-broadening Parameter $\alpha_{1/2}$ for the H_β line of Hydrogen (4861 Å) for Various Temperatures and Electron Densities.

	N_e (cm^{-3})			
T(K)	10^{15}	10^{16}	10^{17}	10^{18}
5000	0.0787	0.0808	0.0765	...
10000	0.0803	0.0840	0.0851	0.0781
20000	0.0815	0.0860	0.0902	0.0896
30000	0.0814	0.0860	0.0919	0.0946

SPECTRAL CONTINUUM RADIATION

Many-Electron Systems

For many-electron systems, only approximate theoretical treatments exist, based on the quantum-defect method (for results of calculations for noble gases). Experimental work is centered on the noble gases. Modifications of the continuum by autoionization processes must also be considered. Near the ionization limit, the f values for bound-bound transitions of a spectral series ($n' \to \infty$) make a smooth connection to the differential oscillator strength distribution $df/d\varepsilon$ in the continuum.

2

Mass Spectrometry

Atomic mass spectrometry is similar to other types of mass spectrometry in that it consists of an ion source, a mass analyzer, and a detector. Atoms' identities are determined by their mass-to-charge ratio (via the mass analyzer) and their concentrations are determined by the number of ions detected. Although considerable research has gone into customizing mass spectrometers for atomic ion sources, it is the ion source that differs most from other forms of mass spectrometry.

These ion sources must also atomize samples, or an atomization step must take place before ionization. Atomic ion sources are generally modifications of atomic optical spectroscopy atom sources.

MASS SPECTROMETER CONVERTS

In order to measure the individuality of individual molecules, a mass spectrometer converts them to ions so that they can be moved about and manipulated by external electric and magnetic fields. The three essential functions of a mass spectrometer, and the associated components, are:

- A small sample is ionized, usually to cations by loss of an electron.
- The ions are sorted and separated according to their mass and charge.
- The separated ions are then measured.

Because ions are very immediate and short-lived, their formation and manipulation must be conducted in a vacuum. Atmospheric pressure is around 760 torr (mm of mercury). The pressure under which ions may be handled is roughly 10^{-5} to 10^{-8} torr (less than a billionth of an atmosphere). Each of the three tasks listed above may be accomplished in different ways. In one common procedure, ionization is effected by a high energy beam of electrons, and ion separation is achieved by accelerating and focusing the ions in a beam, which is then bent by an external magnetic field.

The ions are then detected electronically and the resulting information is stored and analysed in a computer. A mass spectrometer operating in this fashion is outlined in the following diagram.

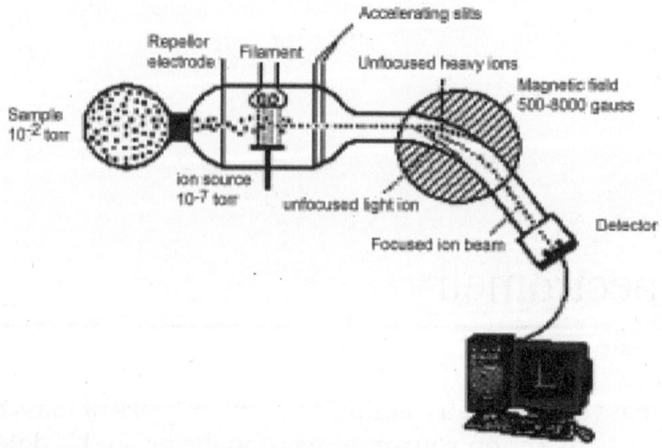

The heart of the spectrometer is the ion source. Here molecules of the sample (black dots) are bombarded by electrons (light blue lines) issuing from a heated filament. This is called an EI (electron-impact) source. Gases and volatile liquid samples are allowed to leak into the ion source from a reservoir (as shown). Non-volatile solids and liquids may be introduced directly. Cations formed by the electron bombardment (red dots) are pushed away by a charged repellor plate (anions are attracted to it), and accelerated toward other electrodes, having slits through which the ions pass as a beam. Some of these ions fragment into smaller cations and neutral fragments. A perpendicular magnetic field deflects the ion beam in an arc whose radius is inversely proportional to the mass of each ion. Lighter ions are deflected more than heavier ions. By varying the strength of the magnetic field, ions of different mass can be focused progressively on a detector fixed at the end of a curved tube (also under a high vacuum).

When a high energy electron collide with a molecule it often ionizes it by knocking away one of the molecular electrons (either bonding or non-bonding). This leaves behind a molecular ion (colored red in the following diagram). Residual energy from the collision may cause the molecular ion to fragment into neutral pieces (colored green) and smaller fragment ions (colored pink and orange). The molecular ion is a radical cation, but the fragment ions may either be radical cations (pink) or carbocations (orange), depending on the nature of the neutral fragment.

M: + e ⟶ 2 e + M + ⟶ M^+ + F* Neutral Fragment

⟶ M^+ + F* Neutral Fragment

The Nature of Mass Spectra

A mass spectrum will usually be presented as a vertical bar graph, in which each bar represent an ion having a specific mass-to-charge ratio (m/z) and the length of the bar indicates the relative abundance of the ion. The most intense ion is assigned an abundance of 100, and it is referred to as the base peak. Most of the ions formed in a mass spectrometer have a single charge, so the m/z value is equivalent to mass itself. Modern mass spectrometers easily distinguish (resolve) ions differing by only a single atomic mass unit, and thus provide completely accurate values for the molecular mass of a compound. The highest-mass ion in a spectrum is normally considered to be the molecular ion, and lower-mass ions are fragments from the molecular ion, assuming the sample is a single pure compound.

Atomic mass is given in terms of the unified atomic mass unit (symbol: μ) or dalton (symbol: Da). In recent years there has been a gradual change towards using the dalton in preference to the unified atomic mass unit. The dalton is classified as a "non-SI unit whose values in SI units must be obtained experimentally". It is defined as one twelfth of the rest mass of an unbound atom of carbon-12 in its nuclear and electronic ground state, and has a value of $1.660538782 \times 10^{-27}$ kg.

The following diagram displays the mass spectra of three simple gaseous compounds, carbon dioxide, propane and cyclopropane. The molecules of these compounds are similar in size, CO_2 and C_3H_8 both have a nominal mass of 44 Da, and C_3H_6 has a mass of 42 Da. The molecular ion is the strongest ion in the spectra of CO_2 and C_3H_6, and it is moderately strong in propane. The unit mass resolution is readily apparent in these spectra (note the separation of ions having m/z=39, 40, 41 and 42 in the cyclopropane spectrum). Even though these compounds are very similar in size, it is a simple matter to identify them from their individual mass spectra.

By clicking on each spectrum in turn, a partial fragmentation analysis and peak assignment will be displayed. Even with simple compounds like these, it should be noted that it is rarely possible to explain the origin of all the fragment ions in a spectrum. Also, the structure of most fragment ions is seldom known with certainty. Since a molecule of carbon dioxide is composed of only three atoms, its mass spectrum is very simple. The molecular ion is also the base peak, and the only fragment ions are CO (m/z=28) and O (m/z=16). The molecular ion of propane also has m/z=44, but it is not the most abundant ion in the spectrum.

Cleavage of a carbon-carbon bond gives methyl and ethyl fragments, one of which is a carbocation and the other a radical. Both distributions are observed, but the larger ethyl cation (m/z=29) is the most abundant, possibly because its size affords greater charge dispersal. A similar bond cleavage in cyclopropane does not give two fragments, so the molecular ion is stronger than in propane, and is in fact responsible for the the base peak. Loss of a hydrogen atom, either before or after ring opening, produces the stable allyl cation (m/z=41). The third strongest ion in the spectrum has m/z=39 (C_3H_3). Its structure is uncertain, but two possibilities are shown in the diagram. The small m/z=39 ion in propane and the absence of a m/z=29 ion in cyclopropane are particularly significant in distinguishing these hydrocarbons. Most stable organic compounds have an even number of total electrons, reflecting the fact that electrons occupy atomic and molecular orbitals in pairs. When a single electron is removed from a molecule to give an ion, the total electron count becomes an odd number, and we refer to such ions as radical cations. The molecular ion in a mass spectrum is always a radical cation, but the fragment ions may either be even-electron cations or odd-electron radical cations, depending on the neutral fragment lost. The simplest and most common fragmentations are bond cleavages producing a neutral radical (odd number of electrons) and a cation having an even number of electrons. A less common fragmentation, in which an even-electron neutral fragment is lost, produces an odd-electron radical cation fragment ion. Fragment ions themselves may fragment further. As a rule, odd-electron ions may fragment either to odd or even-electron ions, but even-electron ions fragment only to other even-electron ions. The masses of molecular and fragment ions also reflect the electron count, depending on the number of nitrogen atoms in the species.

Ions with no nitrogenor an even # N atoms	odd-electron ionseven -number mass	even-electron ionsodd -number mass
Ions having an odd # N atoms	odd-electron ions odd-number mass	even-electron ions even-number mass

This distinction is illustrated nicely by the following two examples. The unsaturated ketone, 4-methyl-3-pentene-2-one, on the left has no nitrogen so the mass of the molecular ion (m/z = 98) is an even number. Most of the fragment ions have odd-numbered masses, and therefore are even-electron cations.

Diethylmethylamine, on the other hand, has one nitrogen and its molecular mass (m/z = 87) is an odd number. A majority of the fragment ions have even-numbered masses (ions at m/z = 30, 42, 56 & 58 are not labeled), and are even-electron nitrogen cations.

The weak even-electron ions at m/z=15 and 29 are due to methyl and ethyl cations (no nitrogen atoms). The fragmentations leading to the chief fragment ions will be displayed by clicking on the appropriate spectrum. Repeated clicks will cycle the display.

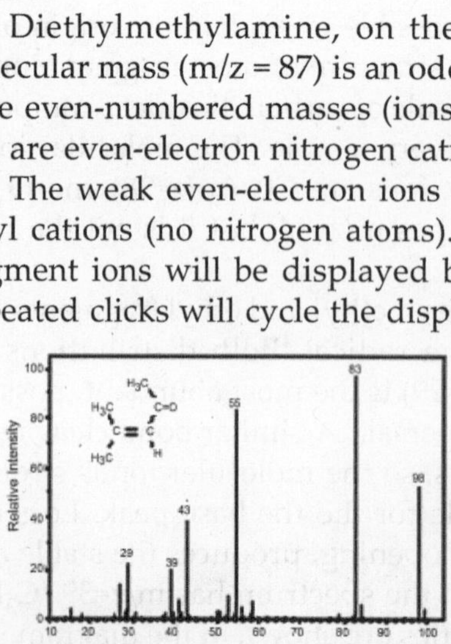

4-methyl-3-pentene-2-one

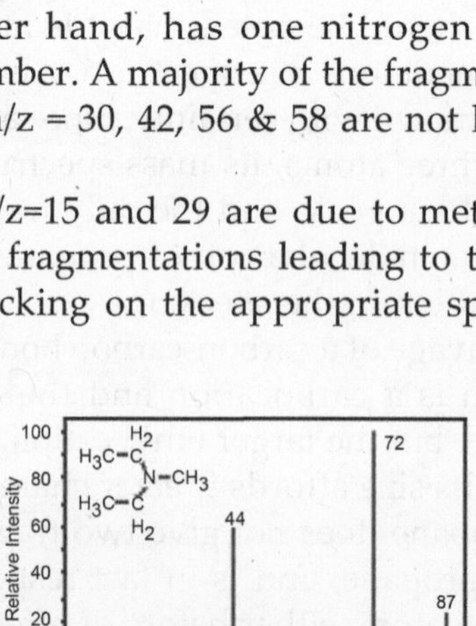

N,N-diethylmethylamine

When non-bonded electron pairs are present in a molecule (*e.g.* on N or O), fragmentation pathways may sometimes be explained by assuming the missing electron is partially localized on that atom. A few such mechanisms are shown above. Bond cleavage generates a radical and a cation, and both fragments often share these roles, albeit unequally.

Isotopes

While a mass spectrometer separates and detects ions of slightly different masses, it with no trouble distinguishes different isotopes of a given element. This is manifested most dramatically for compounds containing bromine and chlorine, as illustrated by the following examples. Since molecules of bromine have only two atoms, the spectrum on the left will come as a surprise if a single atomic mass of 80 Da is assumed for Br. The five peaks in this spectrum demonstrate clearly that natural bromine consists of a nearly 50:50 mixture of isotopes having atomic masses of 79 and 81 Da respectively. Thus, the bromine molecule may be composed of two ^{79}Br atoms (mass 158 Da), two ^{81}Br atoms (mass 162 Da) or the more probable combination of ^{79}Br-^{81}Br (mass 160 Da). Fragmentation of Br_2 to a bromine cation then gives rise to equal sized ion peaks at 79 and 81 Da.

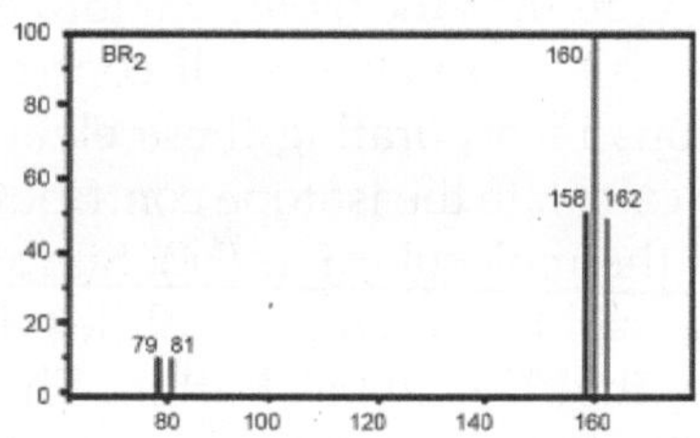

Bromine

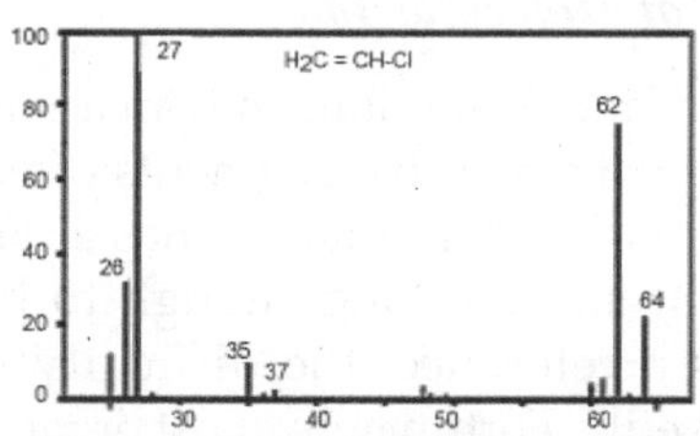

Vinyl chloride

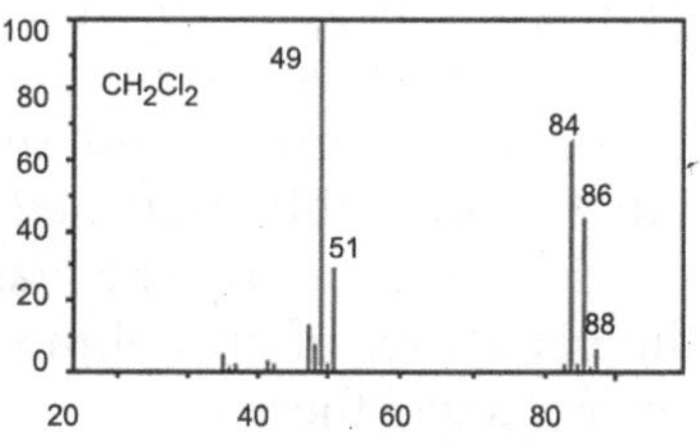

Methylene chloride

The centre and right hand spectra show that chlorine is also composed of two isotopes, the more abundant having a mass of 35 Da, and the minor isotope a mass 37 Da. The precise isotopic composition of chlorine and bromine is:

Chlorine: 75.77 per cent ^{35}Cl and 24.23 per cent ^{37}Cl

Bromine: 50.50 per cent ^{79}Br and 49.50 per cent ^{81}Br

The presence of chlorine or bromine in a molecule or ion is easily detected by noticing the intensity ratios of ions differing by 2 Da. In the case of methylene chloride, the molecular ion consists of three peaks at m/z=84, 86 & 88 Da, and their diminishing intensities may be calculated from the natural abundances given above. Loss of a chlorine atom gives two isotopic fragment ions at m/z=49 & 51 Da, clearly incorporating a single chlorine atom. Fluorine and iodine, by contrast, are monoisotopic, having masses of 19 Da and 127 Da respectively. It should be noted that the presence of halogen atoms in a molecule or fragment ion does not change the odd-even mass rules given above.

Two other common elements having useful isotope signatures are carbon, ^{13}C is 1.1 per cent natural abundance, and sulfur, ^{33}S and ^{34}S are 0.76 per cent and 4.22 per cent natural abundance respectively. For example, the small m/z=99 Da peak in the spectrum of 4-methyl-3-pentene-2-one (above) is due to the presence of a single ^{13}C atom in the molecular ion. Although less significant in this respect, ^{15}N and ^{18}O also make small contributions to higher mass satellites of molecular ions incorporating these elements. The calculator on the right may be used to calculate the isotope contributions to ion abundances 1 and 2 Da greater than the molecular ion (M). Simply enter an appropriate subscript number to the right of each symbol, leaving those elements not present blank, and press the "Calculate" button. The numbers displayed in the M+1 and M+2 boxes are relative to M being set at 100 per cent.

Fragmentation Patterns of Molecular Ions

The fragmentation of molecular ions into an assortment of fragment ions is a mixed blessing. The nature of the fragments often provides a clue to the molecular structure, but if the molecular ion has a lifetime of less than a few microseconds it will not survive long enough to be observed. Without a molecular ion peak as a reference, the difficulty of interpreting a mass spectrum increases markedly. Fortunately, most organic compounds give mass spectra that include a molecular ion, and those that do not often respond successfully to the use of milder ionization conditions.

Among simple organic compounds, the most stable molecular ions are those from aromatic rings, other conjugated pi-electron systems and cycloalkanes. Alcohols, ethers and highly branched alkanes generally show the greatest tendency toward fragmentation. The mass spectrum of dodecane on the right illustrates the behaviour of an unbranched alkane. Since there are no heteroatoms in this molecule, there are no non-bonding valence shell electrons. Consequently, the radical cation character of the molecular ion (m/z = 170) is delocalized over all the covalent bonds. Fragmentation of C-C bonds occurs because they are usually weaker than C-H bonds, and this produces a mixture of alkyl radicals and alkyl carbocations. The positive charge commonly resides on the smaller fragment, so we see a homologous series of hexyl (m/z = 85), pentyl (m/z = 71), butyl (m/z = 57), propyl (m/z = 43), ethyl (m/z = 29) and methyl (m/z = 15) cations. These are accompanied by a set of corresponding alkenyl carbocations (*e.g.* m/z = 55, 41 &27) formed by loss of 2 H. All of the significant fragment ions in this spectrum are even-electron ions. In most alkane spectra the propyl and butyl ions are the most abundant.

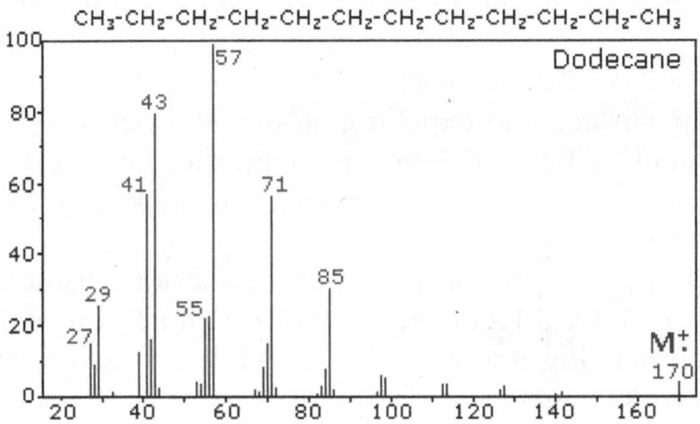

The attendance of a functional group, particularly one having a heteroatom Y with non-bonding valence electrons (Y = N, O, S, X etc.), can dramatically alter the fragmentation pattern of a compound. This influence is thought to occur because of a "localization" of the radical cation component of the molecular ion on the heteroatom. After all, it is easier to remove (ionize) a non-bonding electron than one that is part of a covalent bond. By localizing the reactive moiety, certain fragmentation processes will be favoured. These are summarized in the following diagram, where the green shaded box at the top displays examples of such "localized" molecular ions. The first two fragmentation paths lead to even-electron ions, and the elimination gives an odd-electron ion. Note the use of different curved arrows to show single electron shifts compared with electron pair shifts.

The charge distributions shown above are common, but for each cleavage process the charge may sometimes be carried by the other (neutral) species, and both fragment ions are observed. Of the three cleavage reactions described here, the alpha-cleavage is generally favoured for nitrogen, oxygen and sulfur compounds. Indeed, in the previously displayed spectra of 4-methyl-3-pentene-2-one and N,N-diethylmethylamine the major fragment ions come from alpha-cleavages

Secondary Ion Mass Spectroscopy and Ion Scattering Spectroscopy

For both SIMS and ISS, a primary ion beam with kinetic energy of 0.3–10 keV, usually composed of ions of an inert gas, is directed onto a surface. When an ion strikes the surface, two events can occur. In one scenario the primary ion can be elastically scattered by a surface atom, resulting in a reflected primary ion. It is this ion that is measured in ISS. This is an elastic scattering process, and the kinetic energy of the reflected primary ion will depend on the mass of the surface atom involved in the scattering process, thus providing information about the surface.

In the other possible event, the primary ion can penetrate the surface and become embedded in the solid. When the primary ion penetrates the molecular or atomic lattice, considerable disruption occurs by transfer of momentum to lattice atoms or molecules. This gives rise to an effect called "sputtering." For ~10 keV primary ions, disruption occurs only a few atomic or molecular layers deep. Stopping such energetic ions in such a short distance can be visualized as disrupting the surface and expelling atoms, atom clusters, molecules, or molecular fragments, all of which may be neutral or ionic. The ions, referred to as secondary ions (thus the term *secondary ion mass spectrometry*), are focused by an ion lens into a mass spectrometer.

Secondary ion yields produced in sputtering both organic and inorganic materials are very low, 0.1 percent or less for most materials, with most expelled fragments being neutral particles. However, even for a very low primary ion current density—10^{-9} amperes per square cm (A/cm^2)—the number of secondary ions produced is sufficiently large for mass spectra with reasonable sensitivity to be obtained. The current density of the incident ion beam in ion spectroscopy may be either low (below 10^{-9} A/cm^2; termed the "static" mode) or high (~10^{-6} A/cm^2; termed the "dynamic" mode). In the static mode less than 1 percent of the surface is "hit" by a primary ion in the time required to obtain spectra, and each primary particle therefore encounters an essentially unsputtered surface. In the dynamic mode rapid erosion of the surface occurs in seconds, and the beam bores into the surface. This mode is used for depth profiling.

Secondary Ion Mass Spectroscopy

In SIMS, qualitative analysis for both organic and inorganic materials is obtained through straightforward interpretation of the resulting mass spectra. Quantitative analysis, or determination of absolute ratios of chemical species, however, is difficult because relative ion yields can vary widely with the nature of the surface. The sensitivity variations from sample matrix effects make it impossible to perform absolute quantitative analyses with SIMS. Also, the intensity changes that can occur because of minor changes in the matrix render the relative uncertainties in the measurements, even for calibrated systems, poorer than for the other techniques. SIMS instruments typically use either a quadrupole or a time-of-flight mass spectrometer. Quadrupole instruments typically operate in the mass range of 1–1,000 atomic mass units (amu; 1 amu is 1.66×10^{-24} gram), whereas time-of-flight instruments extend the available mass range to beyond 10,000 amu. Signal intensities in the mass range below 500 amu are typically more intense than for the higher mass range. However, sufficient numbers of large molecular fragments are produced that can be observed with SIMS, which can provide valuable information.

Of the surface analysis techniques, SIMS is the only technique sensitive to all elements. SIMS shows good specificity, although there is some overlap between peaks of different elements. For example, the Si_2^+ peak overlaps with Fe^+, rendering it difficult to detect small amounts of iron in the presence of silicon by using a low-resolution mass spectrometer that can differentiate only unit masses (1 amu). However, a high-resolution instrument readily distinguishes between them ($Si_2^+ = 55.9538$ and $Fe^+ = 55.9349$) because it can typically differentiate 0.001 amu or smaller. Sensitivity variations in SIMS are extremely high. Severe effects from the matrix of the sample give rise to variations in sensitivity of 10^5 between elements for which SIMS is most and least sensitive. As a rule of thumb, the more electropositive elements show higher sensitivity in positive-ion SIMS, and the more electronegative elements show higher sensitivity in negative-ion SIMS.

An significant feature of SIMS is its ability to detect very small amounts of materials on the surface. In the more sensitive cases it is quite possible to achieve a discernible SIMS signal from 10^{-4} percent (1 part per million) of a monolayer. Therefore, although SIMS is the least quantitative of the surface analytical techniques, it can be the most sensitive.

SIMS is equally valuable for both organic and inorganic analysis. Typical organic applications include analysing surface coatings on fabrics, detecting impurities in semiconductors, and studying synthetic polymers. Inorganic applications include characterization of heterogeneous catalysts and determination of the surface composition of alloys. Static SIMS is an ideal method for mapping a surface. Because static SIMS is a fairly nondestructive technique, it is possible to map atoms and molecules of both organic and inorganic species. An ion beam can be focused on a spot with a diameter of less than 0.1 micrometre (μm; 1 μm = 10^{-6} metre), which is the smallest spot size that can contain sufficient material to produce an image. SIMS has been particularly useful in the microelectronics industry for imaging semiconductor devices.

FUNCTION OF SPECTROSCOPIC TECHNIQUES

Spectroscopic techniques function through a "beam in, beam out" device. A beam of photons, electrons, or ions impinges on a material and penetrates to a depth that is dependent on the beam characteristics. A second beam, resulting from the interaction of the first beam with the solid, exits from the surface and is analysed by a spectrometer. The exiting beam carries with it information regarding the composition of the material with which the beam interacted. By varying both the type of particle and the energy of the entering beam, one can generate a large number of surface analytical techniques.

For a surface analytical technique, the information obtained by the spectrometer from the exiting beam should be characteristic of that region of the solid that is defined as the surface. Either the penetration depth of the incident beam, the escape depth of the exiting beam, or both therefore must be limited to the thickness of the surface. This "sampling depth" differs with the type of particle or the energy of the incident beam. For photons, electrons, and ions with an energy of 1,000 electron volts (1 keV), the sampling depths in the energy range used for surface analysis are 1,000, 2, and 1 nm (nanometre; 1 nm is 10^{-9} metre), respectively.

The thickness of a surface is not the same in all materials. The depth into the solid that is of interest therefore depends on the solid being measured. It also depends on the specific application for which the analysis is being carried out. For catalysts, for example, the surface thickness of interest is less than 1 nm, or five atomic layers, which is the thickness of the deposited reactive species. For corrosion studies, on the other hand, the depths of interest are in the 100-nm, or 500-atomic-layer, range. Every technique has a specific sampling profile, and not all techniques are appropriate for all applications. However, different techniques can provide complementary information. Usually no single technique provides sufficient information to solve a complex real-world problem.

Several of these techniques are familiar to the analytical chemist, and each has its own unique capabilities. During the 1970s and '80s, however, four techniques emerged as being most useful for real-world surface analysis because of their general applicability and ease of use. The use of photons in and electrons out provides X-ray photoelectron spectroscopy (XPS, or electron spectroscopy for chemical analysis [ESCA]). Electrons in and out gives Auger electron spectroscopy (AES). The use of ions in and out yields two techniques: secondary ion mass spectroscopy (SIMS) and ion scattering spectroscopy (ISS); the latter includes both low-energy (keV) ISS and high-energy (MeV; 1 million electron volts) ISS, normally called Rutherford back-scattering spectroscopy (RBS).

X-ray Photoelectron Spectroscopy and Auger Electron Spectroscopy

For XPS and AES the primary process is an ionization caused by either a photon or an electron, $m + h\nu \rightarrow m^{+*} + e^-$, or $m + e^- \rightarrow m^{+*} + 2e^-$, where m is an atom in the material. In photoionization an incident photon causes the ejection of an electron with a discrete kinetic energy, which is measured in XPS, leaving an excited ion (m^{+*}). The kinetic energy of the ejected electron is essentially the difference between the energy of the incident photon and the binding energy of the electron. Electron ionization of an atom also produces both an excited ion and a second electron. Unlike in XPS, because of electron-electron interactions, these primary electrons emitted from the atom do not have discrete energies. Excited ions, shaped by either photon or electron ionization, can relax through two mechanisms, or secondary processes. One path involves emission of photons in the form of X-rays, the same process observed in X-ray fluorescence. These photons are usually of such an energy that they have large escape depths (that is, the photons come from deep within the sample) and therefore do not usually contain information relevant to surfaces.

The excited ion can also relax through emission of a secondary electron, called an Auger electron, to form a doubly ionized atom. It is significant to note that XPS can produce electrons with discrete energies through both photoionization and the Auger process, while electron ionization produces electrons with discrete energies only through the Auger process. The kinetic energy of an Auger electron does not depend on the energy of the ionizing photon or electron, whereas that of the XPS electron does.

History of Spectroscopic Techniques

Spectroscopy is a technique that uses the communication of energy with a sample to perform an analysis. A spectrum can be used to obtain information about atomic and molecular energy levels, molecular geometries, chemical bonds, interactions of molecules, and related processes. Often, spectra are used to identify the components of a sample (qualitative analysis). Spectra may also be used to measure the amount of material in a sample (quantitative analysis). There are several instruments that are used to perform a spectroscopic analysis. In simplest terms, spectroscopy requires an energy source (commonly a laser, but this could be an ion source or radiation source) and a device for measuring the change in the energy source after it has interacted with the sample (often a spectrophotometer or interferometer).

There are as many different types of spectroscopy as there are energy sources! Here are some examples: *Atomic Absorption Spectroscopy*: Energy absorbed by the sample is used to assess its characteristics. Sometimes absorbed energy causes light to be released from the sample, which may be measured by a technique such as fluorescence spectroscopy.

- *Astronomical Spectroscopy*: Energy from celestial objects is used to analyse their chemical composition, density, pressure, temperature, magnetic fields, velocity, and other characteristics. There are many energy types (spectroscopies) that may be used in astronomical spectroscopy.
- *Electron Paramagnetic Spectroscopy*: This is a microwave technique based on splitting electronic energy fields in a magnetic field. It is used to determine structures of samples containing unpaired electrons. *Attenuated Total Reflectance Spectroscopy*: This is the study of substances in thin films or on surfaces. The sample is penetrated by an energy beam one or more times and the reflected energy is

analysed. Attenuated total reflectance spectroscopy and the related technique called frustrated multiple internal reflection spectroscopy are used to analyse coatings and opaque liquids.

- *Fourier Transform Spectrosopy*: This is a family of spectroscopic techniques in which the sample is irradiated by all relevant wavelengths simultaneously for a short period of time. The absorption spectrum is obtained by applying a mathematical analysis to the resulting energy pattern.
- *Electron Spectroscopy*: There are several types of electron spectroscopy, all associated with measuring changes in electronic energy levels.
- *Infrared Spectroscopy*: The infrared absorption spectrum of a substance is sometimes called its molecular fingerprint. Although frequently used to identify materials, infrared spectroscopy also may be used to quantify the number of absorbing molecules.
- *Gamma-ray Spectroscopy*: Gamma radiation is the energy source in this type of spectroscopy, which includes activation analysis and Mossbauer spectroscopy.
- *Mass Spectrometry*: A mass spectrometer source produces ions. Information about a sample may be obtained by analysing the dispersion of ions when they interact with the sample, generally using the mass-to-charge ratio.
- *Laser Spectroscopy*: Absorption spectroscopy, fluorescence spectroscopy, Raman spectroscopy, and surface-enhanced Raman spectroscopy commonly use laser light as an energy source. Laser spectroscopies provide information about the interaction of coherent light with matter. Laser spectrocopy generally has high resolution and sensitivity.
- *Multiplex or Frequency-Modulated Spectroscopy*: In this type of spectroscopy, each optical wavelength that is recorded is encoded with an audio frequency containing the original wavelength information. A wavelength analyser can then reconstruct the original spectrum.
- *X-ray Spectroscopy*: This technique involves excitation of inner electrons of atoms, which may be seen as x-ray absorption. An x-ray fluorescence emission spectrum may be produced when an electron falls from a higher energy state into the vacancy created by the absorbed energy.
- *Raman Spectroscopy*: Raman scattering of light by molecules may be used to provide information on a sample's chemical composition and molecular structure.

ADVANCED SPECTROSCOPIC TECHNIQUES

Diffusion is the motion of particles due to Brownian movement. The diffusion coefficient, D, is a measure of the diffusion. The pulsed-gradient spin-echo sequence permits us to measure the diffusion coefficient. The sequence is in theory capable of measuring both the rotational and translational diffusion coefficients, but is used primarily for studying translational diffusion. To understand how the pulsed-gradient spin-echo sequence allows us to measure diffusion, consider the timing diagram for the sequence.

This sequence is very similar to the spin echo sequence, except that two gradient pulses have been applied. These two gradient pulses are identical in amplitude, G, and width, δ. The two gradient pulses are separated by a time Δ and are placed symmetrically about the 180 degree pulse. The function of the gradient pulses is to dephase magnetization from spins which have diffused to a new location in the period Δ. These pulses have no effect on stationary spins. For example, a stationary spin exposed to the first gradient pulse, applied along the Z axis, will acquire a phase in radians given by

$$\phi = 2pg \int z\,Gz\,dt.$$

The spin will acquire an equal but opposite phase from the second pulse since the pulses are on different sides of the 180 degree RF pulse. Thus, their effects cancel each other out.

Consider the following illustration of the effect of the gradient pulses on the phase of stationary and moving spins. The illustration presents the phase of a diffusing spin relative to that of a reference spin and a stationary spin. The reference spin is one which experiences no gradient pulses. The stationary spin is not diffusing during the time illustrated by the sequence. The diffusing spin moves along Z during the sequence. The blue line in the timing diagram represents the time of the 180 degree pulse in the spin echo sequence. When you put the illustration into motion, the stationary spin comes back into phase with the reference one, indicating a positive contribution to the echo. The diffusing spin does not come back into phase with the reference spin so it diminishes the echo height. The relationship between the signal (S) obtained in the presence of a gradient amplitude G_i in the *i* direction and the diffusion coefficient in the same direction is given by the following equation where S_o is the signal at zero gradient.

$$S/S_o = \exp[-(G_i\,\gamma\,\delta)^2\,D_i\,(\Delta-\delta/3)]$$

The diffusion coefficient is typically calculated from a plot of $\ln(S/S_o)$ versus $(G\,\gamma\delta)^2\,(\Delta-\delta/3)$. Diffusion in the x, y, or z direction may be measured by applying the gradient respectively in the x, y, or z direction.

Spin Relaxation Time

The spin-lattice and spin-spin relaxation times, T_1 and T_2 respectively, of the components of a solution are valuable tools for studying molecular dynamics. T_1^{-1} is proportional to the number of molecular motions at the Larmor frequency, while T_2^{-1} is proportional to the number of molecular motions at frequencies less than or equal to the Larmor frequency. When we are dealing with solutions these motions are predominantly rotational motions. There are many pulse sequences which may be used to measure T_1 and T_2. The inversion recovery, 90-FID, and spin-echo sequences may be used to measure T_1. Each technique has its own advantages and disadvantages. The spin-echo sequence may be used to measure T_2.

For accurate measurements with each pulse sequence, it is significant that the sample experience a spatially uniform B_1 magnetic field. When inhomogeneous fields are used, inaccurate rotations result and the spins do not follow the general equation for the pulse sequence. One common cause of an inhomogeneous B_1 magnetic field in a sample is a sample extending beyond the homogeneous bounds of the RF coil. It is best that the sample be confined to the volume of the RF coil. In practice this means filling an NMR tube with sample to a height no greater than the length of the RF coil, and positioning the tube with sample in the centre of the coil. This position will require reshimming the sample if you are concerned about narrow lines widths.

T_1 Measurement: The signal as a function of TI when the sequence is not repeated is

$$S = k\,\rho\,(1\text{-}2e^{TI/T1}).$$

If the curve is well defined (*i.e.* if there is a high density of data points recorded at different TI times), the T_1 value can be determined from the zero crossing of the curve which is T_1 *ln*2. Alternatively the relaxation curve as a function of TI may be fit using the equation

$$S = S_o\,(1\text{-}2e^{-TI/T1}).$$

This approach is favoured when there are fewer data points as a function of TI. T_1 may also be determined from a 90-FID or spin-echo sequence which is repeated at various repetition times (TR). For example, if the 90-FID sequence is repeated many times at TR_1 and then many times at TR_2, TR_3, etc, the plot of signal as a function of TR will be an exponential growth of the form

$$S = k\,(1\text{-}e^{TR/T1}).$$

This data may be fit to obtain T_1. The difficulty with fitting this data and the inversion recovery data is a lack of knowledge of the value of the equilibrium magnetization or signal S_o. Other techniques have been proposed which do not require knowledge of the equilibrium magnetization or signal.
T_2 Measurement: Measurement of the spin-spin relaxation time requires the use of a spin-echo pulse sequence. The echo amplitude, S, as a function of echo time, TE, is exponentially decaying.

Plotting wln(S/S_o) versus TE yields a straight line, the slope of which is-$1/T_2$. A linear least squares algorithm is often used to find the slope and hence T_2 value. This approach can result in lead to large errors in the calculated T_2 values when the data has noise. The later points in the decay curve have poorer signal-to-noise ratio than the earlier points, but are given equal weight by the linear least squares algorithm. The solution to this problem is to use a non-linear least squares procedure.

Solid State: The Magnitude of the Chemical Shift

The magnitude of the chemical shift is related to the extent to which the electron can shield the nucleus from the applied magnetic field. In a spherically symmetric molecule, the chemical shift is independent of molecular orientation. In an asymmetric molecule, the chemical shift is dependent on the orientation. The magnetic field experienced by the nucleus varies as a function of the orientation of the molecule in the magnetic field. The NMR spectrum from a random distribution of fixed orientations, such as in a solid, would look like this. The larger signal at lower field strength is due to the fact that there are more perpendicular orientations. In a nonviscous liquid, the fields at the various orientations average out due to the tumbling of the molecule. The anisotropic chemical shift is one reason why the NMR spectra of solid samples display broad spectral lines. Another reason for broad spectral lines is dipolar broadening. A dipolar interaction is one between two spin 1/2 nuclei. The magnitude of the interaction varies with angle q and distance r. As a function of θ, the magnetic field B experienced by the red nucleus is $(3\cos^2 \theta\text{-}1)$.

A group of dipoles with a random distribution of orientations, as in a solid, gives this spectrum. The higher signal at mid-field strength is due to the larger presence of orientations perpendicular to the direction of the B_o field. This signal is made up of components from the red and blue nuclei in the dipole. In a nonviscous liquid, the interaction averages out due to the presence of rapid tumbling of the molecule. When the angle θ in the above equation is 54.7°, 125.3°, 234.7°, or 305.3°, the dipole interaction vanishes. The angle 54.7° is called the magic angle, θ_m. If all the molecules could be positioned at θ_m, the spectrum would narrow to the fast tumbling limit. Since this is not possible, the next best thing is to cause the average orientation of the molecules to be θ_m.

Even this is not exactly possible, but the closest approximation is to rapidly spin the entire sample at an angle θ_m relative to B_o. In solid state NMR, samples are placed in a special sample tube and the tube is placed inside a rotor. The rotor, and hence the sample, are oriented at an angle θ_m with respect to the B_o magnetic field. The sample is then spun at a rate of thousands of revolutions per second. The spinning rate must be comparable to the solid state line width. The centrifigal force created by spinning the sample tube at a rate of several thousands of revolutions per second is enough to destroy a typical glass NMR sample tube. Specially engineered sample tubes and rotors are needed.

3

Models of Atomic Spectra

The first person to realise that white light was made up of the colours of the rainbow was Isaac Newton, who in 1666 passed sunlight through a narrow slit, then a prism, to project the coloured spectrum on to a wall. This effect had been noticed previously, of course, not least in the sky, but previous attempts to explain it, by Descartes and others, had suggested that the white light became coloured when it was refracted, the colour depending on the angle of refraction.

Newton clarified the situation by using a second prism to reconstitute the white light, making much more plausible the idea that the white light was composed of the separate colours. He then took a monochromatic component from the spectrum generated by one prism and passed it through a second prism, establishing that no further colours were generated. That is, light of a single colour did not change colour on refraction.

He concluded that white light was made up of all the colours of the rainbow, and that on passing through a prism, these different colours were refracted through slightly different angles, thus separating them into the observed spectrum. In 1752, the Scottish physicist Thomas Melvill discovered that putting different substances in flames, and passing the light through a prism, gave differently patterned spectra. Ordinary table salt, for example, generated a "bright yellow". Furthermore, not all the colours of the rainbow appeared - there were dark gaps in the spectrum, in fact for some materials there were just a few patches of light. By the 1820's, Herschel had recognized that spectra provided an excellent way to detect and identify small quantities of an element in a powder put into a flame.

Meanwhile, the white light of the sun was coming in for more detailed scrutiny. In 1802, William Wollaston in England had discovered (perhaps by using a thinner slit or a better prism) that in fact the solar spectrum itself had tiny gaps - there were many thin dark lines in the rainbow of colours. These were investigated much more systematically by Joseph von Fraunhofer, beginning in 1814. He increased the dispersion by using more than one prism. He found an "almost countless number" of lines. He labeled the strongest dark lines A, B, C, D, etc.

BACKGROUND OF THE BOHR ATOM

In 1911, the 26-year-old Niels Bohr earned a Ph. D. at the University of Copenhagen; his dissertation was titled "Studies on the Electron Theory of Metals". He was awarded a postdoctoral fellowship funded by the Carlsberg Brewery Foundation, which enabled him to go to Cambridge in September to study with J. J. Thomson. Bohr was a great admirer of Thomson's many achievements, both experimental and theoretical. In his thesis work, he had closely studied some of the problems covered in Thomson's book *Conduction of Electricity through Gases.*

He had uncovered some apparent errors in Thomson's work, and looked forward to discussing these points with the great man. Unfortunately, by the time Bohr arrived, the Cavendish Laboratory had grown to the point where Thomson as director had more than he could manage. He had no spare time to think about electrons, and was not happy to hear from Bohr that some of his earlier work might be incorrect.

In fact, Thomson went out of his way to avoid theoretical discussions with Bohr. He did assign Bohr an experiment on positive rays, but Bohr was not enthusiastic. Bohr kept himself busy writing a paper on electrons in metals, reading Dickens to improve his English, and playing soccer. In December, Rutherford came down from Manchester for the annual Cavendish dinner. Bohr later said that he was deeply impressed by Rutherford's charm, his force of personality, and his patience to listen to every young man who might have an ideacertainly a refreshing change after J. J.! A little later, Bohr met with Rutherford again when he visited one of his father's friends in Manchester, someone who also knew Rutherford.

Although Rutherford was usually skeptical of theorists, he liked Bohr. For one thing, Rutherford was a soccer fan, and Bohr's brother Harald (only nineteen months younger than Bohr) was famoushe had played in the silver medal winning Danish soccer team at the 1908 Olympics in London. After talking it over with Harald, who visited Cambridge in January, Bohr moved to Manchester in March, and took a six-week lab course, given by Geiger, Marsden and others.

Really, though, his interests were theoretical, and he talked a lot with Charles Galton Darwin"grandson of the real Darwin", as Bohr put it in a letter to Harald. Darwin had just completed a theoretical analysis of the loss of energy of an a -particle going through matterthat is, an a that doesn't get close enough to a nucleus to be scattered. Such a 's gradually lose energy by churning through the electrons, and the rate of loss depends on how many electrons they encounter.

In particular, Bohr concluded, after reviewing and improving on Darwin's work, it seemed clear that the hydrogen atom almost certainly had a *single* electron outside the nucleus.

WHAT DETERMINES ATOMIC SIZE?

A big problem with the nuclear hydrogen atom was: what determined its size? Classical mechanics gives a simple dynamical equation for circular orbits:

$$\frac{mv^2}{r} = \frac{1}{4\pi\varepsilon_0} \cdot \frac{e^2}{r^2}.$$

Now this equation is satisfied by *any* circular orbit centered at the nucleus, however large or small. (Note, by the way, that multiplying both sides by $r/2$ gives that the magnitude of the kinetic energy in the circular orbit is just half the magnitude of the negative potential energy. We need this below.) There is no hint here that the atom in its "natural" ground state should have any particular radius. But it does! This means we're missing *something*. But what?

Bohr (and others) thought that Planck's constant must somehow play a role in determining the size of the orbit. After all, it *did* play a role in restricting allowed orbital changes in the oscillators in black body radiationand these oscillators, although not very clearly understood, were of the same general size as atoms.

So evidently the standard picture of how an oscillating charge radiated couldn't be right at the atomic level. Bohr concluded that in an atom in its natural rest state, the electron must be in a special orbit, he called it a "stationary state" to which the usual rules of electromagnetic radiation didn't apply. In this orbit, which determined the size of the atom, the electron, mysteriously, didn't radiate.

Just how to bring Planck's constant into a discussion of the hydrogen atom was not so clear, though. For the black body oscillators, it related the frequency f of the oscillator with the allowed energy change E by $E = hf$. The obvious parallel approach for the hydrogen atom was to identify the frequency f with the circular frequency of the electron in its orbit. However, in contrast to the simple harmonic oscillator this hydrogen atom frequency *varied* with the size of the orbit. Still, it was the only frequency around, and, dimensionally, multiplying it by h gave an energy. What energy could that be identified with?

Again, the choice was limitedthe electron had a kinetic energy E, the potential energy was $-2E$ and the total energy $-E$. If a hydrogen nucleus captured a passing electron into its ground state, and emitted one quantum of electromagnetic radiation, that quantum would have energy E, the same as the electron kinetic energy in the natural stationary state (called the *ground state*).

Bohr suggested in a note to Rutherford in the summer of 1912 that requiring this energy be some constant (assumed to be of order of magnitude one) multiplied by hf would fix the size of the atom. Actually his argument was a bit more complicated, he considered the several electron atom, and took the electrons to form rings. However, the basic point is the samea condition like this constrains the atomic size, it would be fixed uniquely if we knew the constant. If we assume the constant is 1, for example, we have

$$\frac{1}{2}mv^2$$
$$= hv/2\pi\rho$$

Putting this together with the dynamic equation above determines the atomic radius *r*. It is easy to check that it predicts a radius of 4p e $_0h^2$/p $^2me^2$, which is just four times the "right answer" defined as the Bohr radius. The correct Bohr radius comes out if we choose the constant to be one-half, *E* = ½*hf*, which Bohr used later. Hence the approximate size of the atom follows from *dimensional* arguments alone once one assumes that Planck's constant plays a role! Of course, the nucleus is irrelevant in determining the atomic sizeit just provides a fixed centre of electrostatic attraction. The relevant electronic parameters are the mass *m* and the strength of attraction e^2/4p e $_0$. Together with *h*, these parameters determine a length. It should be mentioned that this assumption explained more than the size of the hydrogen atom. It was believed at the time that in the higher atoms, the electrons formed rings, thought to lie one outside the other, and various stability arguments indicated that there couldn't be more than seven electrons in a ring. The length scale above, $4\pi\varepsilon_0 h^2/2\pi^2 me^2$, would *decrease* for larger atoms, with e^2 replaced by Ze^2 essentially, for nuclear charge Z. Thus as the number of rings increased, the size of the rings would decrease, explaining the observed approximate periodicity in atomic volume with atomic number.

Also, in 1911 Richard Whiddington in Cambridge had found that to cause a substance having atomic number *A* to emit characteristic x-rays by bombarding it with electrons, it was necessary to use electrons of speed approximately $A \times 10^6$ meters per second. Any substance on being bombarded with sufficiently fast electrons emits a continuum of x-ray frequencies up to a maximum frequency *f* given by *hf* = kinetic energy of electron, *plus* some sharply defined linesx-rays at a particular frequency, which does not change as the electron speed is further increased.

The frequency corresponding to these lines was found to increase with atomic number. Applying his length scale argument to the innermost ring of an atom, Bohr found that an electron in that ring would have a speed proportional to the nuclear charge, and hence, at least approximately, to the atomic number. Furthermore, the predicted speed in orbit was of the same order as that of Whiddington's electrons.

NICOLSON: A CLEVER IDEA ABOUT A WRONG MODEL

Meanwhile, in Cambridge one J. W. Nicolson was struggling to incorporate Planck's ideas in a model of the atom, in an attempt to understand some strange sets of spectral lines observed in nebulae and in the sun's corona. He conceived a rather exotic (and quite wrong!) model, in which a ring of electrons, like a necklace, orbited the nucleus. (Actually, many people, including Bohr himself, investigated models like this.

The reason was that the classical radiation from a ring of electrons is a lot less that that from a single orbiting electron, the fields tend to cancel each other.) Oscillations of electrons in this ring gave the spectra. Nicolson predicted the frequencies emitted by a straightforward classical analysis of these oscillation frequencies, in the spirit of earlier work on the plum pudding model.

He did bring in Planck's constant, though. He knew that dimensionally it was a unit of angular momentum, and he suggested that the atom could only lose angular momentum in discrete amountspresumably constant multiples of h. Nicolson felt that, given the dimensionality of Planck's constant, quantization of angular momentum was more plausible than quantization of energy. Of course, for the simple harmonic oscillator they amounted to the same thing, but not for any other system.

BOHR RETURNS TO DENMARK

Bohr left Manchester in July 1912 and was married on the first of August. In the fall, he began work at the University of Copenhagen. At the same time, he began setting down on paper some of his Manchester ideas about atoms. He read Nicolson's work. As he wrote to Rutherford at the end of January 1913, he and Nicolson were really looking at different thingsNicolson was considering atoms in a very hot environment (like the sun's corona, or an electrical discharge tube) and the spectra gave information about how energy was emitted as the atom settled into its ground state.

Bohr himself was only interested in the state in which the system possessed the smallest amount of energy. He went on: " The question of calculation of the frequencies corresponding to the visible part of the spectrum". At that time, Bohr thought of spectra as pretty but peripheral, having as little to do with basic physics as the colours of a butterfly had to do with basic biology.

BOHR CHANGES HIS MIND ABOUT SPECTRA

In February 1913, Bohr was surprised to find out in a casual conversation with the spectroscopist H. R. Hansen that some patterns had been discerned in the apparent chaos of spectral lines. In particular, Hansen (a colleague and former classmate of Bohr) showed him Balmer's formula for hydrogen. They had very likely seen this in class together, but, given Bohr's opinion of the value of spectra, he probably hadn't paid much attention.

Balmer's formula is:

$$\frac{1}{\lambda} = R_H\left(\frac{1}{4}-\frac{1}{n^2}\right)$$

for the sequence of wavelengths of light emitted, with n = 3, 4, 5, 6 being in the visible, the lines used by Balmer in finding the formula. Hansen would doubtless have informed Bohr that the 1/4 could be replaced by $1/m^2$, with m another integer. The constant appearing on the right hand side is called the *Rydberg constant*, $R_H = 109{,}737\ \text{cm}^{-1}$. (This is the modern value-Balmer got it right to one part in 10,000, about the limit of spectral measurements at the time.)

Bohr said later: "As soon as I saw Balmer's formula, the whole thing was immediately clear to me." What he saw was that the set of allowed *frequencies* (proportional to inverse wavelengths) emitted by the hydrogen atom could all be expressed as *differences*.

This immediately suggested to him a generalization of his idea of a "stationary state" lowest energy level, in which the electron did not radiate. There must be a *whole sequence* of these stationary states, with radiation only taking place as the atom jumps from one to another of lower energy, emitting a single quantum of frequency f such that

$$hf = E_n - E_m,$$

the difference between the energies of the two states. Evidently, from the Balmer formula and its extension to general integers m, n, these allowed non-radiating orbits, the stationary states, could be labeled 1, 2, 3,..., n,... and had energies –1, –1/4, –1/9,..., $-1/n^2$,... in units of hcR_H (using l $f = c$ and the Balmer equation above). The energies are of course negative, because these are bound states, and we count energy zero from where the two particles are infinitely far apart. Bohr was very familiar with the dynamics of simple circular orbits in an inverse square field. He knew that if the energy of the orbit was $-hcR_H/n^2$, that meant the kinetic energy of the electron, $\frac{1}{2}mv^2 = hcR_H/n^2$, and the potential energy would be.

$$-(1/4\text{pe}_0)e^2/r = -2hcR_H/n^2.$$

It immediately follows that the *radius* of the nth orbit is proportional to n^2, and the *speed* in that orbit is proportional to $1/n$. It then follows that the angular momentum of the nth orbit is just proportional to n.

Evidently, then the angular momentum in the nth orbit was nKh, where h is Planck's constant and K is some multiplying factor, the same for all the orbits, still to be determined. In fact, the value of K follows from the results above. R_H, m, h, and c are all known quantities (R_H being measured experimentally by observing the lines in the Balmer series) so the above formulas immediately give the electron's speed and distance from the nucleus in the nth orbit, and hence its angular momentum. Therefore, by putting in these experimentally determined quantities, we can find K.

MODELS OF THE ATOM

The first attempt to construct a physical model of an atom was made by William Thomson (later elevated to Lord Kelvin) in 1867. The most striking property of the atom was its *permanence*. It was difficult to imagine any small solid entity that could not be broken, given the right force, temperature or chemical reaction. In contemplating what kinds of physical systems exhibited permanence, Thomson was inspired by a paper Helmholtz had written in 1858 on *vortices*.

This work had been translated into English by a Scotsman, Peter Tait, who showed Thomson some ingenious experiments with smoke rings to illustrate Helmholtz' ideas. The main point was that in an *ideal* fluid, a vortex line is always composed of the same particles, it remains *unbroken*, so it is ring-like. Vortices can also form interesting combinationsA good demonstration is provided by creating two vortex rings one right after the other going in the same direction. They can trap each other, each going through the other in succession. This is probably what Tait showed Thomson, and it gave Thomson the idea that atoms might somehow be vortices in the ether.

Of course, in a non ideal fluid like air, the vortices dissipate after a while, so Helholtz' mathematical theorem about their permanence is only approximate. But Thomson was excited because the ether *was* thought an ideal fluid, so vortices in the ether might last forever!

This was very aesthetically appealing to everybody - "Kirchhoff, a man of cold temperament, can be roused to enthusiasm when speaking of it." In fact, the investigations of vortices, trying to match their properties with those of atoms, led to a much better understanding of the hydrodynamics of vortices - the constancy of the circulation around a vortex, for example, is known as Kelvin's law.

In 1882 another Thomson, J. J., won a prize for an essay on vortex atoms, and how they might interact chemically. After that, though, interest began to wane - Kelvin himself began to doubt that his model really had much to do with atoms, and when the electron was discovered by J. J. in 1897, and was clearly a component of all atoms, different kinds of non-vortex atomic models evolved.

It is fascinating to note that the most exciting theory of fundamental particles at the present time, *string* theory, has a definite resemblance to Thomson's vortex atoms. One of the basic entities is the closed string, a little loop, which has fields flowing around it reminiscent of the swirl of ethereal fluid in Thomson's atom. And it's a very beautiful theory - Kirchhoff would have been enthusiastic!

FLOATING MAGNETS

In 1878, Alfred Mayer, at the University of Maryland, dreamed up a neat demonstration of how he imagined atoms might be arranged in molecules. He took a few equally magnetized needles and stuck them through corks so that they would float with their north poles all at the same height above the water, all repelling each other equally. He then held the south pole of a more powerful magnet some distance above the water, to attract the needles towards this central point. The idea was to see what equilibrium patterns the needles would form for different numbers of needles. He found something remarkable - the needles liked to arrange themselves in shells. Three to five magnets just formed a triangle, square and pentagon in succession. but for six magnets, one went to the centre and the others formed a pentagon. For more magnets, an outer shell began to form.

Kelvin's immediate response to Mayer's publication was that this should give some clues about the vortex atom. Apparently it didn't, but twenty-five years later it guided his thinking on a new model.

PLUM PUDDING

Kelvin, in 1903, proposed that the atom have the newly discovered electrons embedded somehow in a sphere of uniform positive charge, this sphere being the full size of the atom. (Of course, the sphere itself must be held together by unknown non-electrical forces - which is still true of the positive charge in our modern model of the atom.)

This picture was taken up by J. J. Thomson too, and was dubbed the plum pudding model, after traditional English Christmas fare, a large round pudding (rich with suet) with raisins embedded in it. In 1906, J. J. concluded from an analysis of the scattering of X-rays by gases and of absorption of beta-rays by solids, both of which he assumed were effected by electrons, that the number of electrons in an atom was approximately equal to the atomic number. This led to a picture of electron arrangements in an atom reminiscent of Mayer's magnets. Perhaps by analyzing possible modes of vibration of electrons in these configurations, the spectra could be calculated. The simplest case to consider was clearly hydrogen, now assumed (correctly) to contain just one electron.

How Does an Atom's Colour Depend on its Size?

By "colour" we mean here the spectral colours emitted when the atom is excited. In Thomson's plum pudding model, there is a clear relationship between the *size* of the pudding and the *frequency* at which the electron will oscillate, and hence presumably radiate, when excited.

The two are related because the assumption is that the total positive charge - which is uniformly spread throughout the sphere is just equal to the electron's negative charge. At rest in its lowest state, the electron just sits in the middle of this sphere of charge. When bumped somehow, it will oscillate about that point.

If the electron is at distance x from the centre, it will feel a restoring force towards the centre equal to the attraction from that part of the positive charge it is "outside" of - that is, the charge within a sphere of radius x about the centre. Therefore, the *larger* the whole atomthe pudding - the more thinly spread the positive charge is, and the *smaller* the amount of charge within the small sphere of radius x that is attracting the electron back towards the centre. So, the bigger the atom is, the slower the electron's oscillation is, and the lower frequency the radiation emitted. It is straightforward to give a quantitative estimate of the size of the atom based on the observation that when excited it emits radiation in the visible range. Let us assume that the positively charged sphere has radius r_0 (this is then the size of the atom, which we know is about 10^{-10} meters).

If the electron is displaced from the centre of the atom in the x-direction an amount x, it is attracted back by all the charge that is now closer to the centre than itself, that is, an amount of charge equal to ex^3/r_0^3. (Recall e is the total amount of charge on the sphere, and x^3/r_0^3 is the fraction of the sphere closer to the centre than x.)

This charge acts as if it were a point charge at the origin, so the inverse-square law gives a $1/x^2$ factor, and the equation of motion for the electron is therefore:

$$m\frac{d^2x}{dt^2} = -\frac{1}{4\pi\varepsilon_0}\cdot\frac{e^2x}{r_0^3}$$

Provided it stays within the sphere, the electron will execute simple harmonic motion with a frequency,

$$\omega^2 = \frac{1}{4\pi\varepsilon_0}\cdot\frac{e^2}{mr_0^3}.$$

Notice that, as we discussed above, as the size of the atom increases the frequency goes down. And we know the frequency, at least approximately it corresponds to visible light. Therefore, this model will predict a size of the atom, which we can compare with the size from other predictions, such as Brownian motion (plus the assumption that in a liquid, the atoms are fairly close packed - they take up most of the room available).

If we take visible light, say with a frequency 4.10^{15} radians per second, we find r_0 must be about 2.10^{-10} meters, a little on the large side, but encouragingly close to the right answer for a first attempt. Sad to report, though, no real progress was made beyond this in predicting spectra using Thomson's pudding.

Many attempts were made to find stable arrangements of electrons in atoms, not just hydrogen, using models like Mayer's magnets, and also having the electrons going around in circles. It was hoped that if certain numbers of magnets formed a very stable arrangement, that might model a chemically nonreactive atom, etc. - but nobody succeeded in making any real predictions along these lines, the models could not be connected with the properties of real atoms. Evidently, then, the theorists were stuck - and the experimental challenge was to find some way to look *inside* an atom, and see how the electrons were arranged.

SPECTROSCOPE

A spectrum is a range of frequencies or wavelengths. By the process of refraction, a prism can split white light into it's component wavelengths. However this method is rather crude, so a spectroscope is used to analyze the light passing through the prism more accurately. The diagram to the right shows a simple prism spectroscope (click to enlarge). The smaller the difference between distinguishable wavelengths, the higher the resolution of the spectroscope.

The observer (shown as an eye in the diagram) sees the radiation passing through the slit as a spectral line. To obtain accurate measurements of the radiation, and electronic device often takes the place of the observer, the device is then called a spectrophotometer. In more modern Spectrophotometers, a diffraction grating is used instead of a prism to disperse the light.

FOUCAULT CONNECTS MELVILL'S BRIGHT LINES AND FRAUNHOFER'S DARK LINES

In 1849, Foucault (of speed of light and pendulum fame) examined the spectrum of light from a voltaic arc between carbon poles. He saw a bright double yellow line at exactly the same wavelength as Fraunhofer's dark D line in the solar spectrum. Investigating further, Foucault passed the sun's light through the arc, then through a prism. He observed that the D lines in the spectrum were even darker than usual. After testing with other sources, he concluded that the arc, which emitted light at the D line frequency, would also absorb light from another source at that frequency.

This discovery did not surprise Sir George Stokes in Cambridge. He pointed out that any mechanical system with a natural frequency of oscillation will emit at that frequency if disturbed, but will also absorb most readily at that frequency from incoming disturbances, the phenomenon of resonance.

Question: In a total eclipse of the sun, the only sunlight reaching earth comes from the hot gases of the sun's atmosphere, light from the sun's main disc being blocked by our moon. The light from these hot gases was analysed during an eclipse in 1870. How do you think the spectrum observed related to that of full sunlight? The spectrum of *hydrogen,* which turned out to be crucial in providing the first insight into atomic structure over half a century later, was first observed by Anders Angstrom in Uppsala, Sweden, in 1853. His communication was translated into English in 1855. Angstrom, the son of a country minister, was a reserved person, not interested in the social life that centered around the court. Consequently, it was many years before his achievements were recognized, at home or abroad (most of his results were published in Swedish).

Meanwhile, in Freeport, Pennsylvania, in 1855, David Alter described the spectrum of hydrogen and other gases. In the 1840's, Alter had started the first commercial production of bromine from brines. He also found a way to extract oil from coal, but that proved uneconomic after the discovery of oil in Pennsylvania. His work was not widely recognized, either.

BUNSEN AND KIRCHHOFF

The first really systematic investigation of spectra was that of Bunsen and Kirchhoff, in Heidelberg, between 1855 and 1863. They used several techniques. For one thing, they introduced various salts intowhat else?the flame of a Bunsen burner. This was a very effective way of viewing spectra, because the Bunsen burner flame itself gave out practically no light. They also used the cooler flame of alcohol burning mixed with water to generate a vapour to study absorption spectra.

Finally, they studied the spectra of electric arcs between electrodes of different materials. Using iron electrodes gave a spectrum that coincided with dark lines in the sun's spectrum. Copper electrodes did not. They concluded that the sun's atmosphere contained iron, but not much copper, and that, they said, seemed very plausible since there is so much iron in the earth, and in meteors.

(*Cautionary note to philosophers*: In 1835, the French philosopher Auguste Comte (the founder of positivism) wrote: "...Our knowledge concerning the gaseous envelopes [of stars] is necessarily limited to their existence, size and refractive power, we shall not at all be able to determine their chemical composition or even their density. The collaboration of Kirchhoff and Bunsen was a major research effort, even by modern standards. They determined thousands of spectral lines, each to an accuracy of one part in ten thousand. They spectroscopically discovered new elements: rubidium and cesium.

Their method was used to find fifteen more new elements before the end of the century. In 1869, Joseph Lockyer studied the spectra of solar prominences (in eclipses). He found the spectra to be slightly Doppler shifted, so was able to deduce the speeds of the gases whirling around the sunspots. He also found a spectrum never seen before, and conjectured that it came from a new element he named Helium.

In fact, helium was later discovered on earth in 1895, by Ramsay. At that time, it had just become evident that there was an inert component, argon, in the earth's atmosphere. Earlier, an inert gas had been observed to emanate from uranium salts when they were heated. Ramsay assumed this would be the same gas, but decided to check. On heating uranium salts and performing a spectral analysis of the emitted gas, much to his surprise he found it to be helium.

THE BALMER SERIES

It is clear from the above that a tremendous amount of scientific progress was made using spectral lines, yet no-one had the slightest idea why atoms emitted at the frequencies they did. It was appreciated that spectra implied that atoms had structure. In 1852, Stokes had stated that probably the vibrations that produced light were vibrations among the constituent parts of molecules (a term which also included atoms at that time) and in 1875 Maxwell, in enumerating properties atoms must have, included the capability of internal motion or vibration. This worried Maxwell, though. As he said, the spectroscopic evidence forces the conclusion that the atom is quite complex, with many internal degrees of freedom. Yet apparently all these modes of vibration, or almost all of them, are not excited by heat, since if they were this extra capacity of the atom to absorb energy would be reflected in its specific heat.

Obviously, if any pattern could be discerned in the spectral lines for an atom, that might be a clue as to the internal structure of the atom. One might be able to build a model. A great deal of effort went into analyzing the spectral data from the 1860's on. The big breakthrough was made by Johann Balmer, a math and Latin teacher at a girls' school in Basel, Switzerland. Balmer had done no physics before, and made his great discovery when he was almost sixty.

He decided that the most likely atom to show simple spectral patterns was the lightest atom, hydrogen. Angstrom had measured the four visible spectral lines to have wavelengths 6562.10, 4860.74, 4340.1 and 4101.2 in Angstrom units (10^{-10} meters). Balmer concentrated on just these four numbers, and found they were given by the formula:

$$\lambda = b\left(\frac{n^2}{n^2 - 4}\right)$$

where b = 3645.6 Angstroms, and n = 3, 4, 5, 6. Balmer suggested that there would be other lines - in the infrared - corresponding to n = 7, 8, etc., and in fact some of them had already been observed, unbeknownst to Balmer. He further conjectured that the 4 could be replaced by 9, 16, 25, ... and this also turned out to be true - but these lines, further into the infrared, were not detected until the early twentieth century, along with the ultraviolet lines generated by replacing the 4 by 1.

It is instructive to write Balmer's general formula in terms of the inverse wavelength. This is called the wave number - the number of waves that fit in one unit of length.

$$\frac{1}{\lambda} = R_H = \left(\frac{1}{n^2} - \frac{1}{m^2}\right)$$

where n, m are integers, and R_H is the Rydberg constant, 109,737 cm^{-1}.

This constant is named after the Swedish physicist Rydberg who (in 1888) presented a generalization of Balmer's formula, in which the integer n was replaced by n + constant, the constant being less than unity. Rydberg suggested that all atomic spectra formed families with this pattern. (He also said he was unaware of Balmer's work.) It turns out that there *are* families of spectra following Rydberg's pattern, notably in the alkali metals, sodium, potassium, etc., but not with the precision the hydrogen atom lines fit the Balmer formula, and low values of n give lines that deviate considerably.

(Modern footnote: atoms having spectral lines following Rydberg's formula are called Rydberg atoms. These Rydberg atoms have one electron orbiting at a much greater distance from the nucleus than the others. Consequently, Rydberg atoms can only survive in a gas at very low pressure, otherwise that outermost electron gets knocked off. Prof. Tom Gallagher in our Department is a world expert on these atoms, which have proven a rich source of information on the quantum mechanics of atomic structure.)

One pattern that was noticed in the spectra of many atoms is Ritz' Combination Principle: if for a given atom there are spectral lines at two wave numbers, there is sometimes another spectral line at the precise sum of those two wave numbers.It is easy to see from Balmer's formula that this is true for some pairs of lines in the hydrogen spectrum. It also turns out to be true for atoms where the spectral lines have no other discernible pattern.

ATOMIC ORBITALS

Atomic orbitals are in a way that makes them understandable for introductory courses and its equivalents. It explores s and p orbitals in some detail, including their shapes and energies. d orbitals are described only in terms of their energy, and f orbitals only get a passing mention.

ORBITALS AND ORBITS

When the a planet moves around the sun, you can plot a definite path for it which is called an orbit. A simple view of the atom looks similar and you may have pictured the electrons as orbiting around the nucleus. The truth is different, and electrons in fact inhabit regions of space known as *orbitals*.Orbits and orbitals sound similar, but they have quite different meanings. It is essential that you understand the difference between them.

The *Heisenberg Uncertainty Principle* says - loosely - that you can't know with certainty both where an electron is and where it's going next. It is impossible to define with absolute precision, at the same time, both the position and the momentum of an electron. That makes it impossible to plot an orbit for an electron around a nucleus. Is this a big problem? No. If something is impossible, you have to accept it and find a way around it.

Hydrogen's Electron - the 1s Orbital

Suppose you had a single hydrogen atom and at a particular instant plotted the position of the one electron. Soon afterwards, you do the same thing, and find that it is in a new position. You have no idea how it got from the first place to the second.In the hydrogen case, the electron can be found anywhere within a spherical space surrounding the nucleus. The diagram shows a *cross-section* through this spherical space.95 per cent of the time (or any other percentage you choose), the electron will be found within a fairly easily defined region of space quite close to the nucleus. Such a region of space is called an *orbital.* You can think of an orbital as being the region of space in which the electron lives.What is the electron doing in the orbital? We don't know, we can't know, and so we just ignore the problem! All you can say is that if an electron is in a particular orbital it will have a particular definable energy.Each orbital has a name.The orbital occupied by the hydrogen electron is called a *1s orbital.* The *"1"* represents the fact that the orbital is in the energy level closest to the nucleus. The *"s"* tells you about the shape of the orbital. s orbitals are spherically symmetric around the nucleus - in each case, like a hollow ball made of rather chunky material with the nucleus at its centre.

The orbital on the left is a *2s orbital.* This is similar to a 1s orbital except that the region where there is the greatest chance of finding the electron is further from the nucleus - this is an orbital at the second energy level.If you look carefully, you will notice that there is another region of slightly higher electron density (where the dots are thicker) nearer the nucleus. ("Electron density" is another way of talking about how likely you are to find an electron at a particular place.)2s (and 3s, 4s, etc) electrons spend some of their time closer to the nucleus than you might expect. The effect of this is to slightly reduce the energy of electrons in s orbitals. The nearer the nucleus the electrons get, the lower their energy.3s, 4s (etc) orbitals get progressively further from the nucleus.

p Orbitals

Not all electrons inhabit s orbitals (in fact, very few electrons live in s orbitals). At the first energy level, the only orbital available to electrons is the 1s orbital, but at the second level, as well as a 2s orbital, there are also orbitals called *2p orbitals.*A p orbital is rather like 2 identical balloons tied together at the nucleus. The diagram on the right is a cross-section through that 3-dimensional region of space. Once again, the orbital shows where there is a 95 per cent chance of finding a particular electron.Unlike an s orbital, a p orbital points in a particular direction. At any one energy level it is possible to have three absolutely equivalent p orbitals pointing mutually at right angles to each other. These are arbitrarily given the symbols p_x, p_y and p_z. This is simply for convenience - what you might think of as the x, y or z direction changes constantly as the atom tumbles in space.The p orbitals at the second energy level are called $2p_x$, $2p_y$ and $2p_z$. There are similar orbitals at subsequent levels - $3p_x$, $3p_y$, $3p_z$, $4p_x$, $4p_y$, $4p_z$ and so on.All levels except for the first level have p orbitals. At the higher levels the lobes get more elongated, with the most likely place to find the electron more distant from the nucleus.

d and f Orbitals

In addition to s and p orbitals, there are two other sets of orbitals which become available for electrons to inhabit at higher energy levels. At the third level, there is a set of five d orbitals (with complicated shapes and names) as well as the 3s and 3p orbitals ($3p_x$, $3p_y$, $3p_z$). At the third level there are a total of nine orbitals altogether.

At the fourth level, as well the 4s and 4p and 4d orbitals there are an additional seven f orbitals - 16 orbitals in all. s, p, d and f orbitals are then available at all higher energy levels as well.For the moment, you need to be aware that there are sets of five d orbitals at levels from the third level upwards, but you probably won't be expected to draw them or name them. Apart from a passing reference, you won't come across f orbitals at all.

Fitting Electrons into Orbitals

You can think of an atom as a very bizarre house (like an inverted pyramid!) - with the nucleus living on the ground floor, and then various rooms (orbitals) on the higher floors occupied by the electrons. On the first floor there is only 1 room (the 1s orbital); on the second floor there are 4 rooms (the 2s, $2p_x$, $2p_y$ and $2p_z$ orbitals); on the third floor there are 9 rooms (one 3s orbital, three 3p orbitals and five 3d orbitals); and so on. But the rooms aren't very big... Each orbital can only hold 2 electrons. A convenient way of showing the orbitals that the electrons live in is to draw "electrons-in-boxes".

"Electrons-in-Boxes"

Orbitals can be represented as boxes with the electrons in them shown as arrows. Often an up-arrow and a down-arrow are used to show that the electrons are in some way different.

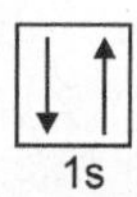

A 1s orbital holding 2 electrons would be drawn as shown on the right, but it can be written even more quickly as $1s^2$. This is read as "one s two" - not as "one s squared".You mustn't confuse the two numbers in this notation:

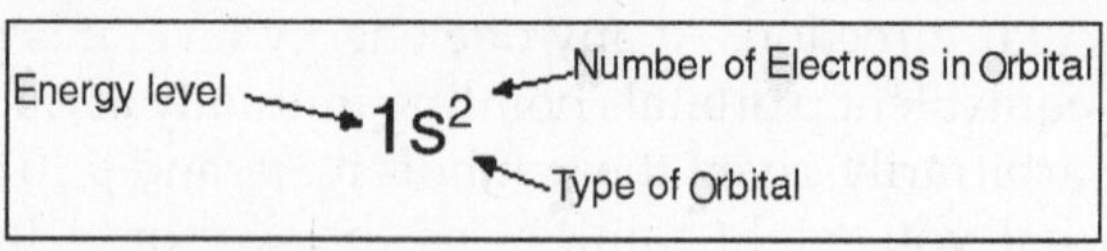

The Order of Filling Orbitals

Electrons fill low energy orbitals (closer to the nucleus) before they fill higher energy ones. Where there is a choice between orbitals of equal energy, they fill the orbitals singly as far as possible.This filling of orbitals singly where possible is known as *Hund's rule*.

It only applies where the orbitals have exactly the same energies (as with p orbitals, for example), and helps to minimise the repulsions between electrons and so makes the atom more stable.

The diagram (not to scale) summarises the energies of the orbitals up to the 4p level.

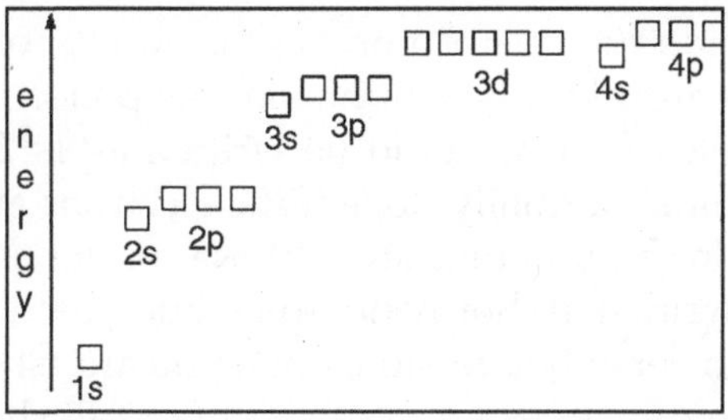

Notice that the s orbital always has a slightly lower energy than the p orbitals at the same energy level, so the s orbital always fills with electrons before the corresponding p orbitals.

The real oddity is the position of the 3d orbitals. They are at a slightly higher level than the 4s - and so it is the 4s orbital which will fill first, followed by all the 3d orbitals and then the 4p orbitals. Similar confusion occurs at higher levels, with so much overlap between the energy levels that the 4f orbitals don't fill until after the 6s, for example.

ELECTRONIC STRUCTURES

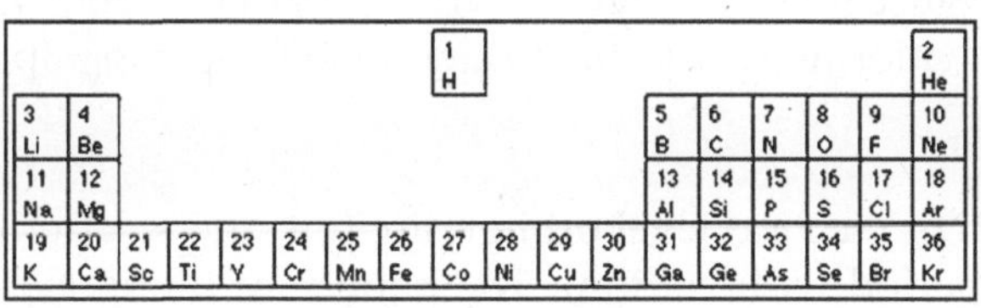

The First Period

Hydrogen has its only electron in the 1s orbital - $1s^1$, and at helium the first level is completely full - $1s^2$.

The Second Period

Now we need to start filling the second level, and hence start the second period. Lithium's electron goes into the 2s orbital because that has a lower energy than the 2p orbitals. Lithium has an electronic structure of $1s^2 2s^1$. Beryllium adds a second electron to this same level - $1s^2 2s^2$.

Now the 2p levels start to fill. These levels all have the same energy, and so the electrons go in singly at first.

B $1s^2 2s^2 2p_x^{\ 1}$

C $1s^2 2s^2 2p_x^{\ 1} 2p_y^{\ 1}$

N $1s^2 2s^2 2p_x^{\ 1} 2p_y^{\ 1} 2p_z^{\ 1}$

The next electrons to go in will have to pair up with those already there.

O $1s^2 2s^2 2p_x^{\ 2} 2p_y^{\ 1} 2p_z^{\ 1}$

F $1s^2 2s^2 2p_x^{\ 2} 2p_y^{\ 2} 2p_z^{\ 1}$

Ne $1s^2 2s^2 2p_x^{\ 2} 2p_y^{\ 2} 2p_z^{\ 2}$

The full electronic structures of atoms as the number of electrons increases. There are two ways around this, All the various p electrons can be lumped together. For example, fluorine could be written as $1s^2 2s^2 2p^5$, and neon as $1s^2 2s^2 2p^6$.This is what is normally done if the electrons are in an inner layer. If the electrons are in the bonding level (those on the outside of the atom), they are sometimes written in shorthand, sometimes in full. Don't worry about this. Be prepared to meet either version, but if you are asked for the electronic structure of something in an exam, write it out in full showing all the p_x, p_y and p_z orbitals in the outer level separately.For example, although we haven't yet met the electronic structure of chlorine, you could write it as $1s^2 2s^2 2p^6 3s^2 3p_x^{\ 2} 3p_y^{\ 2} 3p_z^{\ 1}$.

The 2p electrons are all lumped together whereas the 3p ones are shown in full. The logic is that the 3p electrons will be involved in bonding because they are on the outside of the atom, whereas the 2p electrons are buried deep in the atom and aren't really of any interest. *All* the inner electrons together using, for example, the symbol [Ne]. In this context, [Ne] means *the electronic structure of neon* - in other words: $1s^2 2s^2 2p_x^{\ 2} 2p_y^{\ 2} 2p_z^{\ 2}$ You wouldn't do this with helium because it takes longer to write [He] than it does $1s^2$.On this basis the structure of chlorine would be written $[Ne]3s^2 3p_x^{\ 2} 3p_y^{\ 2} 3p_z^{\ 1}$.

The Third Period

At neon, all the second level orbitals are full, and so after this we have to start the third period with sodium. The pattern of filling is now exactly the same as in the previous period, except that everything is now happening at the 3-level.

For example:

		Short version
Mg	$1s^2 2s^2 2p^6 3s^2$	$[Ne]3s^2$
S	$1s^2 2s^2 2p^6 3s^2 3p_x^{\ 2} 3p_y^{\ 1} 3p_z^{\ 1}$	$[Ne]3s^2 3p_x^{\ 2} 3p_y^{\ 1} 3p_z^{\ 1}$
Ar	$1s^2 2s^2 2p^6 3s^2 3p_x^{\ 2} 3p_y^{\ 2} 3p_z^{\ 2}$	$[Ne]3s^2 3p_x^{\ 2} 3p_y^{\ 2} 3p_z^{\ 2}$

The Beginning of the Fourth Period

At this point the 3-level orbitals aren't all full - the 3d levels haven't been used yet. But if you refer back to the energies of the orbitals, you will see that the next lowest energy orbital is the 4s - so that fills next.

K $1s^2 2s^2 2p^6 3s^2 3p^6 4s^1$

Ca $1s^2 2s^2 2p^6 3s^2 3p^6 4s^2$

There is strong evidence for this in the similarities in the chemistry of elements like sodium ($1s^2 2s^2 2p^6 3s^1$) and potassium ($1s^2 2s^2 2p^6 3s^2 3p^6 4s^1$) The outer electron governs their properties and that electron is in the same sort of orbital in both of the elements. That wouldn't be true if the outer electron in potassium was $3d^1$.

s- and p-block Elements

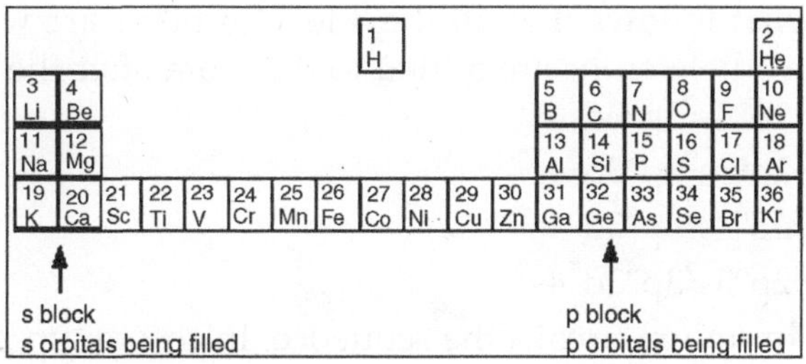

The elements in group 1 of the Periodic Table all have an outer electronic structure of ns^1 (where n is a number between 2 and 7). All group 2 elements have an outer electronic structure of ns^2. Elements in groups 1 and 2 are described as s-block elements.Elements from group 3 across to the noble gases all have their outer electrons in p orbitals. These are then described as p-block elements.

d-block Elements

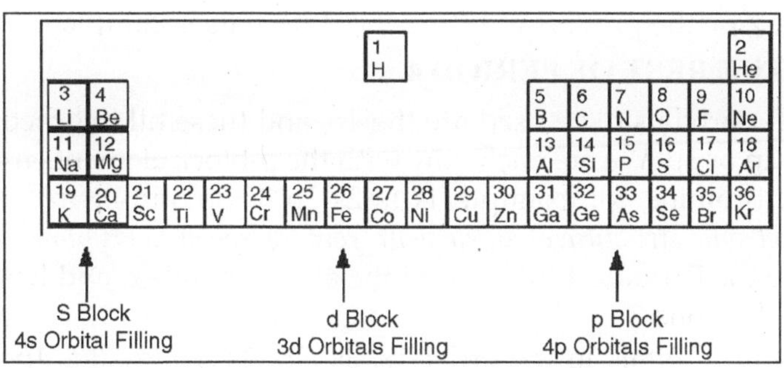

Remember that the 4s orbital has a lower energy than the 3d orbitals and so fills first. Once the 3d orbitals have filled up, the next electrons go into the 4p orbitals as you would expect.d-block elements are elements in which the last electron to be added to the atom is in a d orbital. The first series of these contains the elements from scandium to zinc, which at GCSE you probably called transition elements or transition metals. The terms "transition element" and "d-block element" don't quite have the same meaning, but it doesn't matter in the present context.d electrons are almost always described as, for example, d^5 or d^8 - and not written as separate orbitals.

Remember that there are five d orbitals, and that the electrons will inhabit them singly as far as possible. Up to 5 electrons will occupy orbitals on their own. After that they will have to pair up.

d^5 means	↑	↑	↑	↑	↑
d^8 means	↑↓	↑↓	↑↓	↑↓	↑↓

Notice in what follows that all the 3-level orbitals are written together, even though the 3d electrons are added to the atom after the 4s.

Sc $1s^22s^22p^63s^23p^63d^14s^2$

Ti $1s^22s^22p^63s^23p^63d^24s^2$

V $1s^22s^22p^63s^23p^63d^34s^2$

Cr $1s^22s^22p^63s^23p^63d^54s^1$

Whoops! Chromium breaks the sequence. In chromium, the electrons in the 3d and 4s orbitals rearrange so that there is one electron in each orbital. It would be convenient if the sequence was tidy - but it's not!

Mn $1s^22s^22p^63s^23p^63d^54s^2$ (back to being tidy again)

Fe $1s^22s^22p^63s^23p^63d^64s^2$

Co $1s^22s^22p^63s^23p^63d^74s^2$

Ni $1s^22s^22p^63s^23p^63d^84s^2$

Cu $1s^22s^22p^63s^23p^63d^{10}4s^1$ (another awkward one!)

Zn $1s^22s^22p^63s^23p^63d^{10}4s^2$

And at zinc the process of filling the d orbitals is complete.

FILLING THE REST OF PERIOD 4

The next orbitals to be used are the 4p, and these fill in exactly the same way as the 2p or 3p. We are back now with the p-block elements from gallium to krypton. Bromine, for example, is $1s^22s^22p^63s^23p^63d^{10}4s^24p_x^{\ 2}4p_y^{\ 2}4p_z^{\ 1}$.

The electronic structure of an element from hydrogen to krypton

- Use the Periodic Table to find the atomic number, and hence number of electrons.
- Fill up orbitals in the order 1s, 2s, 2p, 3s, 3p, 4s, 3d, 4p - until you run out of electrons. The 3d is the awkward one - remember that specially. Fill p and d orbitals singly as far as possible before pairing electrons up.
- Remember that chromium and copper have electronic structures which break the pattern in the first row of the d-block.

THE ELECTRONIC STRUCTURE OF BIG S- OR P-BLOCK ELEMENTS

First work out the number of outer electrons. This is quite likely all you will be asked to do anyway.The number of outer electrons is the same as the group number. (The noble gases are a bit of a problem here, because they are normally called group 0 rather then group 8. Helium has 2 outer electrons; the rest have 8.) All elements in group 3, for example, have 3 electrons in their outer level. Fit these electrons into s and p orbitals as necessary. Which level orbitals? Count the periods in the Periodic Table (not forgetting the one with H and He in it). *Iodine* is in group 7 and so has 7 outer electrons. It is in the fifth period and so its electrons will be in 5s and 5p orbitals. Iodine has the outer structure $5s^2 5p_x^{\,2} 5p_y^{\,2} 5p_z^{\,1}$. What about the inner electrons if you need to work them out as well? The 1, 2 and 3 levels will all be full, and so will the 4s, 4p and 4d.

That gives the full structure:

$$1s^2 2s^2 2p^6 3s^2 3p^6 3d^{10} 4s^2 4p^6 4d^{10} 5s^2 5p_x^{\,2} 5p_y^{\,2} 5p_z^{\,1}.$$

When you've finished, count all the electrons to make sure that they come to the same as the atomic number.

Barium is in group 2 and so has 2 outer electrons. It is in the sixth period. Barium has the outer structure $6s^2$.

Including all the inner levels:

$$1s^2 2s^2 2p^6 3s^2 3p^6 3d^{10} 4s^2 4p^6 4d^{10} 5s^2 5p^6 6s^2.$$

It would be easy to include $5d^{10}$ as well by mistake, but the d level always fills *after* the next s level - so 5d fills after 6s just as 3d fills after 4s. As long as you counted the number of electrons you could easily spot this mistake because you would have 10 too many.

Electronic Structures of Ions

Ions are atoms (or groups of atoms) which carry an electric charge because they have either gained or lost one or more electrons. If an atom gains electrons it acquires a negative charge. If it loses electrons, it becomes positively charged. The electronic structure of s- and p-block ions. Write the electronic structure for the neutral atom, and then add (for a negative ion) or subtract electrons (for a positive ion).

To write the electronic structure for Cl^-:

Cl $1s^2 2s^2 2p^6 3s^2 3p_x^{\,2} 3p_y^{\,2} 3p_z^{\,1}$ but Cl^- has one more electron

Cl^- $1s^2 2s^2 2p^6 3s^2 3p_x^{\,2} 3p_y^{\,2} 3p_z^{\,2}$

To write the electronic structure for O^{2-}:

O $1s^2 2s^2 2p_x^{\,2} 2p_y^{\,1} 2p_z^{\,1}$ but O^{2-} has two more electrons

O^{2-} $1s^2 2s^2 2p_x^{\,2} 2p_y^{\,2} 2p_z^{\,2}$

To write the electronic structure for Na^+:

Na $1s^2 2s^2 2p^6 3s^1$ but Na^+ has one less electron

Na^+ $1s^2 2s^2 2p^6$

The electronic structure for Ca^{2+}:

Ca $1s^2 2s^2 2p^6 3s^2 3p^6 4s^2$ but Ca^{2+} has two less electrons

Ca^{2+} $1s^2 2s^2 2p^6 3s^2 3p^6$

The Electronic Structure of d-block Ions

Here you are faced with one of the most irritating facts in A'level chemistry! You will recall that the first transition series (from scandium to zinc) is the result of the 3d orbitals being filled after the 4s orbital. However, once the electrons are established in their orbitals, the energy order changes - and in all the chemistry of the transition elements, the 4s orbital behaves as the outermost, highest energy orbital.

The reversed order of the 3d and 4s orbitals only applies to building the atom up in the first place. In all other respects, the 4s electrons are always the electrons you need to think about first. When d-block elements form ions, the 4s electrons are lost first. Provided you remember that, working out the structure of a d-block ion is no different from working out the structure of, say, a sodium ion.

The electronic structure for Cr^{3+}:

Cr $1s^2 2s^2 2p^6 3s^2 3p^6 3d^5 4s^1$

Cr^{3+} $1s^2 2s^2 2p^6 3s^2 3p^6 3d^3$

The 4s electron is lost first followed by two of the 3d electrons.

The *electronic structure for Zn^{2+}:*

Zn $1s^2 2s^2 2p^6 3s^2 3p^6 3d^{10} 4s^2$

Zn^{2+} $1s^2 2s^2 2p^6 3s^2 3p^6 3d^{10}$

This time there is no need to use any of the 3d electrons.

The electronic structure for Fe^{3+}:

Fe $1s^2 2s^2 2p^6 3s^2 3p^6 3d^6 4s^2$

Fe^{3+} $1s^2 2s^2 2p^6 3s^2 3p^6 3d^5$

The 4s electrons are lost first followed by one of the 3d electrons.The rule is quite simple. Take the 4s electrons off first, and then as many 3d electrons as necessary to produce the correct positive charge.

IONISATION ENERGY

The first ionisation energy is the energy required to remove the most loosely held electron from one mole of gaseous atoms to produce 1 mole of gaseous ions each with a charge of 1+.

This is more easily seen in symbol terms:

$$X_{(g)} \rightarrow X_{(g)} + e.$$

It is the energy needed to carry out this change per mole of X.

The state symbols - (g) - are essential. When you are talking about ionisation energies, everything must be present in the gas state.Ionisation energies are measured in kJ mol^{-1} (kilojoules per mole). They vary in size from 381 (which you would consider very low) up to 2370 (which is very high).All elements have a first ionisation energy - even atoms which don't form positive ions in test tubes. The reason that helium (1st *I.E.* = 2370 kJ mol^{-1}) doesn't normally form a positive ion is because of the huge amount of energy that would be needed to remove one of its electrons.

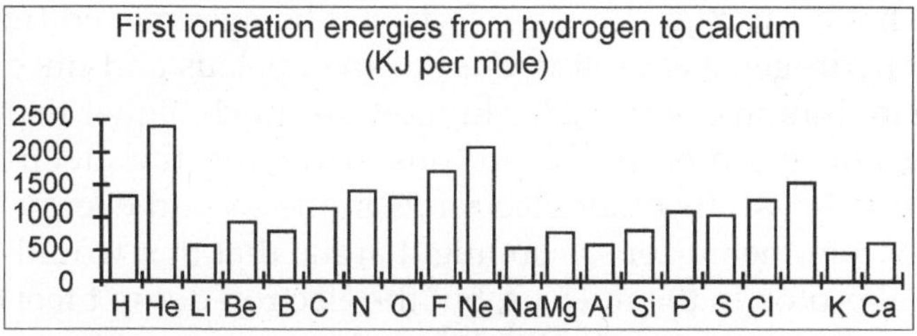

First ionisation energy shows *periodicity*. That means that it varies in a repetitive way as you move through the Periodic Table. For example, look at the pattern from Li to Ne, and then compare it with the identical pattern from Na to Ar.These variations in first ionisation energy can all be explained in terms of the structures of the atoms involved.

Factors Affecting the Size of Ionisation Energy

Ionisation energy is a measure of the energy needed to pull a particular electron away from the attraction of the nucleus. A high value of ionisation energy shows a high attraction between the electron and the nucleus.

The size of that attraction will be governed by:

- The charge on the nucleus.
- The more protons there are in the nucleus, the more positively charged the nucleus is, and the more strongly electrons are attracted to it.
- The distance of the electron from the nucleus.
- Attraction falls off very rapidly with distance. An electron close to the nucleus will be much more strongly attracted than one further away.
- The number of electrons between the outer electrons and the nucleus.
- Consider a sodium atom, with the electronic structure 2,8,1. (There's no reason why you can't use this notation if it's useful!)

If the outer electron looks in towards the nucleus, it doesn't see the nucleus sharply. Between it and the nucleus there are the two layers of electrons in the first and second levels. The 11 protons in the sodium's nucleus have their effect cut down by the 10 inner electrons. The outer electron therefore only feels a net pull of approximately 1+ from the centre. This lessening of the pull of the nucleus by inner electrons is known as *screening* or *shielding*.

Whether the electron is on its own in an orbital or paired with another electron. Two electrons in the same orbital experience a bit of repulsion from each other. This offsets the attraction of the nucleus, so that paired electrons are removed rather more easily than you might expect. Hydrogen has an electronic structure of $1s^1$. It is a very small atom, and the single electron is close to the nucleus and therefore strongly attracted. There are no electrons screening it from the nucleus and so the ionisation energy is high (1310 kJ mol^{-1}).

Helium has a structure $1s^2$. The electron is being removed from the same orbital as in hydrogen's case. It is close to the nucleus and unscreened. The value of the ionisation energy (2370 kJ mol^{-1}) is much higher than hydrogen, because the nucleus now has 2 protons attracting the electrons instead of 1. Lithium is $1s^22s^1$. Its outer electron is in the second energy level, much more distant from the nucleus. You might argue that that would be offset by the additional proton in the nucleus, but the electron doesn't feel the full pull of the nucleus - it is screened by the $1s^2$ electrons.

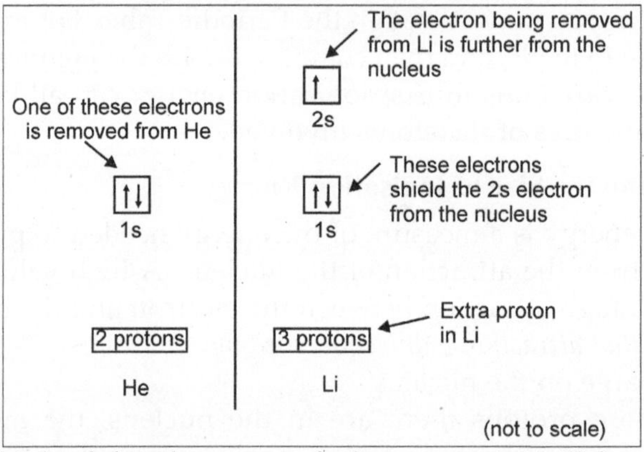

You can think of the electron as feeling a net 1+ pull from the centre (3 protons offset by the two $1s^2$ electrons). If you compare lithium with hydrogen (instead of with helium), the hydrogen's electron also feels a 1+ pull from the nucleus, but the distance is much greater with lithium. Lithium's first ionisation energy drops to 519 kJ mol^{-1} whereas hydrogen's is 1310 kJ mol^{-1}.

The patterns in periods 2 and 3. Talking through the next 17 atoms one at a time would take ages. We can do it much more neatly by explaining the main trends in these periods, and then accounting for the exceptions to these trends. The first thing to realise is that the patterns in the two periods are identical - the difference being that the ionisation energies in period 3 are all lower than those in period 2.

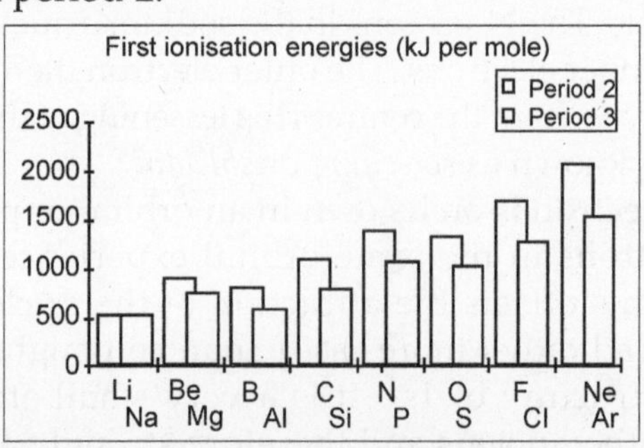

The general trend is for ionisation energies to increase across a period.In the whole of period 2, the outer electrons are in 2-level orbitals – 2s or 2p. These are all the same sort of distances from the nucleus, and are screened by the same $1s^2$ electrons.The major difference is the increasing number of protons in the nucleus as you go from lithium to neon. That causes greater attraction between the nucleus and the electrons and so increases the ionisation energies. In fact the increasing nuclear charge also drags the outer electrons in closer to the nucleus. That increases ionisation energies still more as you go across the period.In period 3, the trend is exactly the same. This time, all the electrons being removed are in the third level and are screened by the $1s^22s^22p^6$ electrons. They all have the same sort of environment, but there is an increasing nuclear charge.

The explanation lies with the structures of boron and aluminium. The outer electron is removed more easily from these atoms than the general trend in their period would suggest.

Be $1s^22s^2$ 1st *I.E.* = 900 kJ mol^{-1}

B $1s^22s^22p_x^{\ 1}$ 1st *I.E.* = 799 kJ mol^{-1}

You might expect the boron value to be more than the beryllium value because of the extra proton. Offsetting that is the fact that boron's outer electron is in a 2p orbital rather than a 2s. 2p orbitals have a slightly higher energy than the 2s orbital, and the electron is, on average, to be found further from the nucleus.

This has two effects:

1. The increased distance results in a reduced attraction and so a reduced ionisation energy.
2. The 2p orbital is screened not only by the $1s^2$ electrons but, to some extent, by the $2s^2$ electrons as well. That also reduces the pull from the nucleus and so lowers the ionisation energy.

The explanation for the drop between magnesium and aluminium is the same, except that everything is happening at the 3-level rather than the 2-level.

Mg $1s^22s^22p^63s^2$ 1st *I.E.* = 736 kJ mol^{-1}

Al $1s^22s^22p^63s^23p_x^{\ 1}$ 1st *I.E.* = 577 kJ mol^{-1}

The 3p electron in aluminium is slightly more distant from the nucleus than the 3s, and partially screened by the $3s^2$ electrons as well as the inner electrons. Both of these factors offset the effect of the extra proton. Once again, you might expect the ionisation energy of the group 6 element to be higher than that of group 5 because of the extra proton. What is offsetting it this time?

N $1s^22s^22p_x^{\ 1}2p_y^{\ 1}2p_z^{\ 1}$ 1st *I.E.* = 1400 kJ mol^{-1}

O $1s^22s^22p_x^{\ 2}2p_y^{\ 1}2p_z^{\ 1}$ 1st *I.E.* = 1310 kJ mol^{-1}

The screening is identical (from the $1s^2$ and, to some extent, from the $2s^2$ electrons), and the electron is being removed from an identical orbital. The difference is that in the oxygen case the electron being removed is one of the $2p_x^{\ 2}$ pair. The repulsion between the two electrons in the same orbital means that the electron is easier to remove than it would otherwise be.The drop in ionisation energy at sulphur is accounted for in the same way. As you go down a group in the Periodic Table ionisation energies generally fall. You have already seen evidence of this in the fact that the ionisation energies in period 3 are all less than those in period 2.Taking Group 1 as a typical example:

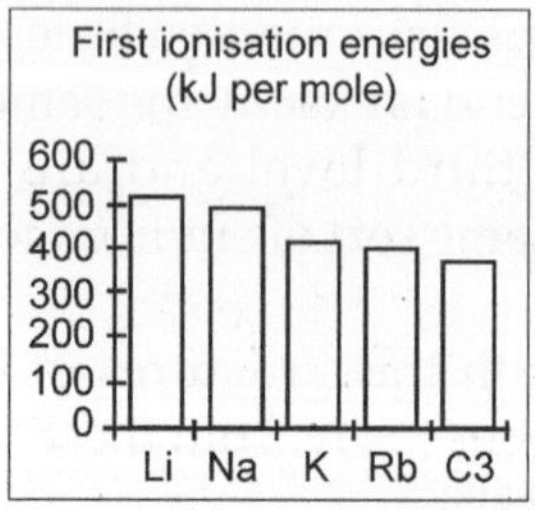

There are 11 protons in a sodium atom but only 3 in a lithium atom, so the nuclear charge is much greater. You might have expected a much larger ionisation energy in sodium, but offsetting the nuclear charge is a greater distance from the nucleus and more screening.

Li	$1s^2 2s^1$	1st *I.E.* = 519 kJ mol^{-1}
Na	$1s^2 2s^2 2p^6 3s^1$	1st *I.E.* = 494 kJ mol^{-1}

Lithium's outer electron is in the second level, and only has the $1s^2$ electrons to screen it. The $2s^1$ electron feels the pull of 3 protons screened by 2 electrons - a net pull from the centre of 1+. The sodium's outer electron is in the third level, and is screened from the 11 protons in the nucleus by a total of 10 inner electrons. The $3s^1$ electron also feels a net pull of 1+ from the centre of the atom. In other words, the effect of the extra protons is compensated for by the effect of the extra screening electrons. The only factor left is the extra distance between the outer electron and the nucleus in sodium's case. That lowers the ionisation energy.Similar explanations hold as you go down the rest of this group - or, indeed, any other group.

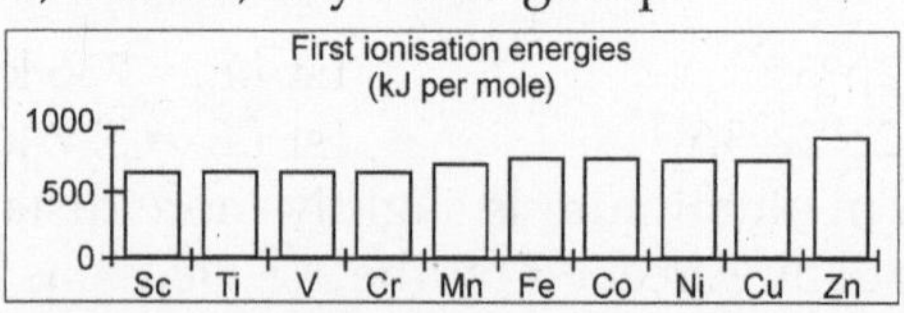

Apart from zinc at the end, the other ionisation energies are all much the same.All of these elements have an electronic structure $[Ar]3d^n 4s^2$ (or $4s^1$ in the cases of chromium and copper). The electron being lost always comes from the 4s orbital.As you go from one atom to the next in the series, the number of protons in the nucleus increases, but so also does the number of 3d electrons. The 3d electrons have some screening effect, and the extra proton and the extra 3d electron more or less cancel each other out as far as attraction from the centre of the atom is concerned.

The rise at zinc is easy to explain:

Cu	$[Ar]3d^{10}4s^1$	1st *I.E.* = 745 kJ mol^{-1}
Zn	$[Ar]3d^{10}4s^2$	1st *I.E.* = 908 kJ mol^{-1}

In each case, the electron is coming from the same orbital, with identical screening, but the zinc has one extra proton in the nucleus and so the attraction is greater. There will be a degree of repulsion between the paired up electrons in the 4s orbital, but in this case it obviously isn't enough to outweigh the effect of the extra proton.

Ionisation Energies and Reactivity

The lower the ionisation energy, the more easily this change happens:

$$X_{(g)} \longrightarrow X^{+}_{(g)} + e^{-}$$

You can explain the increase in reactivity of the Group 1 metals (Li, Na, K, Rb, Cs) as you go down the group in terms of the fall in ionisation energy. Whatever these metals react with, they have to form positive ions in the process, and so the lower the ionisation energy, the more easily those ions will form.The danger with this approach is that the formation of the positive ion is only one stage in a multi-step process.For example, you wouldn't be starting with gaseous atoms; nor would you end up with gaseous positive ions - you would end up with ions in a solid or in solution.

The energy changes in these processes also vary from element to element. Ideally you need to consider the whole picture and not just one small part of it.However, the ionisation energies of the elements are going to be major contributing factors towards the *activation energy* of the reactions. Remember that activation energy is the minimum energy needed before a reaction will take place. The lower the activation energy, the faster the reaction will be - irrespective of what the *overall* energy changes in the reaction are. The fall in ionisation energy as you go down a group will lead to lower activation energies and therefore faster reactions.

ATOMIC AND IONIC RADIUS

Unlike a ball, an atom doesn't have a fixed radius. The radius of an atom can only be found by measuring the distance between the nuclei of two touching atoms, and then halving that distance.

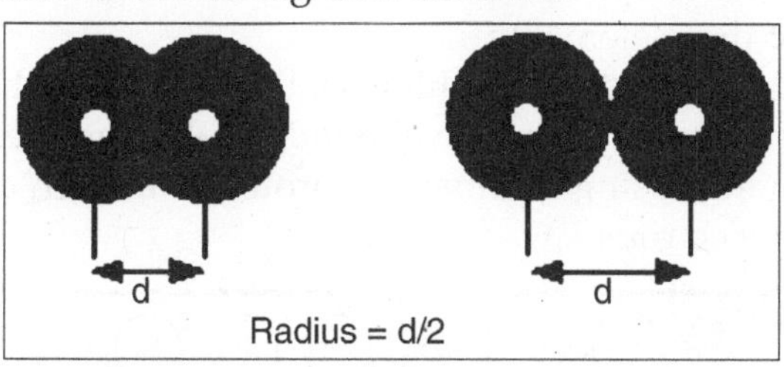

The same atom could be found to have a different radius depending on what was around it.The left hand diagram shows bonded atoms. The atoms are pulled closely together and so the measured radius is less than if they are just touching. This is what you would get if you had metal atoms in a metallic structure, or atoms covalently bonded to each other. The type of atomic radius being measured here is called the *metallic radius* or the *covalent radius* depending on the bonding.The right hand diagram shows what happens if the atoms are just touching. The attractive forces are much less, and the atoms are essentially "unsquashed". This measure of atomic radius is called the *van der Waals radius* after the weak attractions present in this situation.

The exact pattern you get depends on which measure of atomic radius you use - but the trends are still valid.The following diagram uses metallic radii for metallic elements, covalent radii for elements that form covalent bonds, and van der Waals radii for those (like the noble gases) which don't form bonds.

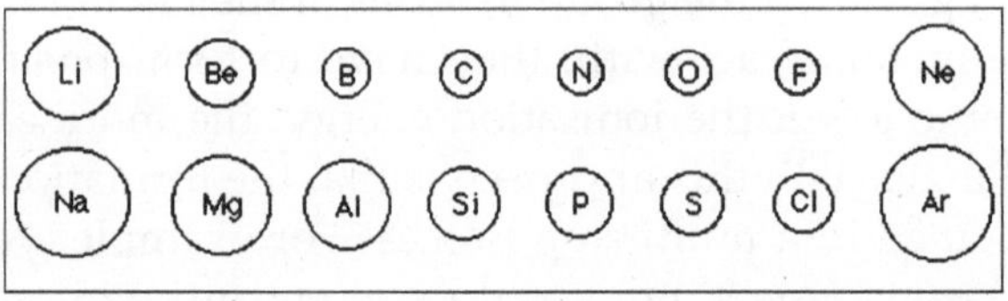

It is fairly obvious that the atoms get bigger as you go down groups. The reason is equally obvious - you are adding extra layers of electrons.

The noble gas at the end of each period. Because neon and argon don't form bonds, you can only measure their van der Waals radius - a case where the atom is pretty well "unsquashed". All the other atoms are being measured where their atomic radius is being lessened by strong attractions. You aren't comparing like with like if you include the noble gases.

Leaving the noble gases out, atoms get smaller as you go across a period.

If you think about it, the metallic or covalent radius is going to be a measure of the distance from the nucleus to the electrons which make up the bond. Look back to the left-hand side of the first diagram, and picture the bonding electrons as being half way between the two nuclei. From lithium to fluorine, those electrons are all in the 2-level, being screened by the $1s^2$ electrons. The increasing number of protons in the nucleus as you go across the period pulls the electrons in more tightly. The amount of screening is constant for all of these elements.

In the period from sodium to chlorine, the same thing happens. The size of the atom is controlled by the 3-level bonding electrons being pulled closer to the nucleus by increasing numbers of protons - in each case, screened by the 1- and 2-level electrons.

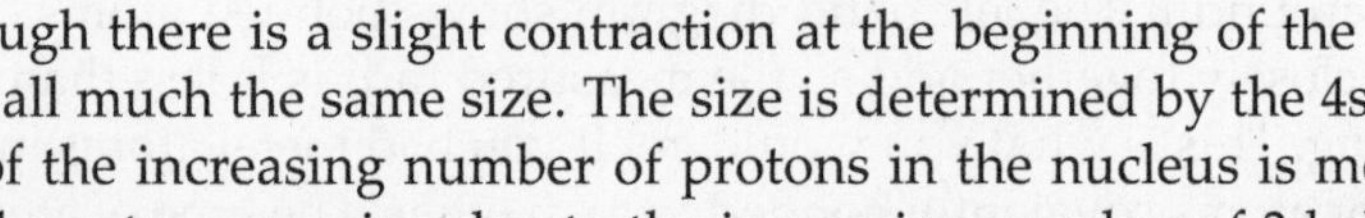

Although there is a slight contraction at the beginning of the series, the atoms are all much the same size. The size is determined by the 4s electrons. The pull of the increasing number of protons in the nucleus is more or less offset by the extra screening due to the increasing number of 3d electrons.

IONIC RADIUS

Ions aren't the same size as the atoms they come from. Compare the sizes of sodium and chloride ions with the sizes of sodium and chlorine atoms.

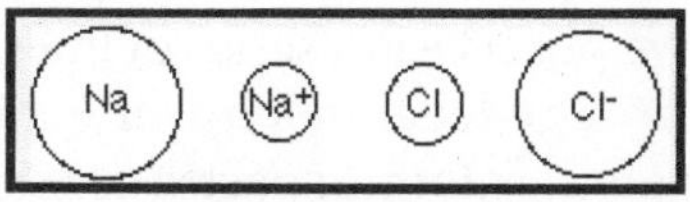

Positive Ions

Positive ions are smaller than the atoms they come from. Sodium is 2,8,1; Na^+ is 2,8. You've lost a whole layer of electrons, and the remaining 10 electrons are being pulled in by the full force of 11 protons.

Negative Ions

Negative ions are bigger than the atoms they come from. Chlorine is 2,8,7; Cl^- is 2,8,8. Although the electrons are still all in the 3-level, the extra repulsion produced by the incoming electron causes the atom to expand. There are still only 17 protons, but they are now having to hold 18 electrons.

THE PHOTOELECTRIC EFFECT

Einstein successfully applied the quantum theory to the photoelectric effect - light shining onto the surface of a metal is capable of ejecting electrons from that metal. The electrons which are emitted come off at energies that depend upon the frequency of the light and (importantly) not its intensity. This can be explained if one assumes that light is composed of photons of energy hv. Einstein suggested that a photon can give up its energy to an electron in the metal. Some of this is used in work, that it needs to remove the electron from the metal, the remainder appears as kinetic energy:

$$hv = Wo + \frac{1}{2} mv^2$$

Hence, light of frequency less than hv will be incapable of ejecting an electron. This shows the dependence of the photoelectric effect on the frequency of light rather than its intensity. Excitation of atoms - electronic spectra Niels Bohr applied the quantum theory to the phenomenon of atomic spectra. When atoms are subjected to high temperatures or electric discharges, they emit electromagnetic radiation which consists of characteristic frequencies.

By subjecting the atoms of a given element to high temperatures or passing an electric discharge through a gas composed of these atoms, the atoms become excited. Electrons within the atoms are raised to higher energy levels. Once at these higher energy levels, they can return to their unexcited, or ground, state by losing this increased energy. This can be achieved by emitting electromagnetic radiation. The frequency of the emitted radiation will be determined by the energy "jump" made by the electron. The electron could return to the ground state in a single stage or it could return by a series of jumps. Rather like being at the top of a staircase and either jumping right to the foot of the stairs or by taking a series of steps.Each jump will be accompanied by a release of radiation, each being characteristic of the magnitude of the jump. Because there are a number of possible routes back to the ground state what we observe is a series of lines, each line corresponding to a jump from a higher energy level to a lower energy level.

In the case of hydrogen the atomic spectra consists of a series of emission lines which have been named after their discoverers - for example the Pfund, Brackett, Paschen and Balmer series.Bohr applied quantum theory to the atomic spectrum of hydrogen and postulated that;electrons could remain in a particular orbit without emitting radiation the angular momentum of electrons in an orbit is quantized These two ideas, at conflict at that time with conventional physics, enabled Bohr to calculate an expression for the energy and radius of an atomic orbital.Although the Bohr theory gives good agreement for hydrogen and hydrogen like (He^{+}, Li^{2+}) it is not satisfactory for other atoms.

On the basis of these ideas, Bohr and later Sommerfeld proposed models of the atom in which electrons had certain allowable energies orbiting the nucleus in circular or elliptical paths.Before going on to develop the ideas of the quantum theory, two important properties of the electron need to be discussed; the dual nature of the electron the Heisenberg Uncertainty Principle when applied to the electron

In the early part of this century (L. V. de Broglie), it was shown that matter has associated wave characteristics and this effect should be particularly marked for a particle of such a small mass as the electron. In other words, the electron can behave as both a particle and a wave.

de Broglie proposed the *wave - matter relationship*

$$\text{lambda} = h / mv$$

Experimental confirmation of the dual nature of the electron was provided by Davidson and Germer who were able to show diffraction of electrons *i.e.* wavelike properties.

HEISENBERG UNCERTAINTY PRINCIPLE

He was one of the greatest physicists of the twentieth century. He is best known as a founder of quantum mechanics, the new physics of the atomic world, and especially for the uncertainty principle in quantum theory. Uncertainty Principle, also called Heisenberg Uncertainty Principle, or Indeterminacy Principle, that the position and the velocity of an object cannot both be measured exactly, at the same time, even in theory. The very concepts of exact position and exact velocity together, in fact, have no meaning in nature.

Uncertainty principle: A principle, central to quantum mechanics, which states that two complementary parameters (such as position and momentum, energy and time, or angular momentum and angular displacement) cannot both be known to infinite accuracy; the more you know about one, the less you know about the other. It can be illustrated in a fairly clear way as it relates to position vs. momentum: To see something (let's say an electron), we have to fire photons at it; they bounce off and come back to us, so we can "see" it. If you choose low-frequency photons, with a low energy, they do not impart much momentum to the electron, but they give you a very fuzzy picture, so you have a higher uncertainty in position so that you can have a higher certainty in momentum. On the other hand, if you were to fire very high-energy photons (*x*-rays or gammas) at the electron, they would give you a very clear picture of where the electron is (higher certainty in position), but would impart a great deal of momentum to the electron (higher uncertainty in momentum).

The more precisely the position is determined, the less precisely the momentum is known in this instant, and vice versa. —Heisenberg, uncertainty Ordinary experience provides no clue of this principle. It is easy to measure both the position and the velocity of, say, an automobile, because the uncertainties implied by this principle for ordinary objects are too small to be observed. The complete rule stipulates that the product of the uncertainties in position and velocity is equal to or greater than a tiny physical quantity, or constant (about 10-34 joule-second, the value of the quantity h (where h is Planck's constant).

Only for the exceedingly small masses of atoms and subatomic particles does the product of the uncertainties become significant. Any attempt to measure precisely the velocity of a subatomic particle, such as an electron, will knock it about in an unpredictable way, so that a simultaneous measurement of its position has no validity. This result has nothing to do with inadequacies in the measuring instruments, the technique, or the observer; it arises out of the intimate connection in nature between particles and waves in the realm of subatomic dimensions. Every particle has a wave associated with it; each particle actually exhibits wavelike behaviour. The particle is most likely to be found in those places where the undulations of the wave are greatest, or most intense.

The more intense the undulations of the associated wave become, however, the more ill defined becomes the wavelength, which in turn determines the momentum of the particle. So a strictly localized wave has an indeterminate wavelength; its associated particle, while having a definite position, has no certain velocity. A particle wave having a well-defined wavelength, on the other hand, is spread out; the associated particle, while having a rather precise velocity, may be almost anywhere. A quite accurate measurement of one observable involves a relatively large uncertainty in the measurement of the other. The uncertainty principle is alternatively expressed in terms of a particle's momentum and position. The momentum of a particle is equal to the product of its mass times its velocity. Thus, the product of the uncertainties in the momentum and the position of a particle equals h/2 or more.

The principle applies to other related pairs of observables, such as energy and time: the product of the uncertainty in an energy measurement and the uncertainty in the time interval during which the measurement is made also equals h/2 or more. The same relation holds, for an unstable atom or nucleus, between the uncertainty in the quantity of energy radiated and the uncertainty in the lifetime of the unstable system as it makes a transition to a more stable state.

$$v = \Delta x \Delta y \Delta z \; \Delta p_x \Delta p_y \Delta p_z = (2\pi\hbar)^3$$

SCHRODINGER'S WAVE EQUATION

Information about a particle in a given energy level is contained in a mathematical expression or function. This is called the wave function and is denoted by the Greek letter psi. The wave function is obtained by solving Schrodinger's Wave Equation. The square of the wave function gives the probability of finding a particle (electron) within a region of space.The Schrodinger's Wave Equation is fundamental to the development of quantum mechanics.

Acceptable solutions to the wave equation must have certain properties, and these give some insight into the significance of the wave equation.

- Psi must be continuous
- Psi must be finite
- Psi must be single valued
- The summation of psi^2 from - infinity to + infinity must = 1

Using Schrodingers equation we can calculate the probability density of the electron - *i.e.* the probability that the electron exists in a given volume of space. For example we can calculate this as a function of distance from the nucleus and we can define the "shape" of an orbital..Before going onto look at where we might "place" a particular electron we need to delve a little deeper into quantum mechanics.There were acceptable solutions to the wave equation.

To obtain such solutions, it is necessary to use four quantum numbers to describe the orbital and the electron within that orbital. By defining the values of each of these four quantum numbers we can identify both the electron and the "orbit" of that electron. In some ways this is rather like giving a map reference for a person and indicating which way that person is facing.

THE ATOMIC HYDROGEN EMISSION SPECTRUM

The atomic hydrogen emission spectrum, showing how it arises from electron movements between energy levels within the atom. It also looks at how the spectrum can be used to find the ionisation energy of hydrogen.

Observing Hydrogen's Emission Spectrum

A hydrogen discharge tube is a slim tube containing hydrogen gas at low pressure with an electrode at each end. If you put a high voltage across this (say, 5000 volts), the tube lights up with a bright pink glow. If the light is passed through a prism or diffraction grating, it is split into its various colours. What you would see is a small part of the hydrogen emission spectrum. Most of the spectrum is invisible to the eye because it is either in the infra-red or the ultra-violet.

Hydrogen's Emission Spectrum

There is a lot more to the hydrogen spectrum than the three lines.It is possible to detect patterns of lines in both the ultra-violet and infra-red regions of the spectrum as well.These fall into a number of "series" of lines named after the person who discovered them. The diagram below shows three of these series, but there are others in the infra-red to the left of the Paschen series shown in the diagram.The diagram is quite complicated, Look first at the Lyman series on the right of the diagram - this is the most spread out one and easiest to see what is happening.

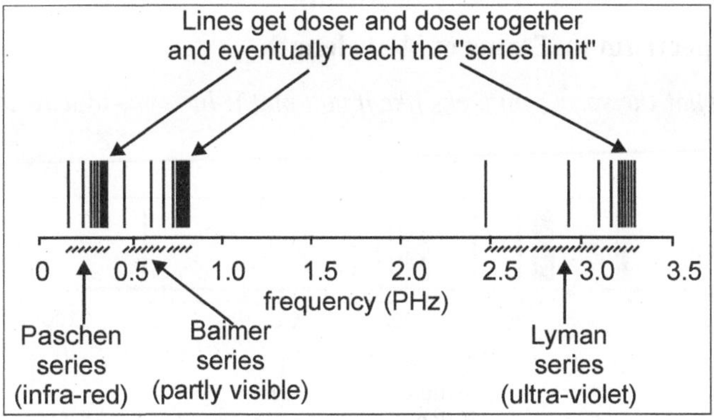

The Lyman series is a series of lines in the ultra-violet. Notice that the lines get closer and closer together as the frequency increases. Eventually, they get so close together that it becomes impossible to see them as anything other than a continuous spectrum. That's what the shaded bit on the right-hand end of the series suggests. Then at one particular point, known as the *series limit,* the series stops.The Balmer series or the Paschen series, you will see that the pattern is just the same, but the series have become more compact. In the Balmer series.

Frequency and Wavelength

The hydrogen spectrum drawn using wavelengths of light rather than frequencies. Unfortunately, because of the mathematical relationship between the frequency of light and its wavelength, you get two completely different views of the spectrum if you plot it against frequency or against wavelength.

The Relationship between Frequency and Wavelength

The mathematical relationship is:

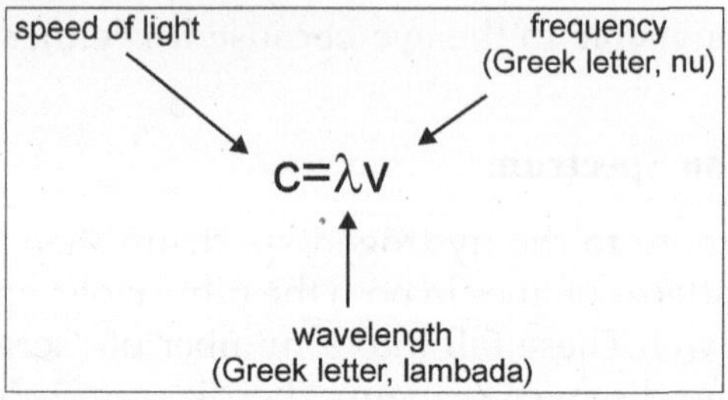

Rearranging this gives equations for either wavelength or frequency.

$$\lambda = \frac{c}{v} \qquad v = \frac{c}{\lambda}$$

What this means is that there is an inverse relationship between the two - a high frequency means a low wavelength and vice versa.

Hydrogen Spectrum in Terms of Wavelength

This is what the spectrum looks like if you plot it in terms of wavelength instead of frequency:

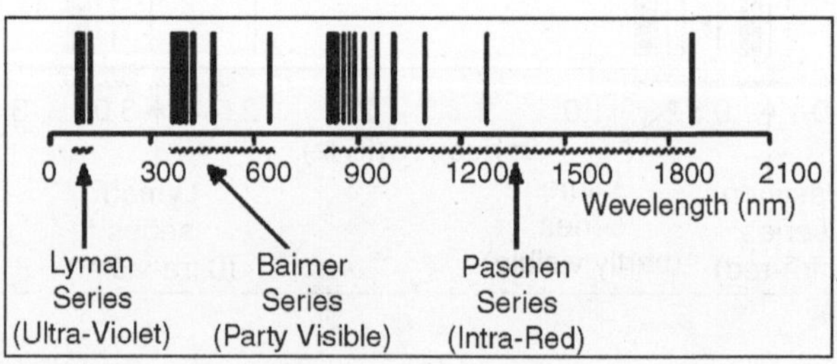

and just to remind you what the spectrum in terms of frequency looks like:

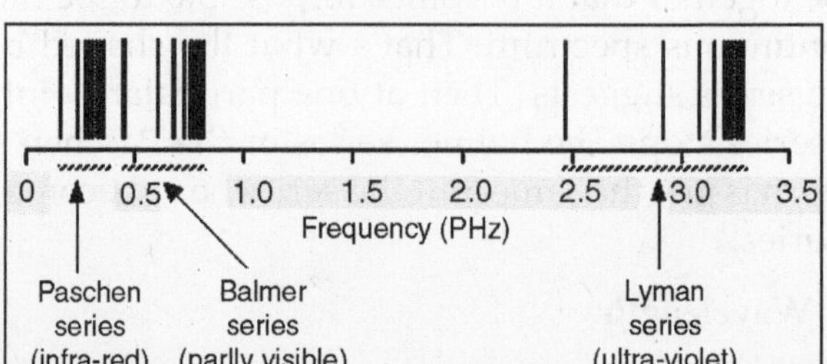

The spectrum plotted against frequency, because it is much easier to relate it to what is happening in the atom. Be aware that the spectrum looks different depending on how it is plotted, but, other than that, ignore the wavelength version unless it is obvious that your examiners want it. If you try to learn both versions, you are only going to get them muddled up!

THE BALMER AND RYDBERG EQUATIONS

By an amazing bit of mathematical insight, in 1885 Balmer came up with a simple formula for predicting the wavelength of any of the lines in what we now know as the Balmer series. Three years later, Rydberg generalised this so that it was possible to work out the wavelengths of any of the lines in the hydrogen emission spectrum.

What Rydberg came up with was:

$$\frac{1}{\lambda} = \mathrm{R_H}\left(\frac{1}{\mathrm{n}_1^2} - \frac{1}{\mathrm{n}_2^2}\right)$$

$\mathrm{R_H}$ is a constant known as the *Rydberg constant.*

n_1 and n_2 are integers (whole numbers). n_2 has to be greater than n_1. In other words, if n_1 is, say, 2 then n_2 can be any whole number between 3 and infinity.

The various combinations of numbers that you can slot into this formula let you calculate the wavelength of any of the lines in the hydrogen emission spectrum - and there is close agreement between the wavelengths that you get using this formula and those found by analysing a real spectrum. You can also use a modified version of the Rydberg equation to calculate the frequency of each of the lines.

$$\nu = \frac{c}{\lambda}$$

$$= c \times \frac{1}{\lambda}$$

$$\text{if} \frac{1}{\lambda} = \mathrm{R_H}\left(\frac{1}{n_1^2} - \frac{1}{n_1^2}\right)$$

$$\nu = c.\mathrm{R_H}\left(\frac{1}{n_1^2} - \frac{1}{n_1^2}\right)$$

The Origin of the Hydrogen Emission Spectrum

The lines in the hydrogen emission spectrum form regular patterns and can be represented by a (relatively) simple equation. Each line can be calculated from a combination of simple whole numbers.When nothing is exciting it, hydrogen's electron is in the first energy level - the level closest to the nucleus. But if you supply energy to the atom, the electron gets excited into a higher energy level - or even removed from the atom altogether.The high voltage in a discharge tube provides that energy.

Hydrogen molecules are first broken up into hydrogen atoms (hence the *atomic* hydrogen emission spectrum) and electrons are then promoted into higher energy levels.Suppose a particular electron was excited into the third energy level.

This would tend to lose energy again by falling back down to a lower level. It could do this in two different ways. It could fall all the way back down to the first level again, or it could fall back to the second level - and then, in a second jump, down to the first level.

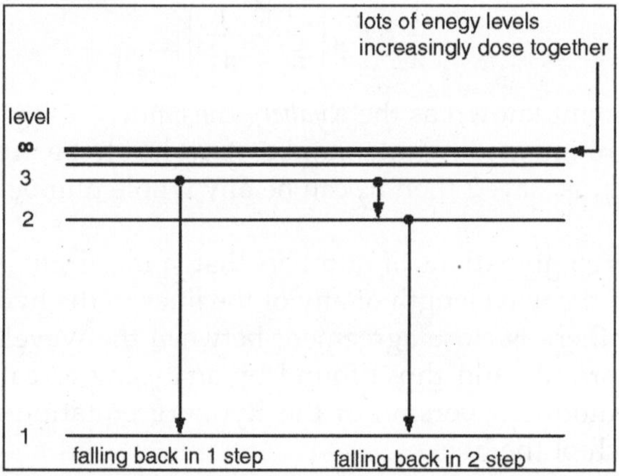

Tying Particular Electron Jumps to Individual Lines in the Spectrum

If an electron falls from the 3-level to the 2-level, it has to lose an amount of energy exactly the same as the energy gap between those two levels. That energy which the electron loses comes out as light (where "light" includes UV and IR as well as visible). Each frequency of light is associated with a particular energy by the equation:

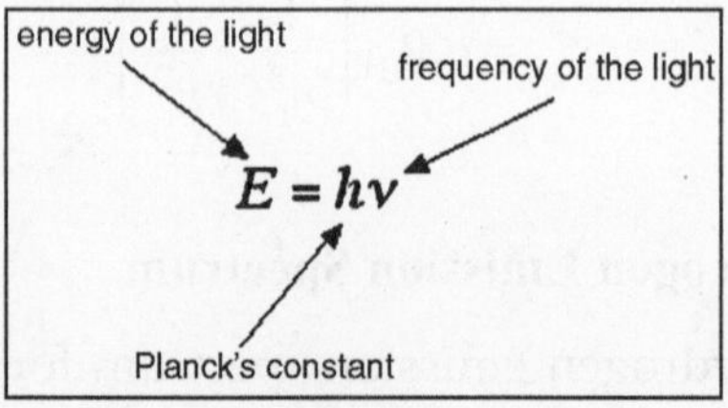

The higher the frequency, the higher the energy of the light. If an electron falls from the 3-level to the 2-level, red light is seen. This is the origin of the red line in the hydrogen spectrum. By measuring the frequency of the red light, you can work out its energy. That energy must be exactly the same as the energy gap between the 3-level and the 2-level in the hydrogen atom.The last equation can therefore be re-written as a measure of the energy gap between two electron levels.

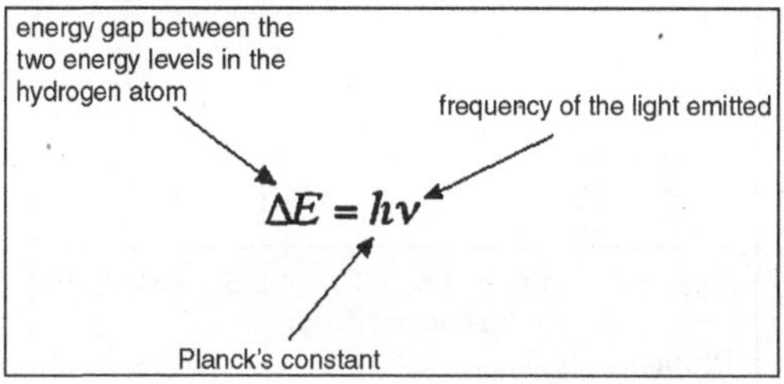

The greatest possible fall in energy will therefore produce the highest frequency line in the spectrum. The greatest fall will be from the infinity level to the 1-level. The next few diagrams are in two parts - with the energy levels at the top and the spectrum at the bottom.

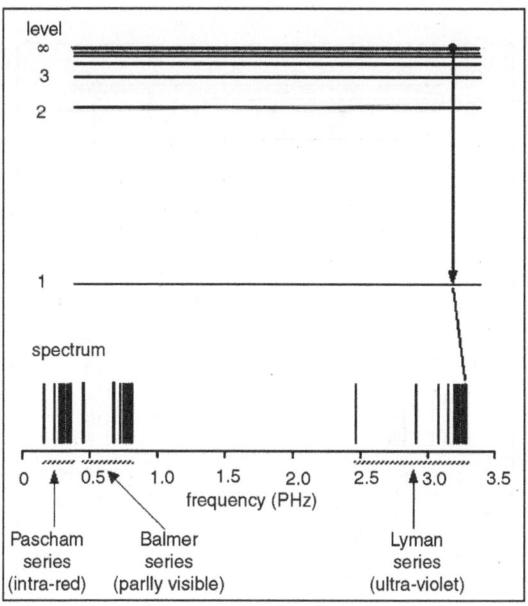

If an electron fell from the 6-level, the fall is a little bit less, and so the frequency will be a little bit lower. (Because of the scale of the diagram, it is impossible to draw in all the jumps involving all the levels between 7 and infinity!)

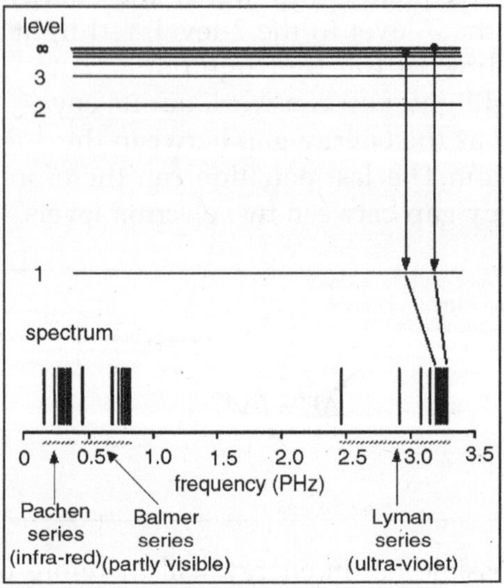

... and as you work your way through the other possible jumps to the 1-level, you have accounted for the whole of the Lyman series. The spacings between the lines in the spectrum reflect the way the spacings between the energy levels change.

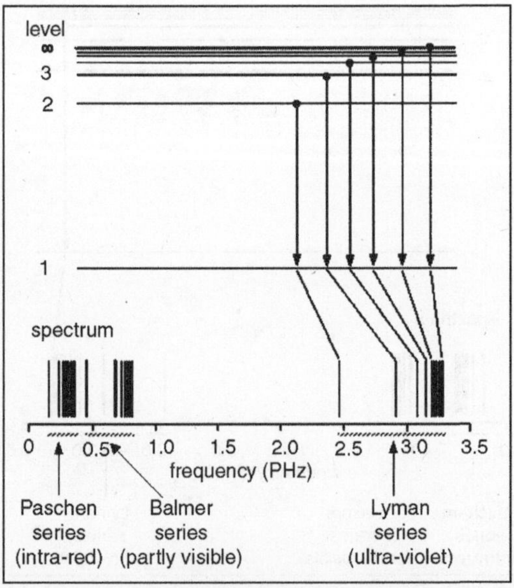

If you do the same thing for jumps down to the 2-level, you end up with the lines in the Balmer series. These energy gaps are all much smaller than in the Lyman series, and so the frequencies produced are also much lower.

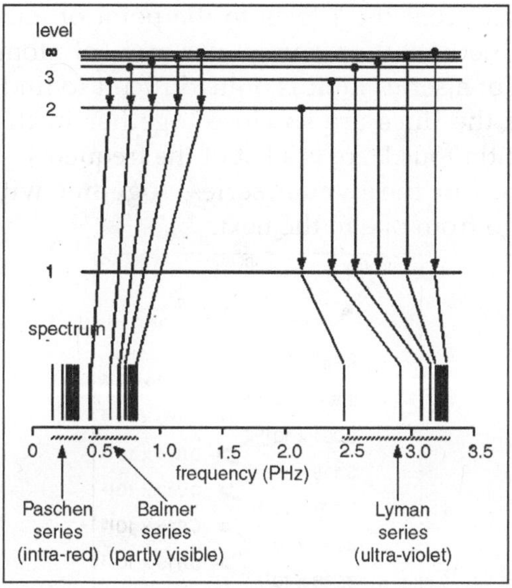

The Paschen series would be produced by jumps down to the 3-level, but the diagram is going to get very messy if I include those as well - not to mention all the other series with jumps down to the 4-level, the 5-level and so on.

RYDBERG EQUATION

n_1 and n_2 in the Rydberg equation are simply the energy levels at either end of the jump producing a particular line in the spectrum. For example, in the Lyman series, n_1 is always 1. Electrons are falling to the 1-level to produce lines in the Lyman series. For the Balmer series, n_1 is always 2, because electrons are falling to the 2-level. n_2 is the level being jumped from. We have already mentioned that the red line is produced by electrons falling from the 3-level to the 2-level.

In this case, then, n_2 is equal to 3. The infinity level represents the highest possible energy an electron can have as a part of a hydrogen atom. So what happens if the electron exceeds that energy by even the tiniest bit? The electron is no longer a part of the atom. The infinity level represents the point at which ionisation of the atom occurs to form a positively charged ion.

Hydrogen's Ionisation Energy

When there is no additional energy supplied to it, hydrogen's electron is found at the 1-level. This is known as its ground state. If you supply enough energy to move the electron up to the infinity level, you have ionised the hydrogen. The ionisation energy *per electron* is therefore a measure of the distance between the 1-level and the infinity level. If you look back at the last few diagrams, you will find that that particular energy jump produces the series limit of the Lyman series.If you can determine the frequency of the Lyman series limit, you can use it to calculate the energy needed to move the electron in one atom from the 1-level to the point of ionisation. From that, you can calculate the ionisation energy per mole of atoms. The problem is that the frequency of a series limit is quite difficult to find accurately from a spectrum because the lines are so close together in that region that the spectrum looks continuous.Here is a list of the frequencies of the seven most widely spaced lines in the Lyman series, together with the increase in frequency as you go from one to the next.

frequency ν (Hz)	increase in frequency ν (Hz)
2.467×10^{15}	
	0.457×10^{15}
2.924×10^{15}	
	0.160×10^{15}
3.084×10^{15}	
	0.074×10^{15}
3.158×10^{15}	
	0.040×10^{15}
3.198×10^{15}	
	0.025×10^{15}
3.223×10^{15}	
	0.015×10^{15}
3.238×10^{15}	

As the lines get closer together, obviously the increase in frequency gets less. At the series limit, the gap between the lines would be literally zero. That means that if you were to plot the increases in frequency against the actual frequency, you could extrapolate (continue) the curve to the point at which the increase becomes zero. That would be the frequency of the series limit.In fact you can actually plot two graphs from the data in the table above.

The frequency *difference* is related to two frequencies. It doesn't matter, as long as you are always consistent - in other words, as long as you *always* plot the difference against either the higher or the lower. At the point you are interested in (where the difference becomes zero), the two frequency numbers are the same. By plotting both of the possible curves on the same graph, it makes it easier to decide exactly how to extrapolate the curves. Because these are curves, they are much more difficult to extrapolate than if they were straight lines.

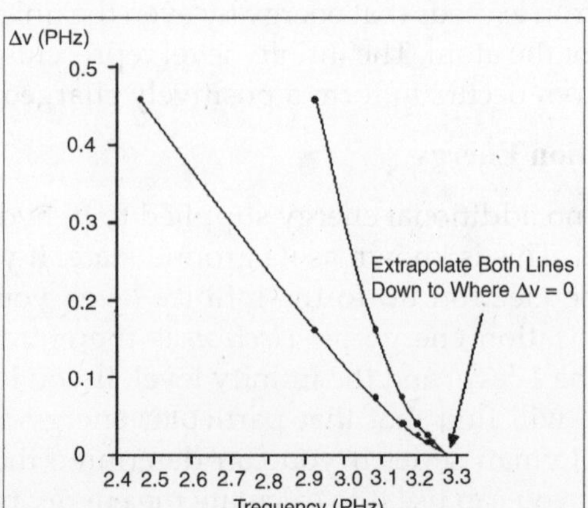

Both lines point to a series limit at about 3.28×10^{15} Hz. So... now we can calculate the energy needed to remove a single electron from a hydrogen atom.

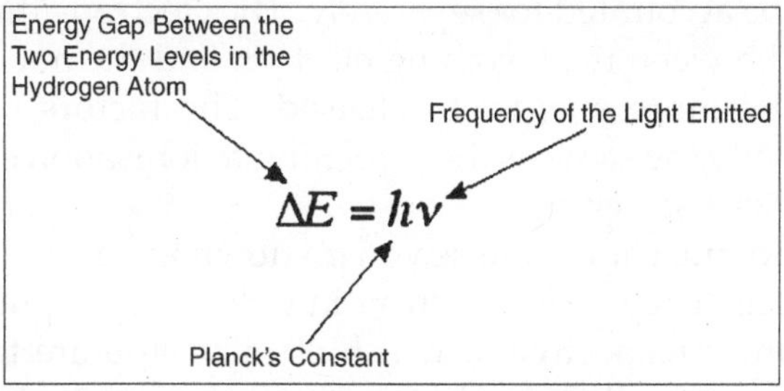

We can work out the energy gap between the ground state and the point at which the electron leaves the atom by substituting the value we've got for frequency and looking up the value of Planck's constant from a data book.

$$\begin{aligned}\Delta E &= h\nu \\ &= 6.626 \times 10^{-34} \times 3.28 \times 10^{15} \\ &= 2.173 \times 10^{-18}\ \mathrm{J}\end{aligned}$$

That gives you the ionisation energy for a single atom. To find the normally quoted ionisation energy, we need to multiply this by the number of atoms in a mole of hydrogen atoms (the Avogadro constant) and then divide by 1000 to convert it into kilojoules.

$$\text{Ionisation energy} = 2.173 \times 10^{-18} \times 6.022 \times 10^{23} \times \frac{1}{1000}$$

$$= 1310\ \mathrm{kJ\,mol^{-1}}$$

This compares well with the normally quoted value for hydrogen's ionisation energy of 1312 kJ mol^{-1}.

ELECTRON AFFINITY

Ionisation energies are always concerned with the formation of positive ions. Electron affinities are the negative ion equivalent, and their use is almost always confined to elements in groups 6 and 7 of the Periodic Table.The first electron affinity is the energy released when 1 mole of gaseous atoms each acquire an electron to form 1 mole of gaseous 1- ions. This is more easily seen in symbol terms $x_{(g)} + e^- \rightarrow x^-_{(g)}$.

It is the energy released (per mole of X) when this change happens. First electron affinities have negative values. For example, the first electron affinity of chlorine is -349 kJ mol^{-1}. By convention, the negative sign shows a release of energy.

The first electron affinities of the group 7 elements:

F -328 kJ mol^{-1}
Cl -349 kJ mol^{-1}
Br -324 kJ mol^{-1}
I -295 kJ mol^{-1}

The group, first electron affinities become less (in the sense that less energy is evolved when the negative ions are formed). Fluorine breaks that pattern, and will have to be accounted for separately. The electron affinity is a measure of the attraction between the incoming electron and the nucleus the stronger the attraction, the more energy is released. The factors which affect this attraction are exactly the same as those relating to ionisation energies - nuclear charge, distance and screening.

The increased nuclear charge as you go down the group is offset by extra screening electrons. Each outer electron in effect feels a pull of 7+ from the centre of the atom, irrespective of which element you are talking about.For example, a fluorine atom has an electronic structure of $1s^22s^22p_x{}^22p_y{}^22p_z{}^1$. It has 9 protons in the nucleus. The incoming electron enters the 2-level, and is screened from the nucleus by the two $1s^2$ electrons. It therefore feels a net attraction from the nucleus of 7+ (9 protons less the 2 screening electrons).By contrast, chlorine has the electronic structure $1s^22s^22p^63s^23p_x{}^23p_y{}^23p_z{}^1$. It has 17 protons in the nucleus.But again the incoming electron feels a net attraction from the nucleus of 7+ (17 protons less the 10 screening electrons in the first and second levels).

The over-riding factor is therefore the increased distance that the incoming electron finds itself from the nucleus as you go down the group. The greater the distance, the less the attraction and so the less energy is released as electron affinity.The incoming electron is going to be closer to the nucleus in fluorine than in any other of these elements, so you would expect a high value of electron affinity.However, because fluorine is such a small atom, you are putting the new electron into a region of space already crowded with electrons and there is a significant amount of repulsion. This repulsion lessens the attraction the incoming electron feels and so lessens the electron affinity.A similar reversal of the expected trend happens between oxygen and sulphur in Group 6. The first electron affinity of oxygen (–142 kJ mol^{-1}) is smaller than that of sulphur (–200 kJ mol^{-1}) for exactly the same reason that fluorine's is smaller than chlorine's.

COMPARING GROUP 6 AND GROUP 7 VALUES

As you might have noticed, the first electron affinity of oxygen (–142 kJ mol^{-1}) is less than that of fluorine (–328 kJ mol^{-1}). Similarly sulphur's (–200 kJ mol^{-1}) is less than chlorine's (-349 kJ mol^{-1}).It's simply that the Group 6 element has 1 less proton in the nucleus than its next door neighbour in Group 7. The amount of screening is the same in both.That means that the net pull from the nucleus is less in Group 6 than in Group 7, and so the electron affinities are less.

FIRST ELECTRON AFFINITY AND REACTIVITY

The reactivity of the elements in group 7 falls as you go down the group - fluorine is the most reactive and iodine the least.Often in their reactions these elements form their negative ions. At GCSE the impression is sometimes given that the fall in reactivity is because the incoming electron is held less strongly as you go down the group and so the negative ion is less likely to form. That explanation looks reasonable until you include fluorine!

An overall reaction will be made up of lots of different steps all involving energy changes, and you cannot safely try to explain a trend in terms of just one of those steps. Fluorine is much more reactive than chlorine (despite the lower electron affinity) because the energy released in other steps in its reactions more than makes up for the lower amount of energy released as electron affinity.

SECOND ELECTRON AFFINITY

You are only ever likely to meet this with respect to the group 6 elements oxygen and sulphur which both form 2- ions.The second electron affinity is the energy required to add an electron to each ion in 1 mole of gaseous 1- ions to produce 1 mole of gaseous 2- ions.

This is more easily seen in symbol terms:

$$x^{-}_{(g)} + e^{-} \rightarrow x^{2-}_{(g)}$$

It is the energy needed to carry out this change per mole of X^-. You are forcing an electron into an already negative ion. It's not going to go in willingly!

$$O_{(g)} + e^{-} \rightarrow O^{-}(g) \text{ 1st EA} = -142 \text{ kJ mol}^{-1}$$

$$O^{-}_{(g)} + e^{-} \rightarrow O^{2-}(g) \text{ 2nd EA} = +844 \text{ kJ mol}^{-1}$$

The positive sign shows that you have to put in energy to perform this change. The second electron affinity of oxygen is particularly high because the electron is being forced into a small, very electron-dense space.

4

Infrared Spectroscopy

Infrared spectroscopy (IR spectroscopy) is the spectroscopy that deals with the infrared region of the electromagnetic spectrum, that is light with a longer wavelength and lower frequency than visible light. It covers a range of techniques, mostly based on absorption spectroscopy. As with all spectroscopic techniques, it can be used to identify and study chemicals. A common laboratory instrument that uses this technique is a Fourier transform infrared (FTIR) spectrometer.

The infrared portion of the electromagnetic spectrum is usually divided into three regions; the near-, mid- and far- infrared, named for their relation to the visible spectrum. The higher-energy near-IR, approximately 14000–4000 cm^{-1} (0.8–2.5 μm wavelength) can excite overtone or harmonic vibrations. The mid-infrared, approximately 4000–400 cm^{-1} (2.5–25 μm) may be used to study the fundamental vibrations and associated rotational-vibrational structure. The far-infrared, approximately 400–10 cm^{-1} (25–1000 μm), lying adjacent to the microwave region, has low energy and may be used for rotational spectroscopy. The names and classifications of these subregions are conventions, and are only loosely based on the relative molecular or electromagnetic properties.

NEAR INFRARED

Near infrared (NIR) spectroscopy is an analytical tool that is still not fully integrated into the pharmaceutical industrial environment. However, its advantages are potentially of considerable benefit for the quality control of herbal medicinal products. Four methods were developed to demonstrate the ability of NIR spectroscopy as a quality control method in the different steps of the manufacturing process of herbal medicinal products.

Qualitative and quantitative methods were established to control the quality of herbal and packaging raw material at reception, and to quantify the active content in the final dosage form. NIR methods showed several clear benefits, such as speed, low analysis costs and environmental friendliness compared to traditional analytical tools. The information obtained from NIR analysis is, however, different to that provided by separative methods such as high.performance liquid chromatography (HPLC). Qualitative NIR techniques can only confirm whether the sample is of required quality or not, and for quantitative measurements NIR intensities have to be calibrated for the sample.s properties and do not give any information about any other property. The most labour.intensive part of the NIR analysis is method development. The results of NIR analysis are obtained in less than one minute for a single sample. In contrast, HPLC analysis is time.consuming but very specific, and provides detailed results about the presence or concentration of identity markers.

The use of chemometric tools and the study of factors affecting the spectra during feasibility studies are highly informative. They were used to optimise the calibration set, the regression model and the sample presentation mode, and were found to be critical steps in the development of specific and robust NIR models. Pharmaceutical guidelines that are currently in force or in preparation were used and compared for the validation of the four NIR methods studied. The validation results proved that the NIR methods were as reliable as the reference analysis methods. NIR spectroscopy is therefore a very suitable analytical tool for the quality control of herbal medicinal products.

The NIR (near infrared) spectroscopy, previously called the "sleeping technique", is being accepted by the pharmaceutical industry. Official instances such as the American Pharmacopea and the European Pharmacopea have recently adopted monographs describing this technique. A draft for the validation of NIR methods is being processed at the European Pharmacopea, and the number of NIR pharmaceutical applications has not stopped increasing since the beginning of the 90′s. It was about time, because NIR spectroscopy was an analytical tool already widely used in the agricultural and food industries in the beginning of the 70′s. The main obstacle to the integration of this technique into the pharmaceutical world has been the regulations governing the introduction of new techniques in quality control labouratories.

In contrast, the agricultural industries are not subjected to such strict regulations. Thanks to the regulatory framework that is now being created around the technique, the pharmaceutical industries will soon be able to fully enjoy the several benefits of NIR - speed, flexibility and low running costs. Herbal medicinal products (HMPs) represent a considerable part of the pharmaceutical market in the world: Europeans are believed to spend more than seven billion US dollars on herbal supplements, and the US market is estimated to grow at about 15 per cent per year. In the domain of herbal medicines, a large part of the costs are attributed to quality testing. The wet chemical, spectroscopic and chromatographic methods that are commonly used as quality control tools are time. and solvent.consuming. The use of NIR was recently investigated for controlling *e.g.* the origin of the drug and quantifying its active or marker substances. It proved to be a very reliable tool compared to traditional methods of analysis. NIR could be more widely used to monitor the complete manufacturing process of the herbal product, *i.e.* from authentication of the plants to the quantification of active compounds in the final dosage form. However, the task still remains to develop methods that fulfil pharmaceutical regulations for HMPs.

Herbal Drug Preparations and Herbal Medicinal Products

Herbal medicinal products are medicinal products containing as active substance exclusively herbal drugs or herbal drug preparations. Herbal drugs are mainly whole, fragmented or cut, plants, parts of plants, algae, fungi or lichens in an unprocessed sate, usually in the dried form but sometimes fresh. Certain exudates may also be considered as herbal drugs. Herbal drug preparations are obtained when herbal drugs are subjected to treatment such as extraction, distillation, expression, fractionation, purification, concentration or fermentation.

They include comminuted or powdered herbal drugs, tinctures, extracts, essential oils, expressed juices and processed exudates. In Europe, herbal drugs are described by monographs from the European Pharmacopea created in 1964. They describe general methods of analysis and the appropriate method of storage. In Germany.Commission E. monographs. have been published by the health authorities since the beginning of the 80′s and describe more than 380 medicinal plants. A third source of monographs is the European Scientific Cooperative on Phytotherapy, ESCOP, which publishes monographs on individual plant drugs. These monographs highlight the clinical and pharmacological properties of the plants in order to represent a statement on efficacy and safety of a medicinal plant and its preparations. Finally, the world Health Organisation (WHO) has also published useful monographs supporting the demonstration of safety and efficacy of herbal medicinal products.

Regulations Applicable to Herbal Medicinal Products

During the last few years, European regulations concerning herbal medicinal products have been submitted to various modifications, and it is worth mentioning the latest issues. Herbal medicinal products (HMP) are, above all, medicines and therefore fall within the scope of the recent European Economic Community Council directive 2001/83/EC concerning medicinal products for human use, adopted in November 2001. Directive 2001/83/EC prescribes that no medicinal product may be placed on the market without having obtained a marketing authorisation. Marketing authorisation as a herbal medicinal product is, in principle, granted on the basis of a.full. dossier in terms of proof of quality, safety and efficacy in all Member States, with the exception of Denmark and Finland. In these countries it is only possible to use bibliographic applications for herbal medicinal products. In Finland, the National Agency of Medicines has defined herbal medicinal products as products traditionally used for medicinal purposes. These products can be derived from plants, animals, bacteria or minerals, and can contain herbal drugs or herbal drug preparations, but not single purified substances. In the United States, herbal products are mostly registered as dietary supplements since the Food and Drug Administration does not accept bibliographic evidence of effectiveness, but prefers randomized controlled trials as evidence of efficacy.

The legal framework concerning medicines, and especially directive 2001/83/EC, is well applicable to certain herbal medicinal products, but is not as suitable for the so called traditional herbal medicinal products. These are herbal medicinal products that have been in use for a long period in the European Community in the form of oral, external and/or inhalation preparations, and are designed for use without the intervention of a medical practitioner and in accordance with specified daily doses.

The efficiency and safety of these traditional medicinal products are plausible on the basis of long-term use. For these products, applying for a "full" marketing authorisation would be irrelevant and too expensive for the pharmaceutical industries. Therefore, in order to simplify the legislation for the traditional HMP, the Commission adopted in January 2002 a proposal for a new directive amending directive 2001/83/EC with respect to traditional herbal medicinal products.

The European Parliament gave a partial agreement to a modified draft in November 2002. This new directive provides a simplified procedure for the registration, and hence marketing, of certain traditional herbal medicinal products. In other words, for the registration of a traditional HMP, it will no longer be necessary to submit a file containing experimental evidence of the pharmacodynamic, pharmaco-kinetic and toxicological aspects or clinical evidence proving their therapeutic effects and tolerability in man.

However, documentation will have to be submitted providing evidence of quality, efficacy and safety on the basis of experience gained from at least 30 years. traditional use. The group of herbal products that is not considered as herbal medicinal products does not have to conform the legislation concerning medicines. These products can be sold as food supplements, as long as no medical claim is made, and their control is in accordance with the food legislation. They are therefore outside the scope of this dissertation.

Process Manufacturing and Distribution

Because HMP are medicinal products, their production must follow the Good Manufacturing Practice (GMP), the principles and guidelines of which are laid down in the European directive 91/356/EEC. A detailed guideline for the manufacture of medicinal products in accordance with these principles was also published recently. Moreover, good agricultural and collection practice should be followed in the handling of starting material of herbal origin.

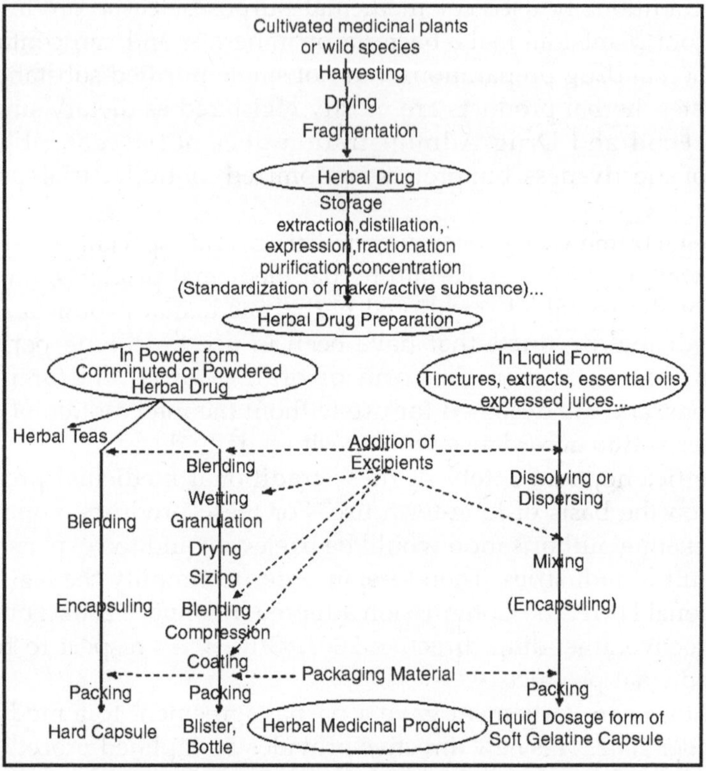

Fig. Basic Steps in Some of the Most Widely Used Manufacturing Processes for Herbal Medicinal Products.

This document is a recent guideline based on a document proposed by the European Herb Growers Association (EUROPAM). It addresses the specific concerns of growing, collecting and primary processing of medicinal plants or herbal drugs. HMP manufacture has different features depending on whether the active substance is in the form of a herbal drug or herbal drug preparation. The basic steps in the most frequently used manufacturing processes for capsule, tablet and liquid dosage forms are described in Figure. Herbal drugs are obtained from cultivated or wild plants, and are produced by the following operations: cultivation, harvesting, drying, fragmentation and storage. These operations must follow the guideline on the manufacture of herbal medicinal products.

For the production of herbal drug preparations, herbal drugs are subjected to treatments such as extraction, distillation, expression, fractionation, purification, concentra-tion or fermentation. They can be in liquid form (*e.g.*, extracts, tinctures, and essential oils) or in powder form. Powdered herbal drug preparations may be supplied in bulk form or as a sachet, *e.g.* for herbal teas, prepared extempo-raneously by the patient. They consist of one or more herbal drugs prepared by means of infusion, decoction or maceration. Infusions are prepared by pouring boiling water over the dried herbal drugs, in chopped form and, after being allowed to draw for 5.10 minutes, strained.

Decoctions are prepared by pouring cold water on the drug placed in a saucepan, bringing it to the boil, simmering for about twenty minutes, and then sieving the suspension. Maceration consists of pouring water onto the herb and leaving it to stand overnight. HMPs are prepared with herbal drugs or herbal drug preparation as active substances. Excipients are added to give the finished product, as for any other medicinal product. The finished products are usually in solid (*i.e.* tablet or capsule), liquid (*i.e.* oral solutions, drops), or semisolid (*i.e.* gel, unguent or cream) forms.

The dosage form of the 84 HMPs marketed in Finland are shown in Figure. In Finland, HMPs are also called semi-medicinal products because they are marketed with a simplified marketing authorisation based on bibliographical application. The retail sale of herbal medicinal products is usually restricted to pharmacies in Belgium, France, Greece, Ireland, Italy, Luxembourg, Portugal and Spain, but it is also permitted in other outlets. at least for certain HMPs. in Austria, Denmark, Finland, Germany, the Netherlands, Sweden and the United Kingdom. In Finland, 44 per cent of the HMPs are sold only in pharmacies, the other 56 per cent being distributed in pharmacies and in general food stores.

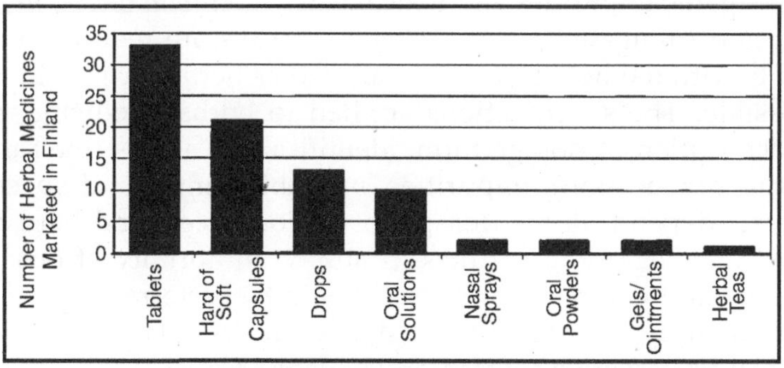

Fig. Distribution of the Dosage forms of the 84 Herbal Medicinal Products Marketed in Finland.

QUALITY CONTROL OF HERBAL MEDICINAL PRODUCTS

In May 1997, a Working Party on herbal medicinal products (HMPWP) was formed at the EMEA to prepare guidance for the mutual recognition in marketing authorisa-tions for herbal medicinal products. These guidelines concern regulations, quality, efficacy and safety of the herbal products, and were prepared because there were assessment differences on these topics in the individual European Union countries. So far, two guidelines under the topic of "Quality" have been adopted. They aim at providing a uniform set of specifications for herbal drugs and HMPs to support marketing authorisations. The specifications consist of a list of tests and acceptance criteria designed to verify the suitability of the herbal drug preparation or the herbal medicinal product for its intended use.

They define the product quality and are therefore useful in ensuring the safety and efficacy of HMPs. The general monographs *Herbal drug* and *Herbal drug preparation* should be used to interpret the specification requirements. Specifications applied to herbal drugs usually include: definition, characterization, identification, tests and assay. Identification testing should consist of three or more of the following tests: macroscopic characters, microscopic characters, chromatographic procedures and chemical reactions. Tests include foreign matter, particle size, water content, inorganic impurities and toxic metals, microbial limits, mycotoxins, pesticides and fumigation agents.

Content assays are applied in the case where the constituents with therapeutic activity are known; otherwise an assay of marker substances is required. Specifications applied to herbal drug preparations include: definition, characterization, identification, tests and assay. Identification testing should be specific and discriminatory with respect to substitutes or adulterants that are likely to occur. A combination of chromatographic tests is recommended for this purpose. The tests include residual solvents, water content, inorganic impurities and toxic metals, microbial limits, mycotoxins, pesticides and fumigation agents. Content assays are required for known constituents with therapeutic activity, as well as herbal drug content assays when possible. The specifications applied to herbal medicinal products include: description of dosage form, identification, assays (the same as for herbal drug preparation), impurities, microbial limits and specific tests. Specific tests depend on the dosage form. For tablets and hard capsules, dissolution/disintegration, hardness/friability, uniformity of dosage units, water content and microbial limit tests should be performed.

For oral liquid: uniformity of dosage units, pH, microbial limits, antimicrobial preservative content, antioxidant preservative content, alcohol content, dissolution, particle size distribution, redispersibility, reconstitution time and water content should be performed among others. The specifications are, nevertheless, only a part of the global quality scheme to be applied to HMPs. The overall quality control procedure should also include control of the raw materials and excipients, in.process testing, process evaluation and validation, stability testing and testing of batch consistency. Analytical procedures not described in a Pharmacopea should be validated according to the ICH guidelines in order to prove that they are suitable for their intended purpose.

The two ICH guidelines are a worldwide basis for both the regulatory authorities and industry. They are called.Validation of analytical methods: Definitions and terminology and.Validation of analytical procedures: Methodology. They prescribe that identification tests should be validated with respect to their specificity, and that assays should be validated with respect to their accuracy, precision, specificity, linearity and range.

Traditional Quality Control Tools

The quality control of bulk herbal drugs and finished products is mainly performed in the quality control labouratory with conventional spectroscopic, chromato-graphic, titrimetric, or other wet chemical analytical methods. However, these procedures are time. Consuming and expensive because they require the use of environmentally unfriendly chemicals and personnel with a relatively high level of training. A sample preparation step is frequently required before performing the analysis of the herbs. Recent sample, preparation techniques for the extraction, clean up, and concentration of analytes from herbal materials include solid, phase microextraction, supercritical. fluid extraction, pressurised, liquid extraction, microwave. assisted extraction, solid, phase extraction, and surfactant. mediated extraction. The most widely applied sample. preparation techniques are nevertheless selective solvent extraction, filtration and precipitation. The usual qualitative analysis techniques are infrared spectroscopy (IR), ultraviolet/visible (UV/VIS) absorption spectrometry, thin layer chromatography (TLC), and microscopic identification.

The most common tools for quantitative analyses are high performance liquid chromatography (HPLC) for a wide variety of compounds, and gas chromatography for volatile organic compounds. HPLC is one of the most widely used analytical tools for qualitative and quantitative measurements. It is also used for the isolation of pure natural compounds from an extract. An HPLC chromatograph consists of the following devices: solvent reservoir, high.pressure pump, packed column, detector and recorder.

The most widely used detectors are UV detectors. The two main modes of HPLC analysis are liquid.solid and liquid.liquid chromatography. They are also called adsorption (normal phase) and partition (reverse phase) modes, respectively. The reverse phase HPLC separates compounds according to their hydrophobicity, and is widely used for all types of biomolecules.

NEAR INFRARED (NIR) SPECTROSCOPY

Historical and Physicochemical Basis

The principles of NIR spectroscopy are linked to the development of Fourier Transform spectroscopy (FTS) as early as in the middle of the 19th century. In 1862, Fizeau constructed the first variable path interferometer for wavelength measurement and it is considered to be the ancestor of the actual Fourier interferometers. Michelson used this idea to develop his two.beam interferometer. Michelson, however, never realised the greatest potentiality of his technique and faced several limitations. For example, the only detector that Michelson used was his own eye. Therefore he was not able to record precisely the fringe intensity. Moreover, he did not realise that multiplexing could be used to measure all frequencies simultaneously, and had no concern about the energy throughput advantage. Multiplexing, a technique also used in telephone engineering to send a large number of messages simultan-eously, was first applied to the interferometric device in the middle 50′s by Fellgett during his doctoral studies. He was also the first to derive a spectrum from the interferogram by using Fourier transformation. At approximately the same time, Jacquinot discovered the throughput advantage.

FTS was seldom used until the advent of computer technology that allowed rapid and economical Fourier transformation of an interferogram into a spectrum. A major step in the acceptance of this technique by industry was the development of an algorithm by Cooley and Tukey in 1965 that significantly increased the speed of the Fourier transformation computation and led to Fast Fourier Transform equipment. Since then, the number of practical applications of the FTIR and FT.NIR spectroscopy has increased dramatically. The NIR region, discovered by William Herschel in 1800, is situated between the visible and the IR region of the electromagnetic spectrum, and ranges from approximately 780 nm to 2500 nm, corresponding to a frequency range of 4000 cm^{-1} to 12 800 cm^{-1}.

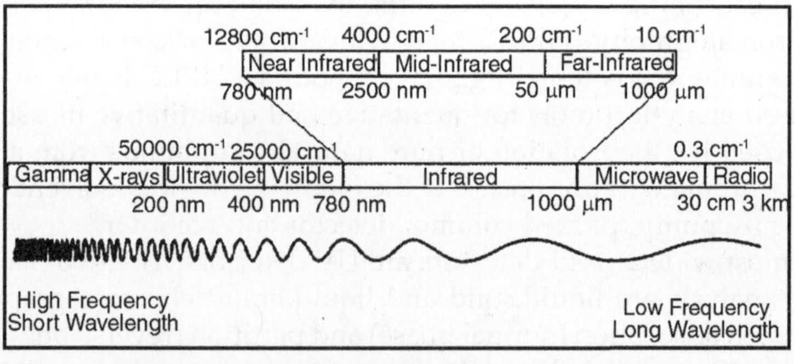

Fig. Electromagnetic Spectrum and Positioning of the Spectral Regions.

At temperatures above absolute zero, all the atoms in molecules are continuously vibrating. Two of the major types of vibration are stretching and bending, as illustrated for a non-linear group CH_2 in Figure. If the fundamental frequency of a specific vibration, ½, is equal to the frequency of the radiation impinging on the molecules, and if the molecule undergoes a change in its dipole moment during the vibration, then the radiation is absorbed and excites a vibrational transition in the molecule. Molecular vibrations are often described by means of the harmonic oscillator model. The basic assumption is that the shift of an atom is directly proportional to the force opposing the shift. For a harmonic model, only transitions between two consecutive vibrational energy levels are allowed. The energy difference between the two levels is *h*½, where ½ is the fundamental frequency. In practice, molecular vibrations tend to follow an anharmonic model. This leads to the appearance of additional allowed transitions with energy differences of two, three *etc.*, times the fundamental frequency.

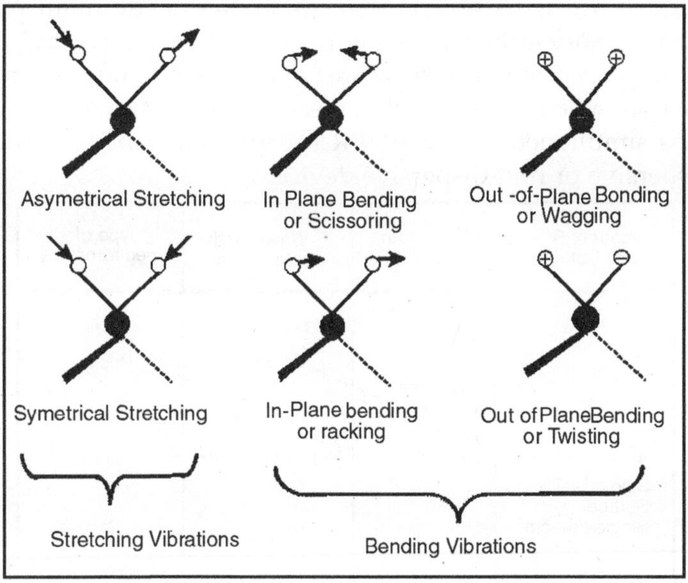

Fig. Main Vibrational Modes of a Nonlinear CH_2 Group.

These higher transitions are called overtones and they are observed beyond the fundamental vibration region in the near-infrared region. In the NIR region also combination vibrations are observed. The most common group frequencies seen in the NIR region are.OH.CH.NH. and SH overtones. Overtone and combination transitions are much less likely than the fundamental transitions. This explains why the intensities of the generally broadly overlapping NIR bands are weaker than the intensities of the fundamental IR bands, by a factor of 10 to 1000. The weakness of absorption bands in the NIR region is the key to the main advanta ge of the technique: it allows for longer path lengths (1 mm.1 cm) to be used than in the Mid. IR region (often < 0.1 mm for liquids). The samples are thus often analysed without sample pretreatment.

NIR SPECTROPHOTOMETERS

The essential features of NIR spectrophotometers are: a source of radiation, an operating contrivance and a detector. The NIR source produces radiation spanning a large or a narrow range of frequencies in the NIR region.

They can be thermal or non.thermal sources. Thermal sources consist of an incandescent filament producing thermal radiation, *e.g.* the Nernst filament, which is a heated ceramic filament containing rare-earth oxides, or quartz. halogen lamps. Non thermal sources usually consist of light. emitting diodes (LED), laser diodes, or lasers that emit much narrower bands of radiation than thermal sources. NIR spectrophotometer can be divided into three groups: those with one source and one detector, those with several sources and one detector, and those with several detectors. One of the main parts of the operating contrivance is the wave-length selection device. It may be a discrete absorption device, *i.e.* only a narrow area of the spectral range is measured at once, or a whole spectrum device that measures the information from several wavelengths simultaneously. The NIR spectrometer optical system can be either a dispersive or non-dispersive device.

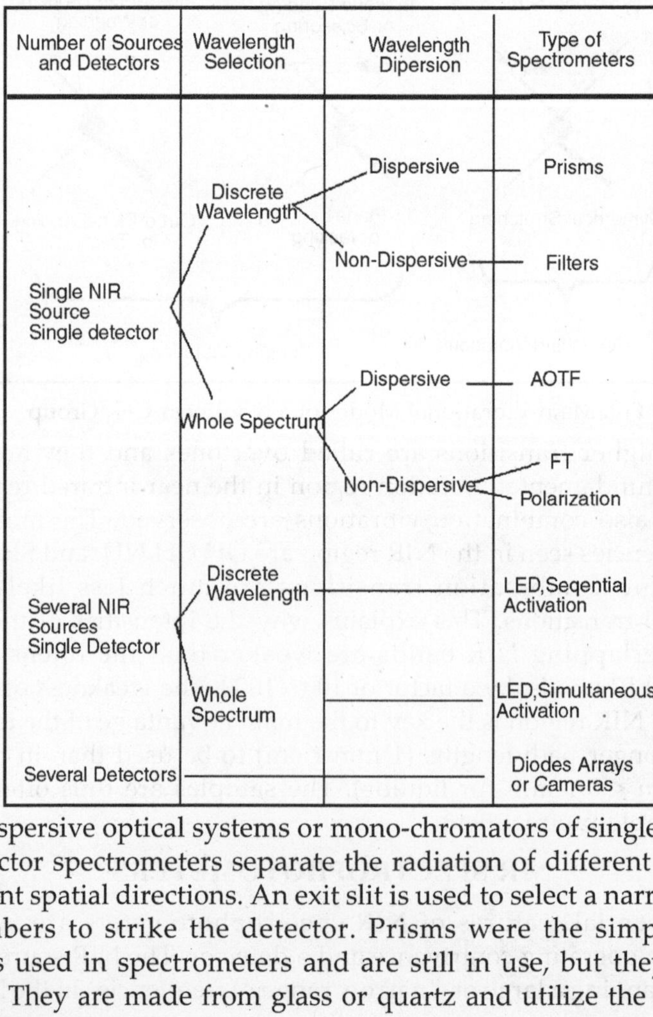

The dispersive optical systems or mono-chromators of single source and single detector spectrometers separate the radiation of different frequencies into different spatial directions. An exit slit is used to select a narrow range of wave numbers to strike the detector. Prisms were the simplest mono-chromators used in spectrometers and are still in use, but they give poor dispersion. They are made from glass or quartz and utilize the variation of the refractive index as a function of the frequency as a separating tool.

Non-dispersive optical systems based on filter devices may include up to 20 filters on a carousel. This type of instrument is robust and still in use for routine analysis. Acousto-Optic Tunable Filters (AOTF) have been incorporated in NIR spectrophotometers in recent years. This technique uses acousto,optic diffraction of light in an anisotropic crystalline medium as the separation device. The absence of moving parts in AOTF ensures good wavelength stability, and provides a rugged, cost.effective instrument with a high signal to noise ratio. The resolution of AOTF instruments is approximately 5 nm. Fourier Transformed instruments are based on interferometers that are widely used in modern spectrometers. The Fourier Transform technique is based on the use of an interferometer (mostly of the Michelson - type) that is able to detect intensities of several spectral frequencies in a composite signal. The Fourier Transform of the recorded interferogram is the infrared spectrum. The purpose of an interferometer is to split a beam of light into two beams and to introduce a difference in their respective travelling distances. The optical path difference is denoted as δ.

The interferometer shown in Figure consists of four arms, one for the source, the second having a moving mirror M2, the third a fixed mirror M1, and the last one is open. The beamsplitter is used to transmit half of the radiation obtained from the source to the moving mirror and to reflect the other half of the radiation to the fixed mirror. The beamsplitter usually consists of a very thin film of germanium covered on both sides by a potassium bromide (KBr) substrate. The two separated beams respectively strike M1 and M2 and are reflected back to the beamsplitter. They are then recombined and exit the interferometer in the direction of the sample and detector. M2 is moving longitudinally back and forth. When $\delta = 0$, both mirrors are equidistant from the beamsplitter. This is called the.zero path difference. (ZPD). When the interferometer is in the position of ZPD ($\delta = 0$) or when $\delta = n\lambda$, the two recombined beams are in phase with each other and the intensity of the detector signal will thus be maximum.

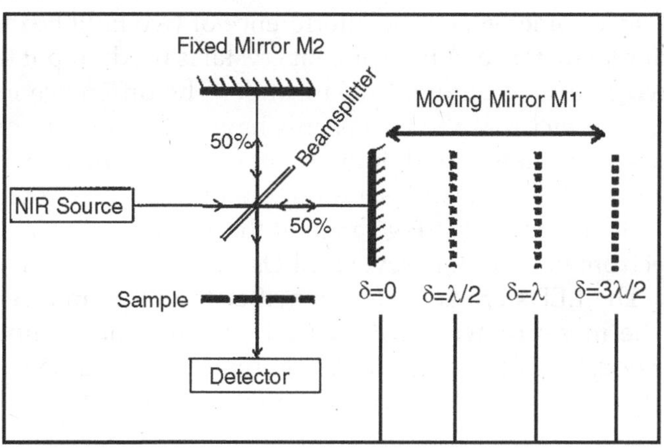

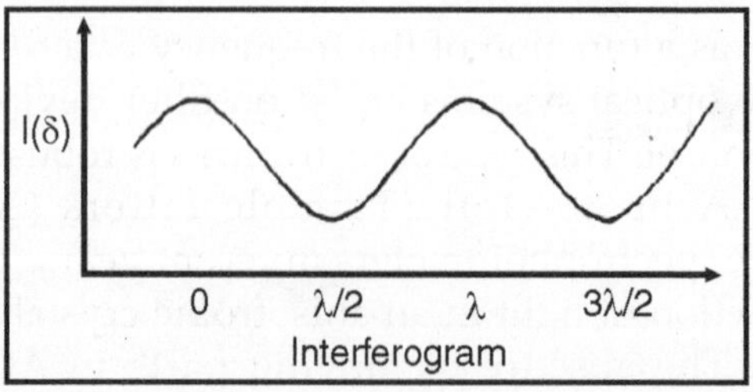

Fig. Principle of a Michelson Interferometer and Example of an Interferogram Obtained from a Monochromatic Source.

These states are called constructive interference. Destructive interferences are obtained when $\delta = (n + 1/2)\lambda$, and in these cases the resulting beam intensity is zero. Intermediate intensities are obtained at intermediate positions of δ. The plot of the intensity versus the optical path difference is called an interferogram. Figure shows an interferogram obtained with a monochromatic source. When the source is polychromatic, radiation of different wavelengths undergoes destructive and constructive interference at different optical path differences. Each wavelength of light leads to an interferogram with a specific path difference, resulting in intensity typical of their frequency that can be measured by the detector.

The signal passing through the sample is the sum of each specific interferogram and therefore contains intensity information about all the wavelengths contained in the band passing the sample. The interferogram, an.intensity versus time. function, is then Fourier transformed to obtain the final NIR spectrum, which is an.intensity versus frequency. function. FT.NIR spectrophotometers can be obtained from several suppliers, including Bomem Inc., Bran + Luebbe, Brücker Instrument, Büchi Labotecknik or Perkin Elmer. The newest features on FT instruments include, for example, an imaging - system providing pictures of the samples showing the chemical distribution at the microscopic level.

Polarization or crystal spectrometers are also.whole spectrum. techniques, but they are not as well known as FT techniques. They are based, as is the case with FT spectrometers, on the interference of two light beams travelling a slightly different distance. A birefringent crystal is used to split the incoming beam into two beams of different polarisation. The difference in the optical path is due to the fact that the two beams have different refraction indices. The group of spectrometers containing several sources includes non.thermal optical designs, such as LED or laser diodes, and selection of the wavelength is inherent in the narrow emitting range of the source.

LED spectrometers contain several LEDs, each coupled to a narrow band optical filter. The LEDs are activated one after the other in a sequence and, because all the measurements are focused on the same channel, only one detector is needed. LEDs can also be activated simultaneously and the instrument functions as a multiwavelength device. The last group of spectrophotometers contains several detectors and they are called multichannel spectrometers. The operating principle is based on diode arrays or cameras that can measure many wavelengths simultaneously. This type of instrument is available from Büchi Labotecknik, Perten Instruments or Multichannel Instruments. However, single detector instruments are normally used. Concerning detector technology, silicon,based photodetectors are recommended for the short,wavelength infrared range (700.1000 nm or 14286.10000 cm^{-1}). For lower energies and longer wavelengths (1100.2500 nm or 9090.4000 cm^{-1}), semiconductors such as lead sulphide (PbS), indium gallium arsenide (InGaAs) or indium arsenide (InAs) can be used as detectors.

SPECTROMETRIC TECHNIQUES COMPARED TO TRADITIONAL ANALYTICAL METHODS

The advantages of FT transform techniques over dispersive instruments have resulted in almost total replacement of the dispersive instruments in spectroscopy.

- *The Multiplex or Fellgett Advantage*: In a dispersive spectrometer, wavenumbers are observed sequenti-ally. In an FT.IR and FT.NIR spectrometer, all the wavenumbers of light are observed simultaneously. Therefore, when spectra are collected under identical conditions the signal.to.noise (S/N) ratio of the FT.IR spectrum will be greater than that of the dispersive IR spectrum.
- *The Throughput or Jacquinot Advantage*: In FT.IR instru-ments there is no need to limit the beam width in order to obtain an adequate resolution. In fact, a circular optical aperture is used in FT instruments, and the beam area is 75 to 100 times larger than the slit area of dispersive instruments. As a consequence, there is an advantage of increased beam intensity going through the sample and therefore a much higher throughput with a FT.IR than with a dispersive instrument.
- *Wavenumber accuracy or Conne's advantage*: In FT instruments, *e.g.* in FT.IR Bomem spectrometers, a wavelength accuracy of 0.04 cm^{-1} can be obtained, which is much higher than the traditional wavelength accuracy obtained with dispersive instruments (only about 1.5 cm^{-1}). This difference is due to the fact that the frequency-stabilised helium. neon laser is used as internal wavelength standard. Therefore the frequency precision is determined by the frequency stability of the laser, which leads to precise and reproducible wavelengths.
- *High and constant resolution*: Spectral resolution is a measure of how well a spectrometer can distinguish closely spaced spectral features. Filter instruments cannot offer high resolution because, in dispersive instruments, resolution decreases as lower frequencies are scanned. In FT.IR, the resolution depends on the optical path difference (OPD)

that can be achieved. Thus it is constant across the scanning range.

- *Practical and powerful data station*: FTIR or FTNIR instruments are normally equipped with a powerful computer capable of carrying out the Fast Fourier Transformation needed to obtain the spectrum. Additionally, the instrument computer uses software that can perform data processing such as baseline correction, smoothing, derivatisation or library searching, and therefore improve data information.

The most attractive advantage of FT.NIR spectroscopy over traditional analytical tools and any other spectroscopic method is probably that the measurements are nondestructive, and noninvasive, and that it is possible to use solid samples without pre. treatment and therefore without solvents. This leads to a large increase in the analysis speed compared to traditional analysis methods, and decreases the risk of errors due to weighing and dilution operations. The variety in sampling technologies is another attractive feature of NIR spectroscopy. Several accessories are adaptable to a number of situations, and can be used with different scanning modes. For instance, fibre optic probes were used already ten years ago for real-time analysis and are nowadays often used in the diffuse reflectance mode for routine qualitative and quantitative applications. Diffuse reflectance is also easily used for off-line analysis for samples contained in simple glass vials. The transmittance mode is more and more widely used for recording spectra from intact tablets. It gives results with a better repeatability and a smaller prediction error than reflectance measurements.

The transflectance mode, which is a variant of the diffuse reflectance mode, has also been investigated recently. In this case, incident light crosses the sample, is reflected by a reflectance material such as stainless steel or PTFE (Polytetrafluoroethylene) located on the opposite side, and travels back through the sample before reaching the detector. An important property of the NIR signal is that, because it depends on both the chemical composition and the physical properties of the sample, analysis of these two characteristics can be performed by the same technique. NIR also has the potential to be used for developing on.line methods, leading to real. time control systems. This advantage can be well used in the pharmaceutical or chemical industry to give real.time information about processes.

On the other hand, there are three main disadvantages of NIR spectroscopy over traditional techniques. First, the development of a NIR method is time.consuming because it is necessary to analyse several representative samples by a time.consuming reference analysis method (HPLC or Karl Fisher titration for example). Secondly, NIR methods lack robustness: calibrations often need to be updated, *e.g.* when a sample is provided by a new supplier, or when the manufacturing process of the sample is modified. This is especially problematic with raw material whose quality may vary from time to time leading to false identification of the material. Furthermore, NIR spectroscopy is not very sensitive and it can usually be satisfactorily applied to major components but not to impurities or low.dose substances. Other minor disadvantages are the following: First, in contrast to IR spectra, NIR raw spectra exhibit low specificity. They do not show clear peaks characteristic to a specific compound of interest.

Thus, extensive statistical calculations are required to extract useful qualitative or quantitative information. Second, NIR spectroscopy methods are developed using the reference analysis results of the calibration samples. Thus, the accuracy of the NIR method cannot be better than the accuracy of the reference method. Furthermore, the transferability of NIR methods from one instrument to another is limited due to the frequent need for updating calibrations after routine maintenance or repair of the instrument. Finally, the absence of NIR training in pharmacy schools is one of the major obstacles to the acceptance of NIR spectroscopy by pharmacists. The specialised vocabulary used in the chemometrics world makes things even less accessible for pharmacists.

USE OF CHEMOMETRICS

Chemometrics is a chemical discipline that utilizes mathematics and statistics to design optimal measurement procedures and experiments and to provide maximum relevant chemical information by analysing chemical data. Traditional applications of chemometrics often involve data pre.processing for enhancing analytical measurements to obtain chemically or physically relevant information from the sample and to reduce the irrelevant variability that arises from the effect of instrument changes over time or physical phenomena, such as temperature, or scattering. Of the number of existing signal.preprocessing techniques, only the most widely used mathematical tools will be described here. In the reflectance mode, NIR spectra are subjected to large baseline shifts introduced by the spectrometer or sample especially in the case of solid powdered samples with a large particle size distribution, because scattering of the light is strong. Baseline effects can also be due to a number of reasons such as detector drift, changing environmental conditions *e.g.* temperature and humidity, and sampling accessories.

One of the best methods for removing baseline effects is to use derivative spectra. A constant background can be removed by transforming the original spectra into first.derivative spectra, while the linear background can be removed by taking second.derivative spectra. The second derivative is more often used because it increases the selectivity of interesting bands and thus simplifies the data interpretation. However, as derivation amplifies the spectral noise, it is necessary to smooth the data before derivation. The most widely used differentiation method is the Savitzky and Golay algorithm, which combines smoothing and differentiation and thus removes the noise.

The use of Standard Normal Variate (SNV) transformation leads to the removal of the major effects of light scattering and particle size. The SNV algorithm normalises each spectrum by dividing the difference between the transmittance and average transmittance by the standard deviation of transmittance.

De.trending is also a baseline correction method. It removes offset and curves linearity, which often occurs in the case of powdered, densely packed samples. The baseline is modelled as a function of wavelength and subtracted from the spectrum. Normally, de.trending is carried out in combination with SNV transformation. Multiplicative signal correction (MSC) can be used to resolve the problem of a varying background due to differences in optical path length and to compensate for different scatter and particle sizes from sample to sample. The principle is that MSC establishes a linear regression between spectral variables and the average spectrum. The slope and offset values of the regression spectrum are then removed from the original spectrum in order to give a corrected spectrum. Multivariate calibration remains, by far, the fastest growing area of chemometrics.

This procedure is used to relate the analyte concentration or the measured value of a physical or chemical property to a measured response. PLS is now dominating the practice of multivariate calibration, because of the quality of the calibration models produced and the ease of their implementation. This algorithm was developed by WOLD and MARTENS in the beginning of the 80's. Industrial problems can frequently be described on the basis of an input/ output system: X are the input variables, and Y the output variables being observed. The PLS regression is a linear regression technique that can be used to understand and explain the relationship between X and Y. The PLS regression principle is to find new variables to estimate the latent or underlying X variables. The new variables are called X.scores and denoted by T. The X.scores are used to model X and to predict Y (response variables). The X.scores are orthogonal and restricted in number and are linear combinations of the original variables X with the coefficients, or weights W. The relationship between the matrices X, Y, T and W are shown in Equations and in Figure.

The explanation of the abbreviations are as follow: X is the (N x K) matrix of the predictor variables, Y is the (N x M) matrix of the response variables, N is the number of observations, k is the index of the X variables, W is the X weight matrix, W* is the matrix of the X weights transformed to be independent between components, A is the number of components in the PLS model, C, is the transposed Y weight matrix, T is the X.score matrix (N x A), U is the Y, score matrix (N x A), P, is the transposed loading matrix, E is the matrix of the X residuals, and F is the matrix of the Y residuals. The notation employs uppercases for the matrix (*e.g.*, X), and lowercases for their corresponding values (*e.g.* xa) T = XW* X.scores, multiplied by the loadings

P, are a good estimation of X providing that the residues E are small X = TP′ + E Y.scores, multiplied by the weights C, are a good estimation of Y providing that the residues G are small. Y = UC′ + G The X.scores are good predictors of Y, providing that the residues F are small. Y = TC′ + F Therefore the summarising equation is in the form of a multiple regression, with XW* as the PLS regression coefficients: Y = XW*C′ + F If the predictive power of this regression is too weak when only one component is calculated, then a second component is calculated.

This iterative procedure can be continued to calculate as many components or factors as are predictively significant. The geometric interpretation of the PLS model is a projection of the K. dimensional X matrix down on an A. dimensional plane (A<K). Each plane has a direction corresponding to a PLS component, A. The direction of each plane is described by its slope, pak (loadings). Each point projected on the plane is characterised by its coordinates, also called scores t. The plane satisfactorily approximates X and, at the same time, the positions of the projected data points on this plane (scores t), are related to the responses, Y.

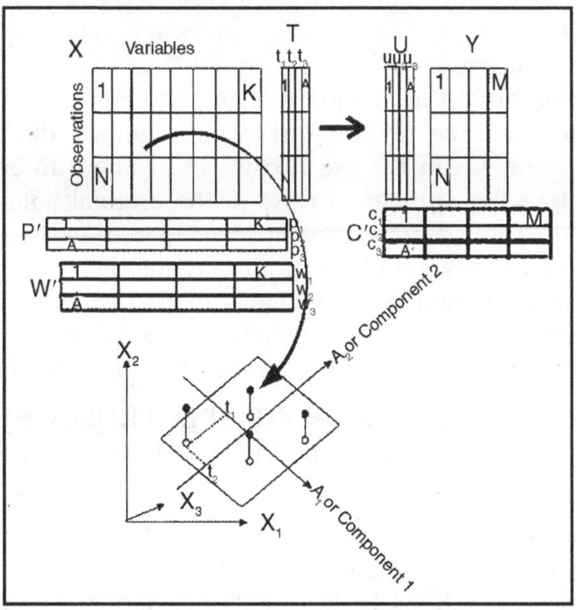

Fig. Matrix and Geometric Representation of a PLS Model.

The number of factors to be retained can be evaluated in several ways, often based on cross.validation. The principle of cross.validation is to remove one sample or a group of samples from the calibration set and then to calculate the model with the remaining samples. The model will be different depending on which sample is removed and on the number of factors included. The removed sample is predicted by each model that includes a successive number of factors. The cross.validation is repeated by omitting another sample, and so on, until each sample from the calibration set has been removed once. Then, the differences between the actual and predicted Y values are calculated for the deleted data. The sum of squares of these differences (Predictive Residual Sum of Squares or PRESS) gives an estimation of the predictive ability of the model. The number of significant PLS components is usually calculated to be the minimum number for which the PRESS value is not significantly different from the lowest PRESS value, as described by HAALAND and THOMAS.

If the number of factors (or components) is too high, the risk of overfitting the model is increased. An overfitted model has little or no predictive power, because it includes factors that are not related to the constituent of interest but instead to the noise. Principal Component Regression is another widely used multivariate calibration method, dominated by the use of a compression technique, Principal Component Analysis (PCA). PCA also allows data visualisation by means of data dimensionality reduction. PCA is the most popular linear projection method.

It projects multidimensional data onto a few directions called principal components (PCs). PCs are a linear combination of the original variables that describe the data variance. They explain successively decreasing amounts of variance in the matrix X. Thus, the first PC is the direction that best approximates (minimised least square error) the original data and explains the maximum variance of the data. The second PC improves the approximation, and so on for the further PCs. The number of extracted components equals the number of rows in the original data matrix, but all components with small eigenvalues are considered as data noise and are eliminated. There are several other important chemometrics methods, such as multiple linear regression (MLR) which is another widely used multivariate calibration method, quantitative structure activity relationship (QSAR), pattern recognition, multivariate process modelling or artificial neural systems.

APPLICATIONS OF NIR SPECTROSCOPY IN PHARMACEUTICAL TECHNOLOGIES

In addition to applications in the food, textile, biological, petroleum and chemical industries, pharmaceutical technology is one of the main application fields for NIR spectroscopy. The current applications concern a large part of the pharmaceutical operations. In the initial stage of pharmaceutical processes, NIR is used for the identification of raw materials: active substances and excipients. The physical properties of the raw material are also determined by near infrared spectroscopy, *e.g.* the particle size of drugs or excipients. In the following step of the manu-facturing process, the monitoring of blending processes can be performed successfully with NIR. NIR can provide real.time information about the blending, which is often not the case with traditional analysis methods.

• When tablets are manufactured, wet granulation is often the next process phase. NIR has frequently been used to monitor the granulation process, for example, to quantify a drug during the different steps of the granulation process, to measure the particle size, or to follow the moisture content during granulation. After granulation, the tablets are sometimes coated. Here, as well, NIR can be applied *e.g.* to monitor the film coating process and to perform a final identification test of the active compound in the final dosage form. The identification of active substances through blister packaging can also be performed *e.g.* for discriminating between active tablets and placebos during clinical trials.

One of the most important steps in tablet quality control is the quantitative assay for the active substance. Currently, mostly other spectroscopy methods, and chromatographic, titrimetric and wet chemical methods, are used to analyse finished products. However, these tools are destructive for the sample, while NIR is not. Quantitative measurements are nevertheless not the major application for NIR spectroscopy in the pharmaceutical industry, probably because it requires more extensive work than qualitative methods, and also because of the lack (nearly fulfilled nowadays!) of adapted guidelines for the validation of such procedures. However, a large number of publications reflect the advantages of the NIR assay of active substances in semi.finished or finished products such as granules, cores or tablets.

Quantitative measurements for the determination of moisture in finished products are also a popular application of NIR spectroscopy. Very few papers have described the use of NIR spectroscopy to monitor the quality of a pharmaceutical product during the different steps of its manufacture. Concerning herbal medicinal applications, NIR spectroscopy has, during the past few years, become a useful tool for the non.destructive analysis of plant species and herbal products. An electronic search using the SciFinder Scholar database showed that there are approximately twenty applications in this field in the literature. In fact NIR spectroscopy has primarily been used to identify or classify herbal drugs and herbal drug preparations, but seldom to control the quality of herbal medicinal products. Literature applications report significant improvements in terms of speed and flexibility of NIR analysis compared to the conventional or traditional analysis of drugs. However, none of these applications were validated according to pharmaceutical regulations.

NIR SPECTROSCOPY IN PHARMACEUTICAL INDUSTRIES

Although there are still no monographs in which the use of near infrared spectroscopy is mandatory, the regulations concerning NIR spectroscopy are nowadays better adapted to its practical use in the industry. The first sign of the acceptance of NIR spectroscopy was the publication of the European Pharmacopoeia monograph on NIR describing the control of instrument performance and the establishment of a spectral reference library for qualitative purposes. Then appeared the revision in 1998 of the USP general chapter of Spectrophotometry and Light scattering, and the publication of a guideline for the qualification of NIR instruments. However, the European Pharmacopoeia monograph on NIR does not give extended details on the way to develop and validate a qualitative application or any information about the development or validation of quantitative methods.

In fact, up until now, quantitative NIR methods were validated according to the ICH guidelines Q2A and Q2B which, nevertheless, address traditional method validation requir-ements more than NIR method validation. The reason for this is that these guidelines were planned for separative techniques such as high.performance liquid chromatography or gas chromatography, and are not as suitable for non.separative procedures such as direct spectroscopic methods. The lack of specific validation guidelines led several authors to attempt to adapt the ICH guidelines to the purposes of their own NIR quantitative application. The Pharmaceutical Analytical Sciences Group (PASG) even proposed its own guideline. This guideline covers the design, development and validation of NIR methods, whilst remaining complementary to ICH Q2A and Q2B.

In the meanwhile, the EMEA published draft guideline for the use of near infrared spectroscopy in the pharmaceutical industries. This document aims at regulating the validation processes of qualitative and quantitative NIR methods, and provides some recommendations about the qualification of NIR spectrometers. The draft is under modification and will greatly contribute to the acceptance of this technique as an analysis tool in the European industries. After acceptance of this draft, the pharmaceutical industries will have the regulations required to implement NIR spectroscopy as a routine analysis tool in their labouratories.

5

Raman Spectroscopy

Raman spectroscopy provides information about molecular vibrations that can be used for sample identification and quantitation. The technique involves shining a monochromatic light source (*i.e.* laser) on a sample and detecting the scattered light. The majority of the scattered light is of the same frequency as the excitation source; this is known as Rayleigh or elastic scattering. A very small amount of the scattered light (ca. 10-5 per cent of the incident light intensity) is shifted in energy from the laser frequency due to interactions between the incident electromagnetic waves and the vibrational energy levels of the molecules in the sample. Plotting the intensity of this "shifted" light versus frequency results in a Raman spectrum of the sample. Generally, Raman spectra are plotted with respect to the laser frequency such that the Rayleigh band lies at 0 cm-1. On this scale, the band positions will lie at frequencies that correspond to the energy levels of different functional group vibrations. The Raman spectrum can thus be interpreted similar to the infrared absorption spectrum.

Spectroscopy is an analytical technique arising from the interaction of a species with electromagnetic radiation - the electromagnetic radiation absorbed, emitted or scattered by the molecule is analyzed. Typically, a beam of radiation from a source such as a laser is passed through a sample, and the radiation exiting the sample is measured. For example, in vibrational absorption spectroscopy, by varying the frequency of the radiation, a spectrum can be produced, showing the intensity of the exiting radiation for each frequency. This spectrum will show which frequencies of radiation have been absorbed by the molecule to raise it to higher vibrational energy states. Each molecule will have its own characteristic spectrum, making spectroscopy indispensible in analytical chemistry.

THEORETICAL BASIS

The Raman effect occurs when light impinges upon a molecule and interacts with the electron cloud and the bonds of that molecule. For the spontaneous Raman effect, which is a form of light scattering, a photon excites the molecule from the ground state to a virtual energy state. When the molecule relaxes it emits a photon and it returns to a different rotational or vibrational state. The difference in energy between the original state and this new state leads to a shift in the emitted photon's frequency away from the excitation wavelength. The Raman effect, which is a light scattering phenomenon, should not be confused with absorption (as with fluorescence) where the molecule is excited to a discrete (not virtual) energy level.

If the final vibrational state of the molecule is more energetic than the initial state, then the emitted photon will be shifted to a lower frequency in order for the total energy of the system to remain balanced. This shift in frequency is designated as a Stokes shift. If the final vibrational state is less energetic than the initial state, then the emitted photon will be shifted to a higher frequency, and this is designated as an anti-Stokes shift. Raman scattering is an example of inelastic scattering because of the energy transfer between the photons and the molecules during their interaction.

A change in the molecular polarization potential — or amount of deformation of the electron cloud — with respect to the vibrational coordinate is required for a molecule to exhibit a Raman effect. The amount of the polarizability change will determine the Raman scattering intensity. The pattern of shifted frequencies is determined by the rotational and vibrational states of the sample. This dependence on the polarizability differs from Infrared spectroscopy where the interaction between the molecule and light is determined by the dipole moment; this contrasting feature allows to analyze transitions that might not be IR active via Raman spectroscopy, as exemplified by the rule of mutual exclusion in centrosymmetric molecules.

APPLICATION OF RAMAN SPECTROSCOPY

The spectroscopic examination of artworks is of high importance for conservators, art-historians and keepers of private or public museum collections. These investigations reveal information, which is of general (art-) historical interest: the knowledge of the artistic materials that were used and were available at a certain period in a particular region and the dissemination of new methods provide information about the interactions between distinct cultures and about trade routes. The knowledge on the artists' materials that were available in particular regions and periods can help in dating artefacts. The retrieval of pigments with a well-known date of invention enables to date the artefact *post quem*.

Other materials are known to have disappeared from the artists' palette, because they were substituted by others and retrieving enables to date artefacts *ante quem*. Finding anachronisms in the materials that were used, is a straightforward way for the exposure of counterfeit masterpieces. Another method consists of the comparison with the materials that were used in known works of the same artist. If it is well-established that a certain artist in a large group of works from a particular period never used certain pigments, finding these materials in a suspect painting deepens the suspicion and invites for further examinations. Besides these reasons for the spectroscopic examination of objects of art, an important purpose of this work is to help conservators in finding the reasons of the deterioration of a certain artwork and helping them in optimising the conditions of conservation. The main purpose of any analytical examination of artefacts should be to gain as much information as possible in a non-destructive way.

Molecular Raman spectroscopy is well-suited for this purpose: it enables the identification of inorganic and organic pigments as well as binding media and varnishes. By focusing a laser on a sample the intensity of the inelastically scattered light is plotted against the Raman wavenumber, which is proportional to the difference in energy between the laser and the scattered light. By using a microscope to focus the laser beam on the sample, *i.e.* micro-Raman spectroscopy, one can get spectra from particles with a diameter down to 1 μm, which is about the usual dimension of artists' pigments.

This high lateral resolution can be used for the examination of:

- Embedded stratigraphic samples, of
- Micro-samples, or even for the
- Direct investigation of artefacts that can be positioned under the microscope. For large artefacts
- Fibre optics can be used for remote investigation. In this case there is loss of spatial resolution to some mm^2.

EXPERIMENTAL OF RAMAN INVESTIGATIONS

Micro-Raman investigations are performed using a dispersive Renishaw System-1000 Ramanspectrometer. The device is fitted with a diode laser operating at 50 mW and 780 nm. An Olympus BH-2 microscope is used to focus the laser on the sample, using objective lenses of 5x, 20x, 50x and 80x magnification. The microscope is equipped with binoculars and a colour video camera, allowing to position the sample and to select a specific region for investigation. The backscattered light passes through a holographic notch filter (HNF), which rejects the abundant elastic Rayleigh scattering, to avoid outshining of the weaker Raman signal. The backscattered light is dispersed by using a 1200 lines/mm grating and is detected on a Peltier-cooled CCD-detector. This configuration allows to record spectra with a spectral resolution of *ca.* 1 datapoint/cm^{-1} in the spectral region between 150 and 2500 cm^{-1}. The fibre-optics Ramanspectra are recorded on a Bruker FT-Raman spectrometer, allowing to record spectra from 250 cm^{-1} onwards.

Direct Analysis of 19th Century Porcelain Cards

Micro-Raman spectroscopy can be used for the non-destructive direct analysis of small objects of art that can be positioned under the microscope optics of the spectrometer. Examples of such small objects of art are 19th

Century porcelain cards. These luxurious products served as status symbols of the *bourgeoisie* and were used as business cards or invitations for parties. The cards were manufactured in a combined process of printing the text and illustration and colouring by hand. In Flanders only a confined number of printers fabricated these cards and the number of colours that were used was limited. It was our purpose to reveal which colorants were used in these cards. Therefore several porcelain cards were analysed directly under the Raman microscope. The geometry for direct micro-Raman analysis is shown together with some Raman spectra from the investigation of a porcelain card of the Ghent University. This porcelain card, which is dated 1843, was released on the occasion of two fancy dress balls that were organised by the students of the university.

The recorded spectra are of sufficient quality to allow identification of the pigment material. The bright blue areas are painted with a Prussian blue ($Fe_4[Fe(CN)_6]_3$) containing paint. This blue iron-containing pigment was for the first time synthesised by Diesbach in Berlin in 1704. It can easily be identified via the band at *ca.* 2150 cm^{-1}, which is caused by the v(CN) stretch. The darker blue areas are painted with ultramarine ($Na_{8..10}Al_6Si_6O_{24}S_{2..4}$), while the spectrum of the red paint is that of vermilion (HgS). The yellow areas are painted with a chrome yellow ($PbCrO_4$) containing paint. The card is whitened with white lead ($2\ PbCO_3 \cdot Pb(OH)_2$).

MASER

Maser, acronym for microwave amplification by stimulated emission of radiation, a device that amplifies or generates microwaves or radio waves. A maser producing radiation in the optical region is called a laser.

Principle of Operation

As in lasers, amplification of radiation in masers is obtained by stimulated emission. This occurs when a photon induces an excited atom or molecule to fall to a lower energy state while emitting a photon of the same frequency as the incoming photon. The emitted photon travels in the same direction and in phase with the incoming photon, which is not absorbed during the interaction. The amplitudes of the two waves add up, and amplification of the incoming wave has taken place. Masers make use of those transitions in molecules or crystals that correspond to the energies of microwave or radio frequencies.

Types of Masers

The first maser oscillator was developed by the American physicists Charles Hard Townes, James P. Gordon, and Herbert J. Zeiger in 1954, Hard Townes and made use of the frequency of the ammonia molecule. This frequency corresponds to the energy of the photon emitted when the nitrogen atom moves from one side to the other of the triangle formed by the three hydrogen atoms in an ammonia molecule. The hydrogen maser makes use of the frequency corresponding to that of the photon released when the spin of the proton in a hydrogen atom flips over with respect to the spin of the atom's electron. Paramagnetic masers use energy transitions corresponding to the orientations of the magnetic moments of paramagnetic ions in crystalline substances placed in an external magnetic field. Different frequencies can be obtained by varying the magnetic field, thus allowing the tuning of a paramagnetic maser from less than a megacycle to several hundreds of megacycles.

PARTICLE OF LIGHT ENERGY

PHOTON

Photon, particle of light energy, or energy that is generated by moving electric charges. Energy generated by moving charges is called electromagnetic radiation. Visible light is one kind of electromagnetic radiation. Other kinds of radiation include radio waves, infrared waves, and X rays. All such radiation sometimes behaves like a wave and sometimes behaves like a particle. Scientists use the concept of a photon to describe the effects of radiation when it behaves like a particle. Most photons are invisible to humans. Humans only see photons with energy levels that fall within a certain range. We describe these visible photons as visible light. Invisible photons include radio and television signals, photons that heat food in microwave ovens, the ultraviolet light that causes sunburn, and the X rays doctors use to view a person's bones. The photon is an elementary particle, or a particle that cannot be split into anything smaller. It carries the electromagnetic force, one of the four fundamental forces of nature, between particles. The electromagnetic force occurs between charged particles or between magnetic materials and charged particles. Electrically charged particles attract or repel each other by exchanging photons back and forth.

Characteristics of Photons

Photons are particles with no electrical charge and no mass, but they do have energy and momentum, a property that allows photons to affect other particles when they collide with them. Photons travel at the speed of light, which is about 300,000 km/sec (about 186,000 mi/sec). Only objects without mass can travel at the speed of light. Objects with mass must travel at slower speeds, and nothing can travel at speeds faster than the speed of light. The energy of a photon is equal to the product of a constant number called Planck's constant multiplied by the frequency, or number of vibrations per second, of the photon. Scientists write the equation for a photon's energy as $E = hv$, where h is Planck's Constant and v is the frequency. Photons with high frequencies, such as X rays, carry more energy than do photons with low frequencies, such as radio waves. Photons that are visible to the human eye have energy levels around one electron volt (eV) and frequencies from 10^{14} to 10^{15} Hz (hertz or cycles per second). The number 10^{14} is a 1 followed by 14 zeros. The frequency of visible photons corresponds to the color of their light. Photons of violet light have the highest frequencies of visible light, while photons of red light have the lowest frequencies. Gamma rays, the highest-energy photons of all, have energies in the 1 GeV range (10^9 eV) and frequencies higher than 10^{18} Hz. Gamma rays are only produced in special experimental devices called particle accelerators and in outer space.

Although momentum is usually considered a property of objects with mass, photons also have momentum. Momentum determines the amount of force, or pressure, that an object exerts when it hits a surface. In classical physics, or physics that deals with the behavior of objects we encounter in everyday life, momentum is equal to the product of the mass of an object multiplied by its velocity (the combination of its speed and direction). While photons do not have mass, scientists have found that they exert extremely small amounts of pressure when they strike surfaces. Scientists have redefined momentum to include the force exerted by photons, called *light pressure* or *radiation pressure.*

ELECTRON

Electron, negatively charged particle found in an atom. Electrons, along with neutrons and protons, comprise the basic building blocks of all atoms. The electrons form the outer layer or layers of an atom, while the neutrons and protons make up the nucleus, or core, of the atom. Electrons, neutrons, and protons are elementary particles—that is, they are among the smallest parts of matter that scientists can isolate. The electron carries a negative electric charge of -1.602×10^{-19} coulomb and has a mass of 9.109×10^{-31} kg. Electrons are responsible for many important physical phenomena, such as electricity and light, and for physical and chemical properties of matter. Electrons form electric currents by flowing in a stream and carrying their negative charge with them. All electrical devices, from flashlights to computers, depend on the movement of electrons. Electrons also are involved in creating light. The electrons in the outer layers of the atom sometimes lose energy, emitting the energy in the form of light. Because electrons form the outer layers of atoms, they are also responsible for many of the physical and chemical properties of the chemical elements. Electrons help determine how atoms of an element behave with respect to each other and how they react with atoms of other elements.

Electrons as Elementary Particles

The electron is one of the most fundamental and most important of elementary particles. The electron is also one of the few elementary particles that is stable, meaning it can exist by itself for a long period of time. Most other elementary particles can exist independently for only a fraction of a second. Electrons are among the smallest of all elementary particles and have no detectable shape or structure. At the same time, they do have a property that scientists can measure called spin, or intrinsic angular momentum. An electron's spin makes it act as a tiny magnet. Electrons can spin clockwise or counterclockwise.

The electron is affected by three of the four fundamental forces that define the nature and interaction of everything in the universe: gravitation, the electromagnetic force, and the weak nuclear force. Gravitation is the attractive force between every object in the universe that has mass. Gravitation affects the electron because the electron has mass. The electromagnetic force affects objects with an electric charge, so the electron's negative electric charge subjects it to the forces of electromagnetism. The electron attracts positively charged particles, such as protons, and repels negatively charged particles, such as other electrons. The electron is also sensitive to the weak nuclear force, a very feeble force that affects certain types of elementary particles and is only important over very short distances. The one fundamental force that does not affect the electron is the strong nuclear force, which is the force that binds protons and neutrons in the atom's nucleus.

Electrons in Atoms

An atom consists of neutrons and protons packed into a dense nucleus with electrons orbiting around the nucleus. The neutrons have no electrical charge, while each proton carries a positive charge that is equal and opposite to the negative charge of the electron. Each chemical element is defined by the number of protons in the nucleus of its atoms; this number is the element's atomic number. The electrons are equal in number to the protons in the atom, balancing the electrical charge of the nucleus. In other words, the atom's net charge is zero, and the atom is said to be neutral.

Electrons and Light Emission

When an atom's energy is at its minimum, it is said to be in a ground state. In this ground state, the atom's electrons occupy the innermost available shells, those closest to the nucleus. When atoms are excited by heat, by an electric current, or by light or some other form of radiation, the atoms' electrons can acquire energy and jump from an inner to an outer shell, leaving a vacancy in the inner shell. The atom seeks to shed this surplus energy, leading the electron in the outer orbit to fall back down to an inner vacancy. As it falls, the electron releases energy in the form of a photon, a tiny flash of light. The color of the light depends on the amount of energy emitted.

When an electron moves to a different shell, it does not gradually go from one shell to another, but instead jumps directly to the other shell. These jumps are like steps on a staircase (and are different from a smooth incline, or hill). The electron also absorbs or emits the energy to make jumps in steps. It cannot gradually build up or lose energy, but must instantly absorb the exact amount of energy needed to make a certain jump, or instantly emit the exact amount needed to fall to a lower shell.

Each element has a different pattern of allowed jumps within its electronic structure, so the element's atoms can only absorb or emit a distinct set of energies, or spectrum of colors. In this way, a scientist can tell which elements are present in a sample by looking at the colors absorbed or emitted when the sample is excited by heat, electricity, or light.

Electron Orbitals

Scientists cannot simultaneously measure both the exact location of an electron and its precise speed and direction, so they cannot measure the path a specific electron takes as it orbits the nucleus. The law of physics governing this phenomenon is called the uncertainty principle. Scientists can, however, determine the area an electron will probably occupy, and the probability of finding the electron at some place inside this area. A map of this area and its probabilities forms a cloudlike pattern known as an orbital. Each orbital can contain two electrons, but these electrons can not have identical properties, so they must spin in opposite directions.

Orbitals are grouped into shells, like the layers of an onion, around the nucleus. Each shell can contain a limited number of orbitals, which means that each shell can contain a limited number of electrons. Each shell corresponds to a certain level of energy, and all the electrons in the shell have this same level of energy. As the shells get farther from the nucleus, they can contain more electrons, and the electrons in the shells have higher energy.

ELECTRONS AND CHEMICAL BONDING

The electrons in the valence, or outermost, shell of atoms determine the chemical behavior of most elements. The atoms of noble gases (helium, neon, argon, krypton, xenon, and radon) have complete, or full, valence shells. The configuration of a complete outer shell is very stable, so the noble gases usually exist as single atoms and rarely react with other elements. Atoms of the other elements attempt to imitate the stable configuration of the noble gases. They do this by donating, accepting, or sharing electrons in chemical reactions with atoms of the same element or atoms of other elements. When atoms donate, accept, or share electrons with other atoms to complete their valence shells, they form chemical bonds. The resulting substance is called a compound. The type of bond depends on whether the electrons are transferred or shared.

An atom with few electrons in its valence shell will tend to donate these electrons to fill an almost complete shell in another atom. For example, an atom of lithium has two electrons filling its inner shell and a lone electron in an outer shell that could accommodate eight electrons. An atom of fluorine, on the other hand, has seven electrons in the outer shell (as well as two in the inner shell). The lithium atom transfers its outer electron to the fluorine atom. Both atoms now have filled outer shells. Fluorine has ten electrons, with eight electrons completing its outer shell. Lithium no longer has a second shell, but has two electrons completing the first shell. Because the lithium atom lost an electron, it now has a positive charge, while the fluorine atom gains a negative charge. Atoms that have an electrical charge are called ions. These oppositely charged ions attract each other, and an *ionic bond* forms between them. The compound created by lithium and fluorine is called lithium fluoride.

A *covalent bond* forms between atoms when the valence electrons of one atom are shared with another atom with no discrete transfer of electrons. For example, two atoms of hydrogen, each with a single electron (and just one shell), can share their electrons. Each hydrogen atom's shell is now complete with two electrons. This covalent bond yields a molecule of hydrogen. In molecules, each valence electron belongs to the molecule, not to the individual atoms. When metal atoms combine with each other, the outermost electrons lose contact with their parent atoms. The remaining positively charged atomic centers form an ordered structure while the outer electrons move freely around the whole sample. These freely moving electrons, called conduction electrons, can carry heat energy and electric charge easily throughout the metal, making metals good conductors of heat and electricity.

Elements with atoms that have similar valence shell structures react in the same way to complete their outer shells. This predictable behavior led scientists to form the periodic law, which states that the physical and chemical properties of the elements tend to repeat at certain intervals as the atomic number (and number of electrons in the atom) increases. Elements that behave similarly are grouped in columns in the periodic table. For example, the valence shells of hydrogen and the alkali metals (lithium, sodium, potassium, rubidium, cesium, and francium) found in column 1 (or Ia) of the periodic table all contain a single electron, which makes them all highly reactive.

Electrons and Electricity

Electricity refers to the group of effects caused by charged particles, such as electrons and protons. Each charged particle creates an electric field around it that attracts or repels other charged particles. A difference in the amount of attraction or repulsion between any two points in an electrical field is known as a potential difference and is usually measured in volts. The two terminals of a working battery hold different charges: the positively charged terminal attracts electrons, the negative terminal repels them. Because of this difference in attraction, there is a voltage between the terminals. When a piece of metal is connected to the positive and negative terminals of a battery, freely moving conduction electrons will be attracted to and move toward the positive terminal. Such a movement of electric charge is an electric current. Insulators are substances that do not normally conduct electricity. Scientists can make these substances conduct, however, by applying a very high electric field to the substance, a field strong enough to overcome the outer electron's attraction to its nucleus and pull the electrons from the atoms. The electrons will move toward the positive terminal and, in a gas, the positive ions (the atoms stripped of their outer electrons) will move toward the negative terminal. Such currents are seen as electrical discharges of light—for example, in neon lamps.

Applications

In addition to using electrons for electrical devices, manufacturers use beams of pure electrons to produce television pictures and X rays, and to illuminate objects in electron microscopes. The electron beam for each of these devices is created by heating a cathode, a negatively charged metal that emits electrons. The electrons accelerate as they are attracted to the anode, a positively-charged piece of metal. Electron beams are used in the cathode-ray tube (or picture tube) of traditional television screens. In the cathode-ray tube, the electrons race toward a hollow anode so that a narrow, fast beam of electrons shoots out through the hole in the anode. The higher the positive charge on the anode, the greater the speed—and thus the energy—of the beam. The tube must be emptied of air to prevent the electrons from being slowed or scattered by collisions with air molecules.

The beam of electrons is focused so that it hits a specific spot on the television screen, which is covered with luminescent material. When the electrons hit this material, they excite its atoms. The excited atoms then lose this extra energy by releasing flashes of light. A changing electromagnetic field inside the picture tube affects the negatively charged electrons and makes the electron beam rapidly scan across the screen, moving horizontally and vertically. The flashes caused by the beam build up a continually changing picture. When a high-powered electron beam hits a metal anode, it can create X rays for medical or industrial purposes. A fast-moving electron can knock an inner-shell electron out of an atom. As an outer-shell electron jumps inward to fill the inner-shell vacancy, the atom emits an X ray, a high-energy photon invisible to the eye. X rays are absorbed by heavier atoms, such as those in bones, but pass through lighter atoms, such as those in flesh.

X rays can also react with chemicals in specialized film to create a picture. If a patient's arm is placed in front of a photographic film, exposing the arm to an X-ray beam will create an image of the bone on the film.

Scientists use powerful X rays created by electrons to probe the structure of atoms and molecules. They produce these X rays by accelerating a beam of electrons, confined by magnets in a circular tube, to a very high energy. Higher and higher energy electrons release radiation with shorter and shorter wavelengths, in this case, X rays. The shorter the wavelength, the finer the detail the X rays reveal. While scientists usually describe the electron as a particle, the electron can also behave like a wave. Scientists use this aspect of electron behavior to illuminate extremely small objects. Ordinary light can only resolve objects that are larger than the wavelength of the light waves illuminating them. For smaller objects, the light waves scatter randomly off the object and do not reveal its shape. The wavelength of visible light is about a millionth of a meter. Electrons can have smaller wavelengths than visible light and thereby reveal objects many times smaller.

Electron microscopes, using beams of electrons instead of light, can create images of objects, such as viruses, too small to be visible by ordinary microscopes. Electron energies are usually measured in electron volts (eV), where 1 eV is the energy acquired by an electron when it is accelerated in a vacuum by 1 volt. Physicists can use electrons they've accelerated to very high energies (giga-electron volts, or 109 eV, which is 1 billion electron volts) to reveal elementary particles such as protons, neutrons, and even quarks.

THE PRINCIPLES OF RAMAN SPECTROSCOPY

Raman spectroscopy differs from the rotational and vibrational spectroscopy considered above in that it is concerned with the scattering of radiation by the sample, rather than an absorption process. It is named after the Indian physicist who first observed it in 1928, C V Raman. Both rotational and vibrational Raman spectroscopy are possible. The energy of the exciting radiation will determine which type of transition occurs - rotational transitions are lower in energy than vibrational transitions.

In addition to this, rotational transitions are around 3 orders of magnitude slower than vibrational transitions. Therefore, collisions with other molecules may occur in the time in which the transition is occurring. A collision is likely to change the rotational state of the molecule, and so the definition of the spectrum obtained will be destroyed. Rotational spectroscopy is therefore carried out on gases at low pressure to ensure that the time between collisions is greater than the time for a transition. The basic set-up of a Raman spectrometer is shown below. Note that the detector is orthogonal to the direction of the incident radiation, so as to observe only the scattered light. The source needs to provide intense monochromatic radiation, and so is usually a laser.

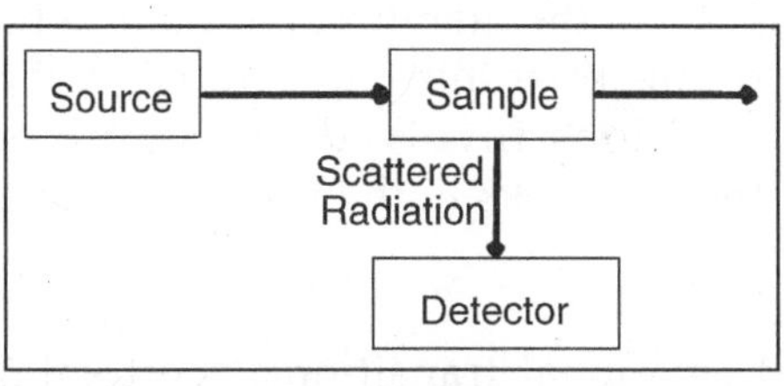

The criteria for a molecule to be Raman active are also different to other types of spectroscopy, which require a permanent dipole moment, at least for diatomic molecules. A molecule will only be Raman active if the following gross selection rule is fulfilled.

SELECTION RULES

Gross Selection Rule

For spectroscopic techniques such as infared spectroscopy, as seen in the first section it is necessary for the molecule being analyzed to have a permanent electric dipole. This is not the case for Raman spectroscopy, rather it is the polarizability (α) of the molecule which is important: The oscillating electric field of a photon causes charged particles (electrons and, to a lesser extent, nuclei) in the molecule to oscillate. This leads to an induced electric dipole moment, μ_{ind}, where

$$\mu_{ind} = \alpha E$$

This induced dipole moment then emits a photon, leading to either Raman or Raleigh scattering. The energy of this interaction is also dependent on the polarizability:

$$\text{Energy of interaction} = -1/2\alpha E^2$$

By comparison with the equation for the interaction energy for other forms of spectroscopy, it can be seen that the energies of Raman transitions are relatively weak. To counter this, a higher intensity of the exciting radiation is used.

For Raman scattering to occur, the polarizability of the molecule must vary with its orientation. One of the strengths of Raman spectroscopy is that this will be true for both heteronuclear and homonuclear diatomic molecules. Homonuclear diatomic molecules do not possess a permanent electric dipole, and so are undetectable by other methods such as infrared.

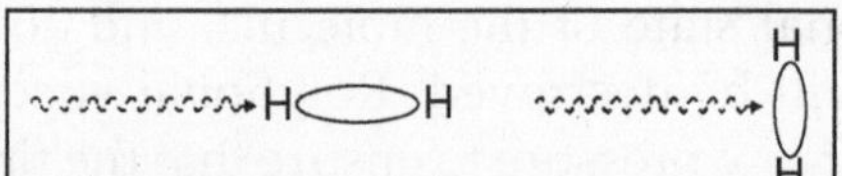

Fig. The Polarizabilities of these Two Homonuclear Diatomic Molecules will be Different.

Selection Rules Arising from Quantum Mechanics

A detailed discussion of the quantum mechanical basis for spectroscopy is beyond the scope of this brief introduction. Here, the quantum numbers and the allowed transitions involved are simply stated. For further information, consult the textbooks in the further reading list.

Rotational

To find allowed rotational transitions, the total angular momentum quantum number for rotational energy states, *J*, must be considered.

Raman spectroscopy involves a 2 photon process, each of which obeys:

$$\Delta J = \pm 1$$

Therefore, for the overall transition,

$$\Delta J = 0, \pm 2$$

where $\Delta J = 0$ corresponds to Raleigh scattering,
$\Delta J = +2$ corresponds to a Stokes transition,
and $\Delta J = -2$ corresponds to an anti-Stokes transition.

Vibrational

To find allowed transitions the vibrational quantum number, õ, must be considered.

For the overall transition,

$$\Delta v = \pm 1$$

Transitions where $\Delta v = \pm 2$ are also possible, but these are of weaker intensity.

Types of Scattering

Consider a light wave as a stream of photons each with energy hí. When each photon collides with a molecule, many things may happen, 2 of which are considered below:

Elastic, or Raleigh scattering

This is when the photon simply 'bounces' off the molecule, with no exchange in energy. The vast majority of photons will be scattered in this way.

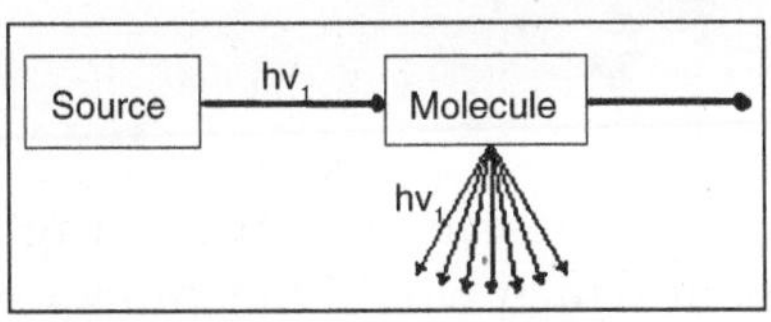

Inelastic, or Raman scattering

This is when there is an exchange of energy between the photon and the molecule, leading to the emission of another photon with a different frequency to the incident photon:

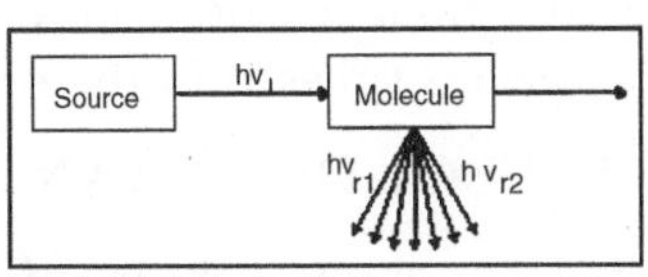

Raman Scattering

The molecule may either gain energy from, or lose energy to, the photon. In the diagram below, 3 energy states of the molecule are shown (E_1, E_2, E_3). The molecule is originally at the E_2 energy state. The photon interacts with the molecule, exciting it with an energy $hí_i$. However, there is no stationary state of the molecule corresponding to this energy, and so the molecule relaxes down to one of the energy levels shown. In doing this, it emits a photon.

If the molecule relaxes to energy state E_1, it will have lost energy, and so the photon emitted will have energy $h\nu_{r1}$, where $h\nu_{r1} > h\nu_i$ These transitions are known as anti-Stokes transitions.If the molecule relaxes to energy state E_3, it will have gained energy, and so the photon emitted will have energy $h\nu_{r2}$, where $h\nu_{r2} < h\nu_i$ These transitions are known as Stokes transitions.

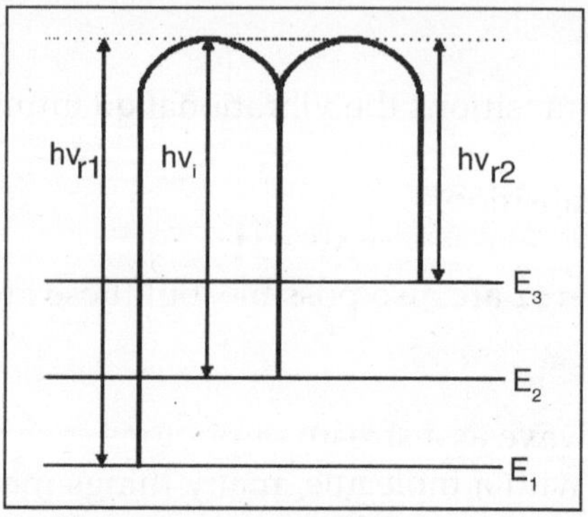

For rotational spectroscopy, both Stokes and anti-Stokes transitions are seen. For vibrational spectroscopy, Stokes transitions are far more common, and so anti-Stokes transitions can effectively be ignored.

Raman Spectra

Uses

Raman spectroscopy is useful for analyzing molecules without a permanent dipole moment which would not show up on an IR spectrum. A useful 'exclusion rule' states that for molecules with an inversion centre, no modes can be both IR and Raman active. It can be used to determine bond lengths in non-polar molecules. It is useful for determining the identity of organic and inorganic species in solution, as the Raman transitions for these species are more characteristic than for IR, where the transitions are much more affected by the other species present in the solution.

EFFECT OF RAMAN

Raman Effect, a change in frequency observed when light is scattered in a transparent material. This phenomenon was discovered by the Indian physicist Sir Chandrasekhara Venkata Raman in 1928. When monochromatic light, such as that obtained from a laser, is passed through a transparent gas, liquid, or solid and is observed with the spectroscope, the spectral line ordinarily produced by the light has associated with it lines of longer and of shorter wavelength, called the Raman spectrum. These lines are caused by photons losing or gaining energy by elastic collisions with the molecules of the transparent substance. The Raman spectrum of a particular spectral line varies with the nature of the material that scatters the light. The Raman effect has practical importance in spectrographic chemical analysis and in the determination of the structure of molecules.

LASER

Laser, a device that produces and amplifies light. The word laser is an acronym for Light Amplification by Stimulated Emission of Radiation. Laser light is very pure in color, can be extremely intense, and can be directed with great accuracy. Lasers are used in many modern technological devices including bar code readers, compact disc (CD) players, and laser printers. Lasers can generate light beyond the range visible to the human eye, from the infrared through the X-ray range. Masers are similar devices that produce and amplify microwaves.

Principles of Operation

Lasers generate light by storing energy in particles called electrons inside atoms and then inducing the electrons to emit the absorbed energy as light. Atoms are the building blocks of all matter on Earth and are a thousand times smaller than viruses. Electrons are the underlying source of almost all light. Light is composed of tiny packets of energy called photons. Lasers produce coherent light: light that is monochromatic (one color) and whose photons are "in step" with one another.

Stimulated Emission

Lasers are different from more familiar sources of light. Excited atoms in lasers collectively emit photons of a single color, all traveling in the same direction and all in step with one another. When two photons are in step, the peaks and troughs of their waves line up. The electrons in the atoms of a laser are first pumped, or energized, to an excited state by an energy source. An excited atom can then be "stimulated" by a photon of exactly the same color (or, equivalently, the same wavelength) as the photon this atom is about to emit spontaneously.

If the photon approaches closely enough, the photon can stimulate the excited atom to immediately emit light that has the same wavelength and is in step with the photon that interacted with it. This stimulated emission is the key to laser operation. The new light adds to the existing light, and the two photons go on to stimulate other excited atoms to give up their extra energy, again in step. The phenomenon snowballs into an amplified, coherent beam of light: laser light.

In a gas laser, for example, the photons usually zip back and forth in a gas-filled tube with highly reflective mirrors facing inward at each end. As the photons bounce between the two parallel mirrors, they trigger further stimulated emissions and the light gets brighter and brighter with each pass through the excited atoms. One of the mirrors is only partially silvered, allowing a small amount of light to pass through rather than reflecting it all. The intense, directional, and single-colored laser light finally escapes through this slightly transparent mirror. The escaped light forms the laser beam.

Albert Einstein first proposed stimulated emission, the underlying process for laser action, in 1917. Translating the idea of stimulated emission into a working model, however, required more than four decades. The working principles of lasers were outlined by the American physicists Charles Hard Townes and Arthur Leonard Schawlow in a 1958 patent application. (Both men won Nobel Prizes in physics for their work, Townes in 1964 and Schawlow in 1981).

The patent for the laser was granted to Townes and Schawlow, but it was later challenged by the American physicist and engineer Gordon Gould, who had written down some ideas and coined the word *laser* in 1957. Gould eventually won a partial patent covering several types of laser. In 1960 American physicist Theodore Maiman of Hughes Aircraft Corporation constructed the first working laser from a ruby rod.

Excited Atoms

At the heart of an atom is a tightly bound cluster of particles called the nucleus. This cluster is made up of two types of particles: protons, which have a positive charge, and neutrons, which have no charge. The nucleus makes up more than 99.9 percent of the atom's mass but occupies only a tiny part of the atom's space. Enlarge an atom up to the size of Yankee Stadium and the equally magnified nucleus is only the size of a baseball.

Electrons, tiny particles that have a negative charge, whirl through the rest of the space inside atoms. Electrons travel in complex orbits and exist only in certain specific energy states or levels. Electrons can move from a low to a high energy level by absorbing energy. An atom with at least one electron that occupies a higher energy level than it normally would is said to be excited.

An atom can become excited by absorbing a photon whose energy equals the difference between the two energy levels. A photon's energy, color, frequency, and wavelength are directly related: All photons of a given energy are the same color and have the same frequency and wavelength. Usually, electrons quickly jump back to the low energy level, giving off the extra energy as light. Neon signs and fluorescent lamps glow with this kind of light as many electrons independently emit photons of different colors in all directions.

Types of Lasers

Lasers are generally classified according to the material, called the medium, they use to produce the laser light. Solid-state, gas, liquid, semiconductor, and free electron are all common types of lasers.

Gas Lasers

The lasing medium of a gas laser can be a pure gas, a mixture of gases, or even metal vapor. The medium is usually contained in a cylindrical glass or quartz tube. Two mirrors are located outside the ends of the tube to form the laser cavity. Gas lasers can be pumped by ultraviolet light, electron beams, electric current, or chemical reactions. The helium-neon laser is known for its color purity and minimal beam spread. Carbon dioxide lasers are very efficient at turning the energy used to excite their atoms into laser light. Consequently, they are the most powerful continuous wave (CW) lasers—that is, lasers that emit light continuously rather than in pulses.

Liquid Lasers

The most common liquid laser media are inorganic dyes contained in glass vessels. They are pumped by intense flash lamps in a pulse mode or by a separate gas laser in the continuous wave mode. Some dye lasers are tunable, meaning that the color of the laser light they emit can be adjusted with the help of a prism located inside the laser cavity.

Solid-State Lasers

Solid-state lasers produce light by means of a solid medium. The most common solid laser media are rods of ruby crystals and neodymium-doped glasses and crystals. The ends of the rods are fashioned into two parallel surfaces coated with a highly reflecting nonmetallic film. Solid-state lasers offer the highest power output. They are usually pulsed to generate a very brief burst of light. Bursts as short as 12×10^{-15} sec have been achieved. These short bursts are useful for studying physical phenomena of very brief duration. One method of exciting the atoms in lasers is to illuminate the solid laser material with higher-energy light than the laser produces. This procedure, called pumping, is achieved with brilliant strobe light from xenon flash tubes, arc lamps, or metal-vapor lamps.

Semiconductor Lasers

Semiconductor lasers are the most compact lasers. Gallium arsenide is the most common semiconductor used. A typical semiconductor laser consists of a junction between two flat layers of gallium arsenide. One layer is treated with an impurity whose atoms provide an extra electron, and the other with an impurity whose atoms are one electron short. Semiconductor lasers are pumped by the direct application of electric current across the junction. They can be operated in the continuous wave mode with better than 50 percent efficiency. Only a small percentage of the energy used to excite most other lasers is converted into light. Scientists have developed extremely tiny semiconductor lasers, called quantum-dot vertical-cavity surface-emitting lasers. These lasers are so tiny that more than a million of them can fit on a chip the size of a fingernail. Common uses for semiconductor lasers include compact disc (CD) players and laser printers. Semiconductor lasers also form the heart of fiber-optics communication systems.

Free Electron Lasers

Free electron lasers employ an array of magnets to excite free electrons (electrons not bound to atoms). First developed in 1977, they are now becoming important research instruments. Free electron lasers are tunable over a broader range of energies than dye lasers. The devices become more difficult to operate at higher energies but generally work successfully from infrared through ultraviolet wavelengths. Theoretically, electron lasers can function even in the X-ray range.

The free electron laser facility at the University of California at Santa Barbara uses intense far-infrared light to investigate mutations in DNA molecules and to study the properties of semiconductor materials. Free electron lasers should also eventually become capable of producing very high-power radiation that is currently too expensive to produce. At high power, near-infrared beams from a free electron laser could defend against a missile attack.

LASER APPLICATIONS

The use of lasers is restricted only by imagination. Lasers have become valuable tools in industry, scientific research, communications, medicine, the military, and the arts.

Industry

Powerful laser beams can be focused on a small spot to generate enormous temperatures. Consequently, the focused beams can readily and precisely heat, melt, or vaporize material. Lasers have been used, for example, to drill holes in diamonds, to shape machine tools, to trim microelectronics, to cut fashion patterns, to synthesize new material, and to attempt to induce controlled nuclear fusion.

Highly directional laser beams are used for alignment in construction. Perfectly straight and uniformly sized tunnels, for example, may be dug using lasers for guidance. Powerful, short laser pulses also make high-speed photography with exposure times of only several trillionths of a second possible.

Communications

Laser light can travel a large distance in outer space with little reduction in signal strength. In addition, high-energy laser light can carry 1,000 times the television channels today carried by microwave signals. Lasers are therefore ideal for space communications. Low-loss optical fibers have been developed to transmit laser light for earthbound communication in telephone and computer systems. Laser techniques have also been used for high-density information recording. For instance, laser light simplifies the recording of a hologram, from which a three-dimensional image can be reconstructed with a laser beam. Lasers are also used to play audio CDs and videodiscs.

Scientific Research

Because laser light is highly directional and monochromatic, extremely small amounts of light scattering and small shifts in color caused by the interaction between laser light and matter can easily be detected. By measuring the scattering and color shifts, scientists can study molecular structures of matter. Chemical reactions can be selectively induced, and the existence of trace substances in samples can be detected. Lasers are also the most effective detectors of certain types of air pollution.

Scientists use lasers to make extremely accurate measurements. Lasers are used in this way for monitoring small movements associated with plate tectonics and for geographic surveys. Lasers have been used for precise determination (to within one inch) of the distance between Earth and the Moon, and in precise tests to confirm Einstein's theory of relativity. Scientists also have used lasers to determine the speed of light to an unprecedented accuracy.

Very fast laser-activated switches are being developed for use in particle accelerators. Scientists also use lasers to trap single atoms and subatomic particles in order to study these tiny bits of matter.

Medicine

Lasers have a wide range of medical uses. Intense, narrow beams of laser light can cut and cauterize certain body tissues in a small fraction of a second without damaging surrounding healthy tissues. Lasers have been used to "weld" the retina, bore holes in the skull, vaporize lesions, and cauterize blood vessels. Laser surgery has virtually replaced older surgical procedures for eye disorders. Laser techniques have also been developed for lab tests of small biological samples.

Military Applications

Laser guidance systems for missiles, aircraft, and satellites have been constructed. Guns can be fitted with laser sights and range finders. The use of laser beams to destroy hostile ballistic missiles has been proposed, as in the Strategic Defense Initiative urged by U.S. president Ronald Reagan and the Ballistic Missile Defense program supported by President George W. Bush. The ability of tunable dye lasers to selectively excite an atom or molecule may open up more efficient ways to separate isotopes for construction of nuclear weapons.

Laser Safety

Because the eye focuses laser light just as it does other light, the chief danger in working with lasers is eye damage. Therefore, laser light should not be viewed either directly or reflected. Lasers sold and used commercially in the United States must comply with a strict set of laws enforced by the Center for Devices and Radiological Health (CDRH), a department of the Food and Drug Administration. The CDRH has divided lasers into six groups, depending on their power output, their emission duration, and the energy of the photons they emit.

The classification is then attached to the laser as a sticker. The higher the laser's energy, the higher its potential to injure. High-powered lasers of the Class IV type (the highest classification) generate a beam of energy that can start fires, burn flesh, and cause permanent eye damage whether the light is direct, reflected, or diffused. Canada uses the same classification system, and laser use in Canada is overseen by Health Canada's Radiation Protection Bureau.

ATTRACTION OF GRAVITATION

Gravitation, the force of attraction between all objects that tends to pull them toward one another. It is a universal force, affecting the largest and smallest objects, all forms of matter, and energy. Gravitation governs the motion of astronomical bodies. It keeps the moon in orbit around the earth and keeps the earth and the other planets of the solar system in orbit around the sun. On a larger scale, it governs the motion of stars and slows the outward expansion of the entire universe because of the inward attraction of galaxies to other galaxies. Typically the term *gravitation* refers to the force in general, and the term *gravity* refers to the earth's gravitational pull. Gravitation is one of the four fundamental forces of nature, along with electromagnetism and the weak and strong nuclear forces, which hold together the particles that make up atoms. Gravitation is by far the weakest of these forces and, as a result, is not important in the interactions of atoms and nuclear particles or even of moderate-sized objects, such as people or cars. Gravitation is important only when very large objects, such as planets, are involved. This is true for several reasons.

First, the force of gravitation reaches great distances, while nuclear forces operate only over extremely short distances and decrease in strength very rapidly as distance increases. Second, gravitation is always attractive. In contrast, electromagnetic forces between particles can be repulsive or attractive depending on whether the particles both have a positive or negative electrical charge, or they have opposite electrical charges. These attractive and repulsive forces tend to cancel each other out, leaving only a weak net force. Gravitation has no repulsive force and, therefore, no such cancellation or weakening. The gravitational attraction of objects for one another is the easiest fundamental force to observe and was the first fundamental force to be described with a complete mathematical theory by the English physicist and mathematician Sir Isaac Newton. A more accurate theory called general relativity was formulated early in the 20th century by the German-born American physicist Albert Einstein. Scientists recognize that even this theory is not correct for describing how gravitation works in certain circumstances, and they continue to search for an improved theory.

EARTH'S GRAVITATION

Gravitation plays a crucial role in most processes on the earth. The ocean tides are caused by the gravitational attraction of the moon and the sun on the earth and its oceans. Gravitation drives weather patterns by making cold air sink and displace less dense warm air, forcing the warm air to rise. The gravitational pull of the earth on all objects holds the objects to the surface of the earth. Without it, the spin of the earth would send them floating off into space.

The gravitational attraction of every bit of matter in the earth for every other bit of matter amounts to an inward pull that holds the earth together against the pressure forces tending to push it outward. Similarly, the inward pull of gravitation holds stars together. When a star's fuel nears depletion, the processes producing the outward pressure weaken and the inward pull of gravitation eventually compresses the star to a very compact size.

Acceleration

If an object held near the surface of the earth is released, it will fall and accelerate, or pick up speed, as it descends. This acceleration is caused by gravity, the force of attraction between the object and the earth. The force of gravity on an object is also called the object's weight. This force depends on the object's mass, or the amount of matter in the object. The weight of an object is equal to the mass of the object multiplied by the acceleration due to gravity. A bowling ball that weighs 16 lb is actually being pulled toward the earth with a force of 16 lb. In the metric system, the bowling ball is pulled toward the earth with a force of 71 newtons (a metric unit of force abbreviated N). The bowling ball also pulls on the earth with a force of 16 lb (71 N), but the earth is so massive that it does not move appreciably.

In order to hold the bowling ball up and keep it from falling, a person must exert an upward force of 16 lb (71 N) on the ball. This upward force acts to oppose the 16 lb (71 N) downward weight force, leaving a total force of zero. The total force on an object determines the object's acceleration. If the pull of gravity is the only force acting on an object, then all objects, regardless of their weight, size, or shape, will accelerate in the same manner. At the same place on the earth, the 16 lb (71 N) bowling ball and a 500 lb (2200 N) boulder will fall with the same rate of acceleration. As each second passes, each object will increase its downward speed by about 9.8 m/sec (32 ft/sec), resulting in an acceleration of 9.8 m/sec/sec (32 ft/sec/sec). In principle, a rock and a feather both would fall with this acceleration if there were no other forces acting. In practice, however, air friction exerts a greater upward force on the falling feather than on the rock and makes the feather fall more slowly than the rock. The mass of an object does not change as it is moved from place to place, but the acceleration due to gravity, and therefore the object's weight, will change because the strength of the earth's gravitational pull is not the same everywhere.

The earth's pull and the acceleration due to gravity decrease as an object moves farther away from the center of the earth. At an altitude of 4000 miles (6400 km) above the earth's surface, for instance, the bowling ball that weighed 16 lb (71 N) at the surface would weigh only about 4 lb (18 N). Because of the reduced weight force, the rate of acceleration of the bowling ball at that altitude would be only one quarter of the acceleration rate at the surface of the earth. The pull of gravity on an object also changes slightly with latitude. Because the earth is not perfectly spherical, but bulges at the equator, the pull of gravity is about 0.5 percent stronger at the earth's poles than at the equator.

NEWTON'S THEORY OF GRAVITATION

To develop his theory of gravitation, Newton first had to develop the science of forces and motion called mechanics. Newton proposed that the natural motion of an object is motion at a constant speed on a straight line, and that it takes a force to slow down, speed up, or change the path of an object. Newton also invented calculus, a new branch of mathematics that became an important tool in the calculations of his theory of gravitation. Newton proposed his law of gravitation in 1687 and stated that every particle in the universe attracts every other particle in the universe with a force that depends on the product of the two particles' masses divided by the square of the distance between them.

The gravitational force between two objects can be expressed by the following equation: $F = GMm/d^2$ where F is the gravitational force, G is a constant known as the universal constant of gravitation, M and m are the masses of each object, and d is the distance between them. Newton considered a particle to be an object with a mass that was concentrated in a small point. If the mass of one or both particles increases, then the attraction between the two particles increases.

For instance, if the mass of one particle is doubled, the force of attraction between the two particles is doubled. If the distance between the particles increases, then the attraction decreases as the square of the distance between them. Doubling the distance between two particles, for instance, will make the force of attraction one quarter as great as it was. According to Newton, the force acts along a line between the two particles. In the case of two spheres, it acts along the line between their centers. The attraction between objects with irregular shapes is more complicated. Every bit of matter in the irregular object attracts every bit of matter in the other object. A simpler description is possible near the surface of the earth where the pull of gravity is approximately uniform in strength and direction. In this case there is a point in an object (even an irregular object) called the center of gravity, at which all the force of gravity can be considered to be acting.

Newton's law affects all objects in the universe, from raindrops in the sky to the planets in the solar system. It is therefore known as the universal law of gravitation. In order to know the strength of gravitational forces in general, however, it became necessary to find the value of G, the universal constant of gravitation. Scientists needed to perform an experiment, but gravitational forces are very weak between objects found in a common laboratory and thus hard to observe. In 1798 the English chemist and physicist Henry Cavendish finally measured G with a very sensitive experiment in which he nearly eliminated the effects of friction and other forces. The value he found was 6.754×10^{-11} N-m^2/kg^2—close to the currently accepted value of 6.670×10^{-11} N-m^2/kg^2 (a decimal point followed by 10 zeros and then the number 6670).

This value is so small that the force of gravitation between two objects with a mass of 1 metric ton each, 1 meter from each other, is about 67 millionths of a newton, or about 15 millionths of a pound. Gravitation may also be described in a completely different way. A massive object, such as the earth, may be thought of as producing a condition in space around it called a gravitational field. This field causes objects in space to experience a force. The gravitational field around the earth, for instance, produces a downward force on objects near the earth surface. The field viewpoint is an alternative to the viewpoint that objects can affect each other across distance. This way of thinking about interactions has proved to be very important in the development of modern physics.

Planetary Motion

Newton's law of gravitation was the first theory to accurately describe the motion of objects on the earth as well as the planetary motion that astronomers had long observed. According to Newton's theory, the gravitational attraction between the planets and the sun holds the planets in elliptical orbits around the sun. The earth's moon and moons of other planets are held in orbit by the attraction between the moons and the planets.

Newton's law led to many new discoveries, the most important of which was the discovery of the planet Neptune. Scientists had noted unexplainable variations in the motion of the planet Uranus for many years. Using Newton's law of gravitation, the French astronomer Urbain Leverrier and the British astronomer John Couch each independently predicted the existence of a more distant planet that was perturbing the orbit of Uranus. Neptune was discovered in 1864, in an orbit close to its predicted position.

Problems with Newton's Theory

Scientists used Newton's theory of gravitation successfully for many years. Several problems began to arise, however, involving motion that did not follow the law of gravitation or Newtonian mechanics. One problem was the observed and unexplainable deviations in the orbit of Mercury (which could not be caused by the gravitational pull of another orbiting body). Another problem with Newton's theory involved reference frames, that is, the conditions under which an observer measures the motion of an object. According to Newtonian mechanics, two observers making measurements of the speed of an object will measure different speeds if the observers are moving relative to each other. A person on the ground observing a ball that is on a train passing by will measure the speed of the ball as the same as the speed of the train.

A person on the train observing the ball, however, will measure the ball's speed as zero. According to the traditional ideas about space and time, then, there could not be a constant, fundamental speed in the physical world because all speed is relative. However, near the end of the 19th century the Scottish physicist James Clerk Maxwell proposed a complete theory of electric and magnetic forces that contained just such a constant, which he called c. This constant speed was 300,000 km/sec (186,000 mi/sec) and was the speed of electromagnetic waves, including light waves. This feature of Maxwell's theory caused a crisis in physics because it indicated that speed was not always relative.

Albert Einstein resolved this crisis in 1905 with his special theory of relativity. An important feature of Einstein's new theory was that no particle, and even no information, could travel faster than the fundamental speed *c*. In Newton's gravitation theory, however, information about gravitation moved at infinite speed. If a star exploded into two parts, for example, the change in gravitational pull would be felt immediately by a planet in a distant orbit around the exploded star. According to Einstein's theory, such forces were not possible.

Though Newton's theory contained several flaws, it is still very practical for use in everyday life. Even today, it is sufficiently accurate for dealing with earth-based gravitational effects such as in *geology* (the study of the formation of the earth and the processes acting on it), and for most scientific work in astronomy.

Only when examining exotic phenomena such as *black holes* (points in space with a gravitational force so strong that not even light can escape them) or in explaining the *big bang* (the origin of the universe) is Newton's theory inaccurate or inapplicable.

EINSTEIN'S THEORY OF RELATIVITY

In 1915 Einstein formulated a new theory of gravitation that reconciled the force of gravitation with the requirements of his theory of special relativity. He proposed that gravitational effects move at the speed of c. He called this theory general relativity to distinguish it from special relativity, which only holds when there is no force of gravitation.

General relativity produces predictions very close to those of Newton's theory in most familiar situations, such as the moon orbiting the earth. Einstein's theory differed from Newton's theory, however, in that it described gravitation as a curvature of space and time. In Einstein's general theory of relativity, he proposed that space and time may be united into a single, four-dimensional geometry consisting of 3 space dimensions and 1 time dimension. In this geometry, called spacetime, the motions of particles from point to point as time progresses are represented by curves called world lines.

If there is no gravity acting, the most natural lines in this geometry are straight lines, and they represent particles that are moving always in the same direction with the same speed—that is, particles that have no force acting on them. If a particle is acted on by a force, then its world line will not be straight. Einstein also proposed that the effect of gravitation should not be represented as the deviation of a world line from straightness, as it would be for an electrical force. If gravitation is present, it should not be considered a force. Rather, gravitation changes the most natural world lines and thereby curves the geometry of spacetime.

In a curved geometry, such as the two-dimensional surface of the earth, there are no straight lines. Instead, there are special curves called geodesics, an example of which are great circles around the earth. These special curves are at each point as straight as possible, and they are the most natural lines in a curved geometry. The effect of gravitation would be to influence the geodesics in spacetime. Near sources of gravitation the space is strongly curved and the geodesics behave less and less like those in flat, uncurved spacetime. In the solar system, for example, the effect of the sun and the earth is to cause the moon to move on a geodesic that winds around the geodesic of the earth 12 times a year.

Testing Einstein's Theory

Einstein's theory required verification, but the level of precision in measurement needed to distinguish between Einstein's theory and Newton's theory was difficult to achieve in the early 20th century and remains so today. There were two predictions, however, that could be tested. One involved deviations in the orbit of Mercury. Astronomers had observed that the ellipse of Mercury's orbit itself rotated—that is, the point nearest the sun, called the perihelion, slowly advanced around the sun. The rate of advancement predicted by Newton's theory differed slightly from what astronomers had measured, but Einstein's theory predicted the correct rate. The second test involved measuring the bending of light as it passed around the sun. Both Newton's and Einstein's theories predicted that light would be deflected by gravitation. But the amount of deflection predicted by the two theories differed. The light to be measured in such a test originates in distant stars. However, under ordinary conditions the sun's brightness prevents scientists from observing the light from these stars.

This problem disappears during an eclipse, when the moon blocks the sun's light. In 1919 a special British expedition took photographs during an eclipse. Scientists measured the deflection of starlight as it passed by the sun and arrived at numbers that agreed with Einstein's prediction. Subsequent eclipse observations also have confirmed Einstein's theory. Other physicists have proposed relativistic theories of gravitation to compete with Einstein's. In these competing theories, almost all of which are geometrical like Einstein's, gravitational effects move at the speed c. They differ mostly in the mathematical details. Even using the technology of the late 20th century, scientists still find it very difficult to test these theories with experiments and observations. But Einstein's theory has passed all tests that have been made so far.

Applications of Einstein's Theory

Einstein's general relativity theory predicts special gravitational conditions. The Big Bang theory, which describes the origin and early expansion of the universe, is one conclusion based on Einstein's theory that has been verified in several independent ways. Another conclusion suggested by general relativity, as well as other relativistic theories of gravitation, is that gravitational effects move in waves. Astronomers have observed a loss of energy in a pair of neutron stars (stars composed of densely packed neutrons) that are orbiting each other. The astronomers theorize that energy-carrying gravitational waves are radiating from the pair, depleting the stars of their energy.

Very violent astrophysical events, such as the explosion of stars or the collision of neutron stars, can produce gravitational waves strong enough that they may eventually be directly detectable with extremely precise instruments. Astrophysicists are designing such instruments with the hope that they will be able to detect gravitational waves by the beginning of the 21st century. Another gravitational effect predicted by general relativity is the existence of black holes.

The idea of a star with a gravitational force so strong that light cannot escape from its surface can be traced to Newtonian theory. Einstein modified this idea in his general theory of relativity. Because light cannot escape from a black hole, for any object—a particle, spacecraft, or wave—to escape, it would have to move past light. But light moves outward at the speed c.

According to relativity, c is the highest attainable speed, so nothing can pass it. The black holes that Einstein envisioned, then, allow no escape whatsoever. An extension of this argument shows that when gravitation is this strong, nothing can even stay in the same place, but must move inward. Even the surface of a star must move inward, and must continue the collapse that created the strong gravitational force. What remains then is not a star, but a region of space from which emerges a tremendous gravitational force.

Other Modern Theories

Einstein's theory of gravitation revolutionized 20th-century physics. Another important revolution that took place was quantum theory. Quantum theory states that physical interactions, or the exchange of energy, cannot be made arbitrarily small. There is a minimal interaction that comes in a packet called the quantum of an interaction.

For electromagnetism the quantum is called the photon. Like the other interactions, gravitation also must be quantized. Physicists call a quantum of gravitational energy a graviton. In principle, gravitational waves arriving at the earth would consist of gravitons. In practice, gravitational waves would consist of apparently continuous streams of gravitons, and individual gravitons could not be detected. Einstein's theory did not include quantum effects. For most of the 20th century, theoretical physicists have been unsuccessful in their attempts to formulate a theory that resembles Einstein's theory but also includes gravitons. Despite the lack of a complete quantum theory, it is possible to make some partial predictions about quantized gravitation. In the 1970s, British physicist Stephen Hawking showed that quantum mechanical processes in the strong gravitational pull just outside of black holes would create particles and quanta that move away from the black hole, thereby robbing it of energy.

Theory of Everything

An important trend in modern theoretical physics is to find a theory of everything (TOE), in which all four of the fundamental forces are seen to be really different aspects of the same single universal force. Physicists already have unified electromagnetism and the weak nuclear force and have made progress in joining these two forces with the strong nuclear force.

However, relativistic gravitation, with its geometric and mathematically complex character, poses the most difficult challenge. Einstein spent most of his later years searching for an all-encompassing theory by trying to make electromagnetism geometrical like gravitation. The modern approach is to try to make gravitation fit the pattern of the other fundamental forces. Much of this work involves looking for mathematical patterns. A TOE is difficult to test using experiments because the effects of a TOE would be important only in the rarest circumstances.

6

Applications of Spectrum Analysis

CURRENT RESEARCH AND APPLICATIONS

At the turn of the 21st century, physicists were still finding new problems to study with quantum theory and new applications for quantum theory. This research will probably continue for many decades. Quantum theory is technically a fully formulated theory—any questions about the physical world can be calculated using quantum mechanics, but some are too complicated to solve in practice.

The attempt to find quantum explanations of gravitation and to find a unified description of all the forces in nature are promising and active areas of research. Researchers try to find out why quantum theory explains the way nature works—they may never find an answer, but the effort to do so is underway. Physicists also study the complicated area of overlap between classical physics and quantum mechanics and work on the applications of quantum mechanics.

Studying the intersection of quantum theory and classical physics requires developing a theory that can predict how quantum systems will behave as they get larger or as the number of particles involved approaches the size of problems described by classical physics. The mathematics involved is extremely difficult, but physicists continue to advance in their research. The constantly increasing power of computers should continue to help scientists with these calculations.

New research in quantum theory also promises new applications and improvements to known applications. One of the most potentially powerful applications is quantum computing. In quantum computing, scientists make use of the behavior of subatomic particles to perform calculations. Making calculations on the atomic level, a quantum computer could theoretically investigate all the possible answers to a query at the same time and make many calculations in parallel. This ability would make quantum computers thousands or even millions of time faster than current computers. Advancements in quantum theory also hold promise for the fields of optics, chemistry, and atomic theory.

CHEMICAL ANALYSIS

The spectrum of a particular element is absolutely characteristic of that element. Different elements, however, sometimes give rise to lines that are quite close together, leading to the possibility of error or misinterpretation. The Fraunhofer C line at 430.8 nm, for example, is caused by two different lines, one formed by calcium with a wavelength of 430.7749 nm and the other formed by iron with a wavelength of 430.7914 nm. With an ordinary spectroscope, distinguishing between these two would be difficult.

The other lines of calcium, however, are very different from those of the other lines of iron. Thus, the comparison of the entire spectrum of an element with a known spectrum simplifies its identification. When the spectrum of an unknown substance is excited by flame, as in the flame test, or by an arc, spark, or other suitable method, a quick analysis with a spectrograph is usually sufficient to determine the presence or absence of any particular element. Absorption spectra are frequently useful in identifying chemical compounds.

Spectra beyond the ultraviolet region, of X rays and gamma rays, are detected by suitable ionization detectors. Gamma-ray spectra are useful in neutron-activation analysis. In this technique, a specimen is irradiated with neutrons in a nuclear reactor and becomes radioactive, emitting gamma rays. The spectra of these gamma rays serve to identify minute quantities of certain chemical elements in the specimen. Along with more conventional types of spectroscopy, this technique is valuable in crime detection. Raman spectroscopy, discovered in 1928 by Indian physicist Sir Chandrasekhara Venkata Raman, has had widespread recent application in theoretical chemistry. Raman spectra are formed when, under certain conditions, light in the visible or ultraviolet region is first absorbed, then is reemited at a lower frequency after causing molecules to rotate or vibrate.

Two magnetic methods of spectroscopy at the radio-frequency region of the spectrum, longer than the infrared band, are valuable in providing chemical information on molecules and showing their detailed structure. These methods are nuclear-magnetic resonance (nmr) and electron-paramagnetic resonance (epr), the latter also being called electron-spin resonance (esr). These methods depend on the fact that electrons and protons spin like little tops. To align the spins, the specimen is placed in a magnetic field. Electrons or protons in the specimen "flip" over, reversing their spin axes, when the proper amount of radio-frequency power is supplied.

ASTROPHYSICAL APPLICATIONS

The distance at which a spectroscope may be placed from the source of light is unlimited. Thus, spectroscopic analysis of the light of the Sun permits an accurate chemical analysis of the constituents of the Sun. The Fraunhofer lines were discovered and named early in the 19th century after their discovery as absorption lines in the spectrum of the Sun; a secondary discovery was that these same lines could be produced on Earth. The element helium was discovered on the Sun and named many years before its presence on Earth was detected. More recently, spectroscopic study of the Sun has given strong indirect evidence for the existence of a negative hydrogen ion. Thus, spectroscopic study of the stars has provided scientists with valuable theoretical knowledge, and is continuing to do so because the stars provide laboratories in which conditions unattainable on Earth are maintained, such as extremely high temperatures and extremely high and low pressures. Certain lines, for example, found in the spectra of nebulas were long thought to be due to an element, tentatively called nebulium, undiscovered on Earth. Scientists now know that these lines are produced by common elements under exceedingly high vacuum conditions.

Late in 1969, for example, the Lunar and Planetary Laboratory at the University of Arizona announced that the spectral analysis of the rings surrounding the planet Saturn showed them to be largely formed of ammonia ice. Scientists also utilized spectroscopy to analyze the composition of the planet Jupiter and its atmosphere after fragments of Comet Shoemaker-Levy 9 crashed into the planet in July 1994. The collisions brought heated interior gases to the surface of the planet's atmosphere, where telescopes recorded them in detail.

A shift in the position of the spectrum lines occurs when the source of the radiation is moving toward or away from the observer. This shift in wavelength, known as the Doppler effect discussed above, provides a fairly accurate value for the relative speed of any source of radiation. In general, if all the lines in the spectrum of a star are shifted toward the red, that star is moving away from Earth, and the velocity of recession can be calculated from the amount of the shift. Conversely, if the star is moving toward Earth, the spectrum is shifted toward the violet.

The Doppler shifts observed in the spectra of exterior galaxies indicate that the universe is expanding. The spectra of a few distant stars periodically split up; the doublets then combine into single lines again. This phenomenon is due to the presence of two stars called double stars, or spectroscopic binaries, that are revolving about each other so close together that a telescope cannot resolve them. When one of the stars is moving toward Earth and the other away, all the lines from one star are shifted toward the violet and all the lines of the other star are shifted toward the red.

When both stars are moving transverse to the line of sight from Earth, the lines from the two stars coincide. All the molecules of a gas are in constant motion, so that at any instant some are moving toward a spectroscope and some away from it. The wavelengths of some of the photons are smaller, and those of others larger, than if all the atoms were at rest. Because of this variability of wavelength, each spectrum line is broadened slightly. If the temperature is raised, the average speed of the molecules is increased, and the lines are still broader. Thus, measurement of the width of certain spectrum lines gives an indication of the temperature of the source, such as the Sun. In many cases, the interior of a source is at a higher temperature than the exterior. An emission spectrum of broad lines will then arise from the interior, and an absorption spectrum will be produced in the exterior; the exterior, however, being cooler, produces narrower lines, and the result for each line is a bright region with a dark center. This phenomenon is called self-reversal.

Related to the Doppler effect is the Mössbauer effect, the discovery of which was announced in 1958 by German physicist Rudolf Ludwig Mössbauer. In a Mössbauer-effect experiment, the recoil-free emission and absorption of gamma rays from one nucleus to another is measured. For absorption to occur, the energy spectrum of gamma rays from the emitter must nearly match the spectrum of possible energies of excitation in the absorber. The slightest change in the motion of the absorber relative to the emitter causes the apparent energy of gamma rays "seen" by the absorber to change. By moving the source or the absorber, scientists may sort out the energies of the gamma rays with high precision. This information is valuable in studies of the electronic and magnetic fields at the nuclei of a solid. The effect also provides an accurate picture of relative motion for use in applications such as the docking of space vehicles.

High-resolution spectroscopy is employed in nuclear physics to study the influence of nuclear size and shape on outer atomic structure. Also, when a light source is placed in magnetic or electric fields, spectral lines are often split or widened, thus revealing important information about the atomic structure of the source, or about the fields, not otherwise available. Dutch physicist Pieter Zeeman discovered in 1896 that, when a ray of light from a source placed in a magnetic field is examined spectroscopically, the spectral line is widened, or even doubled. This phenomenon was named Zeeman effect. The so-called Stark effect was named after German physicist Johannes Stark, who succeeded in splitting spectral lines into several components with a strong electric field in 1913.

CHEMICAL ELEMENTS

Chemical Elements, substance made up only of atoms that all have the same atomic number. The atomic number is the number of protons in the atom's nucleus: element atomic number 1 hydrogen has one proton in the nucleus and the element atomic number 92 uranium has 92 protons. Ninety-four natural elements have been detected in the universe. More than 110 elements have been identified, with some created only in laboratories as artificial elements. Under certain conditions, one element may be changed into another element through processes that add or remove protons from a nucleus. Although the number of protons in the nucleus of a particular chemical element is always the same, the number of neutrons can vary, creating isotopes of that element that have different atomic masses and physical properties. The chemical properties of an element are mainly determined by the number of electrons in the outermost shell of an atom of that element. The number of electrons is the same as the number of protons in the nucleus of an element. The unit for atomic weight of the elements is one-twelfth of the weight of the carbon-12 atom, which is arbitrarily set at 12.

Atoms of a single element may combine to form molecules of that element—two atoms of oxygen combine to form oxygen molecules, eight atoms of sulfur combine to form sulfur molecules. Atoms of different elements may combine to form molecules of chemical compounds—two hydrogen atoms combine with one oxygen atom to form molecules of water. Unlike chemical compounds, elements cannot be broken down into simpler substances by ordinary heat, light, electricity, or chemical reactions. Like other forms of matter, most elements can exist as gases, liquids, or solids, depending on the pressure and temperature.

Scientists use one- and two-letter symbols for each element. In some cases these symbols are based on old Greek or Latin names for the elements and don't correspond to their common English names. For example, hydrogen is H, but gold is Au, from Latin *aurum*.

ORIGIN OF ELEMENTS

The process of building up the nucleus of an element from protons and neutrons is called nucleosynthesis. This process only takes place under extreme conditions of pressure and temperature. The first period of nucleosynthesis occurred in the early moments of the universe following the big bang, creating hydrogen and helium. When the first stars formed, another type of nucleosynthesis began in the cores of stars, fusing hydrogen into helium, and releasing energy. The fusion processes continued to create the elements carbon (6 protons), nitrogen (7 protons), and oxygen (8 protons) on up to iron (26 protons). The fusion of iron does not release energy and stars that reach this stage suffer a core collapse, usually exploding as supernovas. When supernovas explode, the extreme conditions can create even heavier elements, up to uranium.

Other processes can also change the number of protons in a nucleus, transforming one element into another. Radioactive decay can break a heavy element into lighter elements. In space impacts of high energy protons called cosmic rays can break off nuclei of lithium (3 protons), beryllium (4 protons), and boron (5 protons) from atoms of carbon, nitrogen and oxygen. Only tiny amounts of elements heavier than uranium (92 protons)—neptunium (93 protons) and plutonium (94 protons)—are known to exist in nature. All other transuranium elements have been made artificially in laboratories by bombarding heavy atoms with neutrons or with charged particles and atomic nuclei accelerated to high energy.

ABUNDANCE OF ELEMENTS

The relative abundance of elements in the universe reflects how the elements were formed. The big bang created the protons that form all the hydrogen nuclei in the universe. Hydrogen accounts for about 73 percent of the ordinary matter in the universe by mass, helium about 25 percent, and all other elements less than 2 percent. By number of atoms, about 90 percent of the atoms in the universe are hydrogen, about 9 percent are helium, and all the other elements account for less than 1 percent. Processes in stars, and particularly supernova explosions, determined the abundance of other elements. Oxygen (8 protons) is the most abundant, followed by carbon (6 protons), nitrogen (7 protons), neon (10 protons), silicon (14 protons), magnesium (12 protons), and iron (26 protons).

The particular abundance of elements found on Earth reflects the processes that formed our planet from material that became the Sun and the solar system. For example, Earth was not massive enough to retain helium from the original material that formed the solar system. The helium now found in Earth is the result of radioactive decay of minerals that release alpha particles, which have 2 protons and 2 neutrons, the same as helium nuclei.

CHEMICAL ANALYSIS

Chemical Analysis, body of procedures and techniques used to identify and quantify the chemical composition of a sample of a substance. A chemist executing a qualitative analysis seeks to identify the substances in the sample. A quantitative analysis is an attempt to determine the quantity or concentration of a specific substance in the sample. Thus, for example, determining whether a sample of salt contains the element iodine is a qualitative analysis; measuring the percentage by weight of any iodine in the sample is a quantitative analysis. The measurement of chemical composition is necessary throughout commerce, regulatory government, and many fields of science. Chemical analysis thus takes on many specialized forms.

Preparation for Analysis

Chemists are commonly asked to analyze such diverse materials as stainless steel, beer, a fingernail, a rose petal, smoke, aspirin, and paper. The determination of the identity or quantity of a constituent of such materials is preceded by a sampling step—the selection of the amount and uniformity of material required for the analysis—and by the separation from the sample of either the desired constituent or the undesired, interfering constituents. The appropriate separation method depends on the nature of the constituent sought and of the overall sample.

Chromatography is the most generally applicable of the separation methods and has many variants according to the nature of the column packing and the sample-constituent interaction. The two most important types of chromatography are gel permeation chromatography, in which large molecules separate according to their size; and ion exchange chromatography, in which charged, or ionic, constituents are separated. Gas chromatography separates the volatile constituents of a sample, and liquid/liquid chromatography separates small, neutral molecules in solution. The goal in conducting a separation is to produce a purified or partly purified form of the desired constituent for analytical measurement, or to eliminate other constituents that would interfere with the measurement, or both. Separation is often unnecessary when the method is highly specific, or selective, and responds to the desired constituent while ignoring others. Measuring the pH, or hydrogen ion content, of blood with a glass electrode is an example of a measurement that does not require a separation step. Another step preparatory to both qualitative and quantitative analyses is standardization, or calibration. The response of the analytical method and the sensitivity of the mechanical and electronic equipment to the desired constituent must be calibrated, or standardized, using a pure constituent or a sample containing a known amount of constituent. A task of the National Institute of Standards and Technology is to develop and provide such standard samples.

Presentation of Results

The numerical result of a quantitative analysis may state the absolute quantity of the constituent or some percentage of it in the sample. The latter can be expressed as weight percent, molar concentration (moles of dissolved constituent per liter of solution), or ppm (parts per million by weight), among others. The accuracy of the analytical result is reflected by how well it agrees with the true quantity of constituent. The precision of the result is reflected by its reproducibility, or repeatability. Results from repeated measurements are called precise if they all lie within a narrow range of values. Such results are termed highly reproducible. Precision does not necessarily mean that the results are accurate, however, because some part of the measuring process may bias the results toward values that are higher or lower than the true value. Standardization of the analysis often uncovers such systematic errors.

Random errors in a measurement tend to cancel one another out. Accuracy is nearly always improved by averaging multiple determinations. Depending on the method used, measurements may need to be repeated only three or four times. For procedures in which computers are connected to the analytical instruments, as many as 100,000 measurements may be made very quickly. This technique is referred to as signal averaging.

An actual analysis of a sample is commonly based on a chemical reaction of the constituent that produces an easily identifiable quality such as color, heat, or insolubility. Gravimetric analysis, which hinges on measuring the mass of precipitates of the constituent, and titrimetric analysis, which depends on measuring the volumes of solutions that react with the constituent, are referred to as "wet methods"; these are more labor intensive and less versatile than newer methods. Instrumental methods of analysis, or analyses that rely on electronic instruments, became important in the 1950s, and today most analytical measurements are conducted with the aid of such devices.

Qualitative Organic Analysis

Organic analysis relies on certain chemical reactions to detect particular functional groups, such as alcohol, amine, aldehyde, olefin, ester, carboxylic acid, and ether. The test reactions are usually employed without prior separation. As an example, olefins (compounds containing carbon-carbon double bonds) can be identified by the bleaching effect they have on a colored bromine solution. For both organic and inorganic qualitative analysis, instrumental methods are currently preferred because they are more sensitive and specific.

Qualitative Inorganic Analysis

A systematic "wet method" qualitative analysis of inorganic ions proceeds by separating the ions into groups by selective precipitation reactions, isolating individual ions in the groups by an additional precipitation reaction, and confirming the identity of the ion by a reaction test that gives a specific precipitate or color. Several schemes exist for doing this, with cations (positively charged ions) and with anions (negatively charged ions). Table 3 is an abbreviated scheme for the analysis of environmentally important cations of metallic elements.

Quantitative Wet Methods

These are mainly gravimetric and titrimetric procedures for inorganic substances. An example of a gravimetric analysis is the determination of chloride ion concentration in a solution by causing the precipitation of insoluble silver chloride ($AgCl$). The precipitate is then collected and weighed. The analysis yields very accurate results. Titrimetric procedures are commonly based on acid-base reactions such as the titration of acetic acid with a solution of sodium hydroxide. Another common reaction employed is that of a complexing agent, such as ethylenediaminetetraacetic acid (EDTA), with solutions of metal ions, such as lead or mercury. Reactions suitable for titrations must proceed rapidly to completion, without side reactions that tend to obscure the results. This requirement is more often satisfied by inorganic reactions than by organic functional group chemistry.

SPECTROSCOPIC TECHNIQUES

Spectroscopy, or the study of the interactions of electromagnetic radiation with matter, is the largest and most nearly accurate class of instrumental methods used in chemical analysis and indeed in all of chemistry. The electromagnetic radiation (emr) spectrum is divided into the following wavelength regions: X ray, ultraviolet, visible, infrared, microwave, and radiowave. Emr interactions with matter involve absorption or emission of emr energy by means of transitions between quantized, or discrete, levels of energy for electrons, bond vibrations, molecular rotations, and electron and nuclear spins in atoms and molecules. The matter-emr interactions take place in devices called spectrometers, spectrophotometers, or spectroscopes. The spectra produced in these devices are recorded graphically or photographically on spectrograms or spectrographs that permit convenient study of the wavelengths and intensities of the emr absorbed or emitted by the sample being analyzed.

Absorption spectrophotometry in the visible and ultraviolet portions of the emr spectrum is a common quantitative spectral method for both organic and inorganic substances. This technique measures the relative transparency of a solution both before and after the solution has been made to react with a color-forming reagent. The resulting decrease in transparency of the solution is proportional to the concentration of the constituent being analyzed. Infrared absorption spectrophotometry is useful for organic analysis because bonds for olefins, esters, alcohols, and other functional groups have very different strengths and therefore absorb infrared radiation of very different frequencies, or energies. Such absorption spectra appear as peaks when plotted on a spectrograph.

Nuclear magnetic resonance (nmr) spectroscopy depends on transitions between nuclear-spin energy states by absorption of radio-frequency emr energy. In nmr spectra of hydrogen, for example, chemically different hydrogen states absorb emr at different energies. For example, the organic groups 8CH_3 and 8CH_2Cl give very different, well-resolved peaks. Accordingly, nmr is a powerful qualitative analysis tool to deduce the structure of organic molecules. Fluorescence spectroscopy is the reverse of absorption spectrophotometry. With this technique, molecules are induced to emit light, which they do at energies characteristic of their structure, and at intensities proportional to the sample concentration. This method yields extremely sensitive quantitative results for certain molecules.

In atomic emission and atomic absorption spectrophotometry the sample is heated to a high temperature and thereby decomposed into atoms and ions that absorb or emit visible or ultraviolet emr at energies characteristic of the elements involved. The yellowing of a flame by the addition of salt, for example, occurs because the sodium in salt emits strongly in the yellow portion of the visible light emr spectrum. These methods are especially useful for low concentrations of metallic elements in both qualitative and quantitative analysis. In mass spectroscopy, the sample of an organic compound is placed in a vacuum, vaporized, ionized, and given extra energy, all of which cause the individual molecules to fragment. These molecular fragments are then sorted out according to their weight by the electric and magnetic fields in a mass analyzer. The spectral pattern, or mass spectrum, produced is a "fingerprint" of the molecule, in that organic molecules display unique fragmentation patterns.

X-ray fluorescence spectroscopy is useful for both qualitative and quantitative analyses of metallic elements, which emit X rays at characteristic energies when bombarded by a high energy X-ray source.

RADIOCHEMICAL TECHNIQUES

These methods rely on the detection of radioactivity in the form of alpha and beta particles and gamma rays that result from nuclear disintegrations. Radioactivity can be induced in the sample by bombarding it with neutrons. Such a procedure, called neutron activation analysis, is commonly used in industry to identify certain metals in a sample. Neutron activation analysis has the advantage of being rapid and highly automated, and it does not destroy the sample.

Electrochemical Techniques

When a positive and a negative electrode are placed in a solution containing ions, and an electric potential is applied to the electrodes, the positively charged ions (cations) move toward the negative electrode, or cathode, and the negatively charged ions (anions) to the positive electrode, or anode. As a result, electric current flows between the electrodes. The strength of the current depends on the electric potential between the electrodes and the concentration of ions in the solution. Hence, this instrumental quantitative method, called conductometry, is often used to measure the ion concentration in a solution. In a related technique, electrodes specially constructed to accept only specific ions are used to determine the sodium ion or calcium ion concentration or the pH of the solution being analyzed. Such ion-selective electrodes are important in several types of clinical analysis.

THE QUANTUM ATOM

The first great achievement of quantum theory was to explain how atoms work. Physicists found explaining the structure of the atom with classical physics to be impossible. Atoms consist of negatively charged electrons bound to a positively charged nucleus. The nucleus of an atom contains positively charged particles called protons and may contain neutral particles called neutrons. Protons and neutrons are about the same size but are much larger and heavier than electrons are. Classical physics describes a hydrogen atom as an electron orbiting a proton, much as the Moon orbits Earth. By the rules of classical physics, the electron has a property called inertia that makes it want to continue traveling in a straight line. The attractive electrical force of the positively charged proton overcomes this inertia and bends the electron's path into a circle, making it stay in a closed orbit. The classical theory of electromagnetism says that charged particles (such as electrons) radiate energy when they bend their paths. If classical physics applied to the atom, the electron would radiate away all of its energy. It would slow down and its orbit would collapse into the proton within a fraction of a second. However, physicists know that atoms can be stable for centuries or longer.

Quantum theory gives a model of the atom that explains its stability. It still treats atoms as electrons surrounding a nucleus, but the electrons do not orbit the nucleus like moons orbiting planets. Quantum mechanics gives the location of an electron as a probability instead of pinpointing it at a certain position. Even though the position of an electron is uncertain, quantum theory prohibits the electron from being at some places. The easiest way to describe the differences between the allowed and prohibited positions of electrons in an atom is to think of the electron as a wave. The wave-particle duality of quantum theory allows electrons to be described as waves, using the electron's de Broglie wavelength.

If the electron is described as a continuous wave, its motion may be described as that of a standing wave. Standing waves occur when a continuous wave occupies one of a set of certain distances. These distances enable the wave to interfere with itself in such a way that the wave appears to remain stationary. Plucking the string of a musical instrument sets up a standing wave in the string that makes the string resonate and produce sound.

The length of the string, or the distance the wave on the string occupies, is equal to a whole or half number of wavelengths. At these distances, the wave bounces back at either end and constructively interferes with itself, which strengthens the wave. Similarly, an electron wave occupies a distance around the nucleus of an atom, or a circumference, that enables it to travel a whole or half number of wavelengths before looping back on itself.

The electron wave therefore constructively interferes with itself and remains stable:

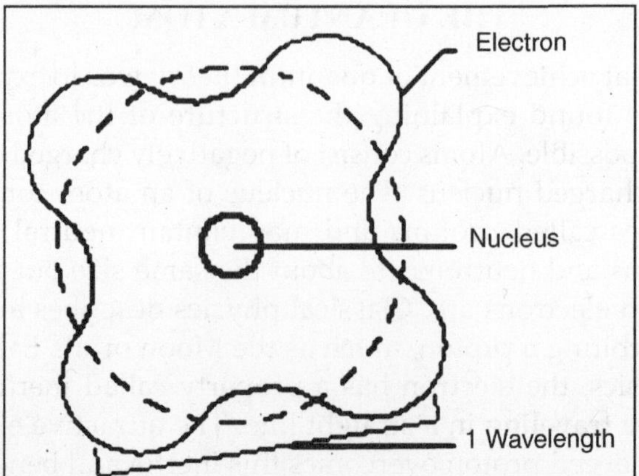

An electron wave cannot occupy a distance that is not equal to a whole or half number of wavelengths.

In a distance such as this, the wave would interfere with itself in a complicated way, and would become unstable:

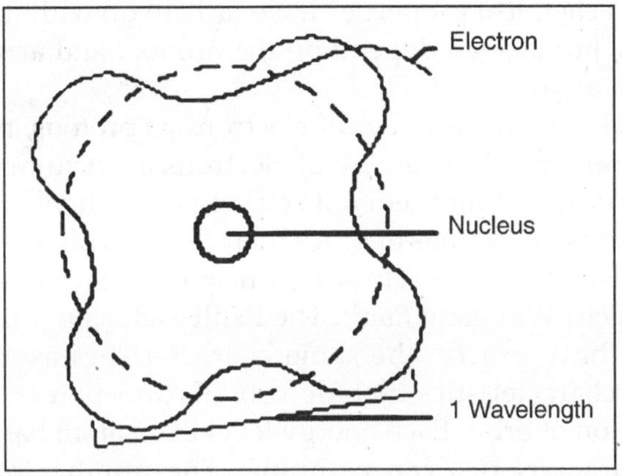

An electron has a certain amount of energy when its wave occupies one of the allowed circumferences around the nucleus of an atom. This energy depends on the number of wavelengths in the circumference, and it is called the electron's energy level. Because only certain circumferences, and therefore energy levels, are allowed, physicists say that the energy levels are quantized. This quantization means that the energies of the levels can only take on certain values.

The regions of space in which electrons are most likely to be found are called orbitals. Orbitals look like fuzzy, three-dimensional shapes. More than one orbital, meaning more than one shape, may exist at certain energy levels. Electron orbitals are also quantized, meaning that only certain shapes are allowed in each energy level. The quantization of electron orbitals and energy levels in atoms explains the stability of atoms. An electron in an energy level that allows only one wavelength is at the lowest possible energy level. An atom with all of its electrons in their lowest possible energy levels is said to be in its ground state. Unless it is affected by external forces, an atom will stay in its ground state forever.

The quantum theory explanation of the atom led to a deeper understanding of the periodic table of the chemical elements. The periodic table of elements is a chart of the known elements. Scientists originally arranged the elements in this table in order of increasing atomic number (which is equal to the number of protons in the nuclei of each element's atoms) and according to the chemical behavior of the elements. They grouped elements that behave in a similar way together in columns. Scientists found that elements that behave similarly occur in a periodic fashion according to their atomic number. For example, a family of elements called the noble gases all share similar chemical properties. The noble gases include neon, xenon, and argon. They do not react easily with other elements and are almost never found in chemical compounds. The atomic numbers of the noble gases increase from one element to the next in a periodic way. They belong to the same column at the far right edge of the periodic table. Quantum theory showed that an element's chemical properties have little to do with the nucleus of the element's atoms, but instead depend on the number and arrangement of the electrons in each atom.

An atom has the same number of electrons as protons, making the atom electrically neutral. The arrangement of electrons in an atom depends on two important parts of quantum theory. The first is the quantization of electron energy, which limits the regions of space that electrons can occupy. The second part is a rule called the Pauli exclusion principle, first proposed by Austrian-born Swiss physicist Wolfgang Pauli. The Pauli exclusion principle states that no electron can have exactly the same characteristics as those of another electron. These characteristics include orbital, direction of rotation (called spin), and direction of orbit. Each energy level in an atom has a set number of ways these characteristics can combine. The number of combinations determines how many electrons can occupy an energy level before the electrons have to start filling up the next level.

An atom is the most stable when it has the least amount of energy, so its lowest energy levels fill with electrons first. Each energy level must be filled before electrons begin filling up the next level. These rules, and the rules of quantum theory, determine how many electrons an atom has in each energy level, and in particular, how many it has in its outermost level. Using the quantum mechanical model of the atom, physicists found that all the elements in the same column of the periodic table also have the same number of electrons in the outer energy level of their atoms. Quantum theory shows that the number of electrons in an atom's outer level determines the atom's chemical properties, or how it will react with other atoms.

The number of electrons in an atom's outer energy level is important because atoms are most stable when their outermost energy level is filled, which is the case for atoms of the noble gases. Atoms imitate the noble gases by donating electrons to, taking electrons from, or sharing electrons with other atoms. If an atom's outer energy level is only partially filled, it will bond easily with atoms that can help it fill its outer level. Atoms that are missing the same number of electrons from their outer energy level will react similarly to fill their outer energy level.

Quantum theory also explains why different atoms emit and absorb different wavelengths of light. An atom stores energy in its electrons. An atom with all of its electrons at their lowest possible energy levels has its lowest possible energy and is said to be in its ground state. One of the ways atoms can gain more energy is to absorb light in the form of photons, or particles of light. When a photon hits an atom, one of the atom's electrons absorbs the photon. The photon's energy makes the electron jump from its original energy level up to a higher energy level. This jump leaves an empty space in the original inner energy level, making the atom less stable. The atom is now in an excited state, but it cannot store the new energy indefinitely, because atoms always seek their most stable state. When the atom releases the energy, the electron drops back down to its original energy level. As it does, the electron releases a photon.

Quantum theory defines the possible energy levels of an atom, so it defines the particular jumps that an electron can make between energy levels. The difference between the old and new energy levels of the electron is equal to the amount of energy the atom stores. Because the energy levels are quantized, atoms can only absorb and store photons with certain amounts of energy. The photon's energy is related to its frequency, or color. As the frequency of photons increases, their energy increases. Atoms can only absorb certain amounts of energy, so only certain frequencies of light can excite atoms. Likewise, atoms only emit certain frequencies of light when they drop to their ground state. The different frequencies available to different atoms help astronomers, for example, determine the chemical makeup of a star by observing which wavelengths are especially weak or strong in the star's light.

DEVELOPMENT OF QUANTUM THEORY

The development of quantum theory began with German physicist Max Planck's proposal in 1900 that matter can emit or absorb energy only in small, discrete packets, called quanta. This idea introduced the particle nature of light. In 1905 German-born American physicist Albert Einstein used Planck's work to explain the photoelectric effect, in which light hitting a metal makes the metal emit electrons. British physicist Ernest Rutherford proved that atoms consisted of electrons bound to a nucleus in 1911. In 1913 Danish physicist Niels Bohr proposed that classical mechanics could not explain the structure of the atom and developed a model of the atom with electrons in fixed orbits. Bohr's model of the atom proved difficult to apply to all but the simplest atoms.

In 1923 French physicist Louis de Broglie suggested that matter could be described as a wave, just as light could be described as a particle. The wave model of the electron allowed Austrian physicist Erwin Schrödinger to develop a mathematical method of determining the probability that an electron will be at a particular place at a certain time. Schrödinger published his theory of wave mechanics in 1926. Around the same time, German physicist Werner Heisenberg developed a way of calculating the characteristics of electrons that was quite different from Schrödinger's method but yielded the same results. Heisenberg's method was called matrix mechanics. In 1925 Austrian-born Swiss physicist Wolfgang Pauli developed the Pauli exclusion principle, which allowed physicists to calculate the structure of the quantum atom for the first time. In 1926 Heisenberg and two of his colleagues, German physicists Max Born and Ernst Pascual Jordan, published a theory that combined the principles of quantum theory with the classical theory of light (called electrodynamics). Heisenberg made another important contribution to quantum theory in 1927 when he introduced the Heisenberg uncertainty principle. Since these first breakthroughs in quantum mechanical research, physicists have focused on testing and refining quantum theory, further connecting the theory to other theories, and finding new applications. In 1928 British physicist Paul Dirac refined the theory that combined quantum theory with electrodynamics. He developed a model of the electron that was consistent with both quantum theory and Einstein's special theory of relativity, and in doing so he created a theory that came to be known as quantum electrodynamics, or QED. In the early 1950s Japanese physicist Tomonaga Shin' ichirô and American physicists Richard Feynman and Julian Schwinger each independently improved the scientific community's understanding of QED and made it an experimentally testable theory that successfully predicted or explained the results of many experiments.

QUANTUM THEORY

Quantum Theory, in physics, description of the particles that make up matter and how they interact with each other and with energy. Quantum theory explains in principle how to calculate what will happen in any experiment involving physical or biological systems, and how to understand how our world works. The name "quantum theory" comes from the fact that the theory describes the matter and energy in the universe in terms of single indivisible units called *quanta* (singular *quantum*). Quantum theory is different from classical physics. Classical physics is an approximation of the set of rules and equations in quantum theory. Classical physics accurately describes the behavior of matter and energy in the everyday universe. For example, classical physics explains the motion of a car accelerating or of a ball flying through the air. Quantum theory, on the other hand, can accurately describe the behavior of the universe on a much smaller scale, that of atoms and smaller particles.

The rules of classical physics do not explain the behavior of matter and energy on this small scale. Quantum theory is more general than classical physics, and in principle, it could be used to predict the behavior of any physical, chemical, or biological system. However, explaining the behavior of the everyday world with quantum theory is too complicated to be practical. Quantum theory not only specifies new rules for describing the universe but also introduces new ways of thinking about matter and energy. The tiny particles that quantum theory describes do not have defined locations, speeds, and paths like objects described by classical physics. Instead, quantum theory describes positions and other properties of particles in terms of the chances that the property will have a certain value. For example, it allows scientists to calculate how likely it is that a particle will be in a certain position at a certain time. Quantum description of particles allows scientists to understand how particles combine to form atoms. Quantum description of atoms helps scientists understand the chemical and physical properties of molecules, atoms, and subatomic particles. Quantum theory enabled scientists to understand the conditions of the early universe, how the Sun shines, and how atoms and molecules determine the characteristics of the material that they make up. Without quantum theory, scientists could not have developed nuclear energy or the electric circuits that provide the basis for computers.

Quantum theory describes all of the fundamental forces—except gravitation—that physicists have found in nature. The forces that quantum theory describes are the electrical, the magnetic, the weak, and the strong. Physicists often refer to these forces as interactions, because the forces control the way particles interact with each other. Interactions also affect spontaneous changes in isolated particles.

Waves and Particles

One of the striking differences between quantum theory and classical physics is that quantum theory describes energy and matter both as waves and as particles. The type of energy physicists study most often with quantum theory is light. Classical physics considers light to be only a wave, and it treats matter strictly as particles. Quantum theory acknowledges that both light and matter can behave like waves and like particles. It is important to understand how scientists describe the properties of waves in order to understand how waves fit into quantum theory. A familiar type of wave occurs when a rope is tied to a solid object and someone moves the free end up and down. Waves travel along the rope. The highest points on the rope are called the crests of the waves. The lowest points are called troughs.

One full wave consists of a crest and trough. The distance from crest to crest or from trough to trough—or from any point on one wave to the identical point on the next wave—is called the wavelength. The frequency of the waves is the number of waves per second that pass by a given point along the rope. If waves traveling down the rope hit the stationary end and bounce back, like water waves bouncing against a wall, two waves on the rope may meet each other, hitting the same place on the rope at the same time. These two waves will interfere, or combine. If the two waves exactly line up—that is, if the crests and troughs of the waves line up—the waves interfere constructively. This means that the trough of the combined wave is deeper and the crest is higher than those of the waves before they combined. If the two waves are offset by exactly half of a wavelength, the trough of one wave lines up with the crest of the other. This alignment creates destructive interference—the two waves cancel each other out and a momentary flat spot appears on the rope.

Light as a Wave and as a Particle

Like classical physics, quantum theory sometimes describes light as a wave, because light behaves like a wave in many situations. Light is not a vibration of a solid substance, such as a rope. Instead, a light wave is made up of a vibration in the intensity of the electric and magnetic fields that surround any electrically charged object. Like the waves moving along a rope, light waves travel and carry energy. The amount of energy depends on the frequency of the light waves: the higher the frequency, the higher the energy. The frequency of a light wave is also related to the color of the light. For example, blue light has a higher frequency than that of red light. Therefore, a beam of blue light has more energy than an equally intense beam of red light has.

Unlike classical physics, quantum theory also describes light as a particle. Scientists revealed this aspect of light behavior in several experiments performed during the early 20th century. In one experiment, physicists discovered an interaction between light and particles in a metal. They called this interaction the photoelectric effect. In the photoelectric effect, a beam of light shining on a piece of metal makes the metal emit electrons. The light adds energy to the metal's electrons, giving them enough energy to break free from the metal. If light was made strictly of waves, each electron in the metal could absorb many continuous waves of light and gain more and more energy. Increasing the intensity of the light, or adding more light waves, would add more energy to the emitted electrons. Shining a more and more intense beam of light on the metal would make the metal emit electrons with more and more energy.

Scientists found, however, that shining a more intense beam of light on the metal just made the metal emit more electrons. Each of these electrons had the same energy as that of electrons emitted with low intensity light. The electrons could not be interacting with waves, because adding more waves did not add more energy to the electrons. Instead, each electron had to be interacting with just a small piece of the light beam. These pieces were like little packets of light energy, or particles of light. The size, or energy, of each packet depended only on the frequency, or color, of the light—not on the intensity of the light. A more intense beam of light just had more packets of light energy, but each packet contained the same amount of energy. Individual electrons could absorb one packet of light energy and break free from the metal. Increasing the intensity of the light added more packets of energy to the beam and enabled a greater number of electrons to break free—but each of these emitted electrons had the same amount of energy.

Scientists could only change the energy of the emitted electrons by changing the frequency, or color, of the beam. Changing from red light to blue light, for example, increased the energy of each packet of light. In this case, each emitted electron absorbed a bigger packet of light energy and had more energy after it broke free of the metal. Using these results, physicists developed a model of light as a particle, and they called these light particles photons. In 1922 American physicist Arthur Compton discovered another interaction, now called the Compton effect, that reveals the particle nature of light. In the Compton effect, light collides with an electron. The collision knocks the electron off course and changes the frequency, and therefore energy, of the light. The light affects the electron in the same way a particle with momentum would: It bumps the electron and changes the electron's path. The light is also affected by the collision as though it were a particle, in that its energy and momentum changes.

Momentum is a quantity that can be defined for all particles. For light particles, or photons, momentum depends on the frequency, or color, of the photon, which in turn depends on the photon's energy. The energy of a photon is equal to a constant number, called Planck's constant, times the frequency of the photon. Planck's constant is named for German physicist Max Planck, who first proposed the relationship between energy and frequency. The accepted value of Planck's constant is 6.626×10^{-34} joule-second. This number is very small—written out, it is a decimal point followed by 33 zeroes, followed by the digits 6626. The energy of a single photon is therefore very small. The dual nature of light seems puzzling because we have no everyday experience with wave-particle duality. Waves are everyday phenomena; we are all familiar with waves on a body of water or on a vibrating rope. Particles, too, are everyday objects—baseballs, cars, buildings, and even people can be thought of as particles. But to our senses, there are no everyday objects that are both waves and particles. Scientists increasingly find that the rules that apply to the world we see are only approximations of the rules that govern the unseen world of light and subatomic particles.

Matter as Waves and Particles

In 1923 French physicist Louis de Broglie suggested that all particles—not just photons—have both wave and particle properties. He calculated that every particle has a wavelength (represented by λ, the Greek letter *lambda*) equal to Planck's constant (h) divided by the momentum (p) of the particle: $\lambda = h/p$. Electrons, atoms, and all other particles have de Broglie wavelengths. The momentum of an object depends on its speed and mass, so the faster and heavier an object is, the larger its momentum (p) will be. Because Planck's constant (h) is an extremely tiny number, the de Broglie wavelength (h/p) of any visible object is exceedingly small. In fact, the de Broglie wavelength of anything much larger than an atom is smaller than the size of one of its atoms. For example, the de Broglie wavelength of a baseball moving at 150 km/h (90 mph) is 1.1×10^{-34} m (3.6×10^{-34} ft). The diameter of a hydrogen atom (the simplest and smallest atom) is about 5×10^{-11} m (about 2×10^{-10} ft), more than 100 billion trillion times larger than the de Broglie wavelength of the baseball. The de Broglie wavelengths of everyday objects are so tiny that the wave nature of these objects does not affect their visible behavior, so their wave-particle duality is undetectable to us. De Broglie wavelengths become important when the mass, and therefore momentum, of particles is very small. Particles the size of atoms and electrons have demonstrable wavelike properties.

One of the most dramatic and interesting demonstrations of the wave behavior of electrons comes from the double-slit experiment. This experiment consists of a barrier set between a source of electrons and an electron detector. The barrier contains two slits, each about the width of the de Broglie wavelength of an electron. On this small scale, the wave nature of electrons becomes evident, as described in the following paragraphs. Scientists can determine whether the electrons are behaving like waves or like particles by comparing the results of double-slit experiments with those of similar experiments performed with visible waves and particles. To establish how visible waves behave in a double-slit apparatus, physicists can replace the electron source with a device that creates waves in a tank of water.

The slits in the barrier are about as wide as the wavelength of the water waves. In this experiment, the waves spread out spherically from the source until they hit the barrier. The waves pass through the slits and spread out again, producing two new wave fronts with centers as far apart as the slits are. These two new sets of waves interfere with each other as they travel toward the detector at the far end of the tank. The waves interfere constructively in some places (adding together) and destructively in others (canceling each other out). The most intense waves—that is, those formed by the most constructive interference—hit the detector at the spot opposite the midpoint between the two slits. These strong waves form a peak of intensity on the detector. On either side of this peak, the waves destructively interfere and cancel each other out, creating a low point in intensity.

Further out from these low points, the waves are weaker, but they constructively interfere again and create two more peaks of intensity, smaller than the large peak in the middle. The intensity then drops again as the waves destructively interfere. The intensity of the waves forms a symmetrical pattern on the detector, with a large peak directly across from the midpoint between the slits and alternating low points and smaller and smaller peaks on either side. To see how particles behave in the double-slit experiment, physicists replace the water with marbles. The barrier slits are about the width of a marble, as the point of this experiment is to allow particles (in this case, marbles) to pass through the barrier. The marbles are put in motion and pass through the barrier, striking the detector at the far end of the apparatus. The results show that the marbles do not interfere with each other or with themselves like waves do. Instead, the marbles strike the detector most frequently in the two points directly opposite each slit.

When physicists perform the double-slit experiment with electrons, the detection pattern matches that produced by the waves, not the marbles. These results show that electrons do have wave properties. However, if scientists run the experiment using a barrier whose slits are much wider than the de Broglie wavelength of the electrons, the pattern resembles the one produced by the marbles. This shows that tiny particles such as electrons behave as waves in some circumstances and as particles in others.

Uncertainty Principle

Before the development of quantum theory, physicists assumed that, with perfect equipment in perfect conditions, measuring any physical quantity as accurately as desired was possible. Quantum mechanical equations show that accurate measurement of both the position and the momentum of a particle at the same time is impossible. This rule is called Heisenberg's uncertainty principle after German physicist Werner Heisenberg, who derived it from other rules of quantum theory. The uncertainty principle means that as physicists measure a particle's position with more and more accuracy, the momentum of the particle becomes less and less precise, or more and more uncertain, and vice versa.

Heisenberg formally stated his principle by describing the relationship between the uncertainty in the measurement of a particle's position and the uncertainty in the measurement of its momentum. Heisenberg said that the uncertainty in position (represented by Δx) times the uncertainty in momentum (represented by Δp;) must be greater than a constant number equal to Planck's constant (h) divided by 4π (π is a constant approximately equal to 3.14). Mathematically, the uncertainty principle can be written as $\Delta x \, \Delta p > h / 4\pi$. This relationship means that as a scientist measures a particle's position more and more accurately—so the uncertainty in its position becomes very small—the uncertainty in its momentum must become large to compensate and make this expression true. Likewise, if the uncertainty in momentum, Δp, becomes small, Δx must become large to make the expression true.

One way to understand the uncertainty principle is to consider the dual wave-particle nature of light and matter. Physicists can measure the position and momentum of an atom by bouncing light off of the atom. If they treat the light as a wave, they have to consider a property of waves called diffraction when measuring the atom's position. Diffraction occurs when waves encounter an object—the waves bend around the object instead of traveling in a straight line. If the length of the waves is much shorter than the size of the object, the bending of the waves just at the edges of the object is not a problem. Most of the waves bounce back and give an accurate measurement of the object's position. If the length of the waves is close to the size of the object, however, most of the waves diffract, making the measurement of the object's position fuzzy. Physicists must bounce shorter and shorter waves off an atom to measure its position more accurately. Using shorter wavelengths of light, however, increases the uncertainty in the measurement of the atom's momentum. Light carries energy and momentum, because of its particle nature (described in the Compton effect). Photons that strike the atom being measured will change the atom's energy and momentum. The fact that measuring an object also affects the object is an important principle in quantum theory. Normally the affect is so small it does not matter, but on the small scale of atoms, it becomes important. The bump to the atom increases the uncertainty in the measurement of the atom's momentum. Light with more energy and momentum will knock the atom harder and create more uncertainty.

The momentum of light is equal to Planck's constant divided by the light's wavelength, or $p = h/\lambda$ Physicists can increase the wavelength to decrease the light's momentum and measure the atom's momentum more accurately. Because of diffraction, however, increasing the light's wavelength increases the uncertainty in the measurement of the atom's position. Physicists most often use the uncertainty principle that describes the relationship between position and momentum, but a similar and important uncertainty relationship also exists between the measurement of energy and the measurement of time.

Probability and Wave Functions

Quantum theory gives exact answers to many questions, but it can only give probabilities for some values. A probability is the likelihood of an answer being a certain value. Probability is often represented by a graph, with the highest point on the graph representing the most likely value and the lowest representing the least likely value.

For example, the graph that shows the likelihood of finding the electron of a hydrogen atom at a certain distance from the nucleus looks like the following:

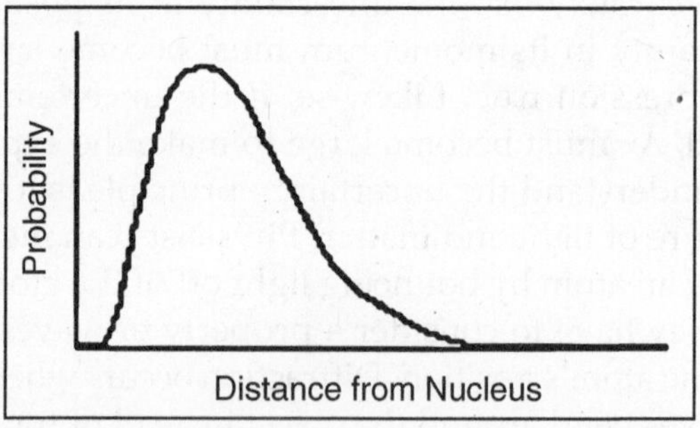

Fig. Probability Density for Electron in Hydrogen Atom

The nucleus of the atom is at the left of the graph. The probability of finding the electron very near the nucleus is very low. The probability reaches a definite peak, marking the spot at which the electron is most likely to be. Scientists use a mathematical expression called a wave function to describe the characteristics of a particle that are related to time and space—such as position and velocity. The wave function helps determine the probability of these aspects being certain values. The wave function of a particle is not the same as the wave suggested by wave-particle duality. A wave function is strictly a mathematical way of expressing the characteristics of a particle. Physicists usually represent these types of wave functions with the Greek letter *psi*, ψ.

The wave function of the electron in a hydrogen atom is:

$$\psi = \frac{1}{\sqrt{\pi a^3}} e^{\frac{-y}{a}}$$

The symbol π and the letter *e* in this equation represent constant numbers derived from mathematics. The letter *a* is also a constant number called the Bohr radius for the hydrogen atom. The square of a wave function, or a wave function multiplied by itself, is equal to the probability density of the particle that the wave function describes. The probability density of a particle gives the likelihood of finding the particle at a certain point. The wave function described above does not depend on time. An isolated hydrogen atom does not change over time, so leaving time out of the atom's wave function is acceptable. For particles in systems that change over time, physicists use wave functions that depend on time. This lets them calculate how the system and the particle's properties change over time. Physicists represent a time-dependent wave function with $\psi(t)$, where t represents time.

The wave function for a single atom can only reveal the probability that an atom will have a certain set of characteristics at a certain time. Physicists call the set of characteristics describing an atom the state of the atom. The wave function cannot describe the actual state of the atom, just the probability that the atom will be in a certain state. The wave functions of individual particles can be added together to create a wave function for a system, so quantum theory allows physicists to examine many particles at once. The rules of probability state that probabilities and actual values match better and better as the number of particles (or dice thrown, or coins tossed, whatever the probability refers to) increases. Therefore, if physicists consider a large group of atoms, the wave function for the group of atoms provides information that is more definite and useful than that provided by the wave function of a single atom. An example of a wave function for a single atom is one that describes an atom that has absorbed some energy.

The energy has boosted the atom's electrons to a higher energy level, and the atom is said to be in an excited state. It can return to its normal ground state (or lowest energy state) by emitting energy in the form of a photon. Scientists call the wave function of the initial exited state ψ_i (with "i" indicating it is the initial state) and the wave function of the final ground state ψ_f (with "f" representing the final state). To describe the atom's state over time, they multiply ψ_i by some function, a(t), that decreases with time, because the chances of the atom being in this excited state decrease with time. They multiply ψ_f by some function, b(t), that increases with time, because the chances of the atom being in this state increase with time. The physicist completing the calculation chooses a(t) and b(t) based on the characteristics of the system.

The complete wave equation for the transition is the following:

$$\psi = a(t)\psi_i + b(t)\psi_f$$

At any time, the rules of probability state that the probability of finding a single atom in either state is found by multiplying the coefficient of its wave function (a(t) or b(t)) by itself. For one atom, this does not give a very satisfactory answer. Even though the physicist knows the probability of finding the atom in one state or the other, whether or not reality will match probability is still far from certain. This idea is similar to rolling a pair of dice. There is a 1 in 6 chance that the roll will add up to seven, which is the most likely sum. Each roll is random, however, and not connected to the rolls before it. It would not be surprising if ten rolls of the dice failed to yield a sum of seven. However, the actual number of times that seven appears matches probability better as the number of rolls increases. For one million or one billion rolls of the dice, one of every six rolls would almost certainly add up to seven. Similarly, for a large number of atoms, the probabilities become approximate percentages of atoms in each state, and these percentages become more accurate as the number of atoms observed increases. For example, if the square of a(t) at a certain time is 0.2, then in a very large sample of atoms, 20 percent (0.2) of the atoms will be in the initial state and 80 percent (0.8) will be in the final state. One of the most puzzling results of quantum mechanics is the effect of measurement on a quantum system. Before a scientist measures the characteristics of a particle, its characteristics do not have definite values. Instead, they are described by a wave function, which gives the characteristics only as fuzzy probabilities. In effect, the particle does not exist in an exact location until a scientist measures its position. Measuring the particle fixes its characteristics at specific values, effectively "collapsing" the particle's wave function. The particle's position is no longer fuzzy, so the wave function that describes it has one high, sharp peak of probability.

If the position of a particle has just been measured, the graph of its probability density looks like the following:

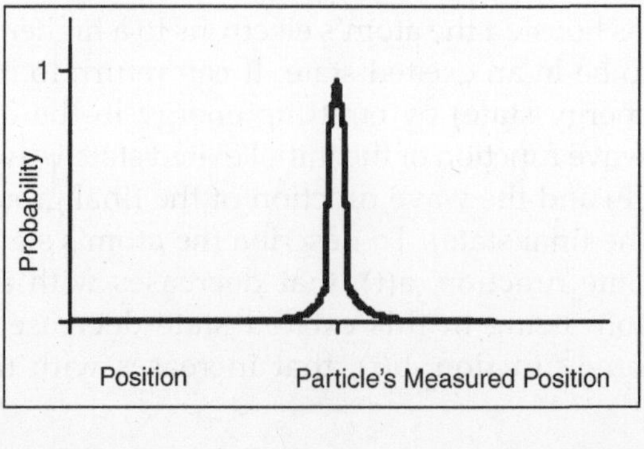

In the 1930s physicists proposed an imaginary experiment to demonstrate how measurement causes complications in quantum mechanics. They imagined a system that contained two particles with opposite values of *spin,* a property of particles that is analogous to angular momentum. The physicists can know that the two particles have opposite spins by setting the total spin of the system to be zero. They can measure the total spin without directly measuring the spin of either particle. Because they have not yet measured the spin of either particle, the spins do not actually have defined values. They exist only as fuzzy probabilities. The spins only take on definite values when the scientists measure them.

In this hypothetical experiment the scientists do not measure the spin of each particle right away. They send the two particles, called an entangled pair, off in opposite directions until they are far apart from each other. The scientists then measure the spin of one of the particles, fixing its value. Instantaneously, the spin of the other particle becomes known and fixed. It is no longer a fuzzy probability but must be the opposite of the other particle, so that their spins will add to zero. It is as though the first particle communicated with the second.

This apparent instantaneous passing of information from one particle to the other would violate the rule that nothing, not even information, can travel faster than the speed of light. The two particles do not, however, communicate with each other. Physicists can instantaneously know the spin of the second particle because they set the total spin of the system to be zero at the beginning of the experiment. In 1997 Austrian researchers performed an experiment similar to the hypothetical experiment of the 1930s, confirming the effect of measurement on a quantum system.

7

Auger Electron Spectroscopy

Auger Electron Spectroscopy (*Auger spectroscopy* or AES) was developed in the late 1960's, deriving its name from the effect first observed by Pierre Auger, a French Physicist, in the mid-1920's. It is a surface specific technique utilising the emission of low energy electrons in the *Auger process* and is one of the most commonly employed surface analytical techniques for determining the composition of the surface layers of a sample.

Auger electron spectroscopy is a common analytical technique used specifically in the study of surfaces and, more generally, in the area of materials science. Underlying the spectroscopic technique is the Auger effect, as it has come to be called, which is based on the analysis of energetic electronsemitted from an excited atom after a series of internal relaxation events. The Auger effect was discovered independently by both Lise Meitner and Pierre Auger in the 1920s. Though the discovery was made by Meitner and initially reported in the journal *Zeitschrift für Physik* in 1922, Auger is credited with the discovery in most of the scientific community. Until the early 1950s Auger transitions were considered nuisance effects by spectroscopists, not containing much relevant material information, but studied so as to explain anomalies in x-ray spectroscopy data. Since 1953 however, AES has become a practical and straightforward characterization technique for probing chemical and compositional surface environments and has found applications in metallurgy, gas-phase chemistry, and throughout the microelectronics industry.

PHYSICAL BASIS

Auger spectroscopy can be considered as involving three basic steps:

1. Atomic ionization (by removal of a core electron)
2. Electron emission (the Auger process)
3. Analysis of the emitted Auger electrons

The last stage is simply a technical problem of detecting charged particles with high sensitivity, with the additional requirement that the kinetic energies of the emitted electrons must be determined.

We will therefore concern ourselves only with the first two processes - before starting, however, it is useful to briefly review the electronic structure of atoms and solids, and associated nomenclature.

Electronic Structure - Isolated Atoms

The diagram below schematically illustrates the energies of the various electron energy levels in an isolated, multi-electron atom, with the conventional chemical nomenclature for these orbitals given on the right hand side.

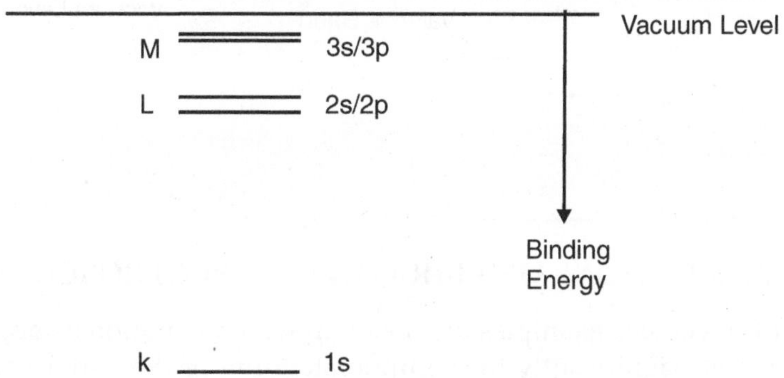

However, scientists working with x-rays tend to use the alternative nomenclature on the left and it is this nomenclature that is used in Auger spectroscopy. The designation of levels to the K,L,M,... shells is based on their having principal quantum numbers of 1,2,3,... respectively. It is convenient to expand the part of the energy scale close to the vacuum level in order to more clearly distinguish between the higher levels.

The numerical component of the KLM.. style of nomenclature is usually written as a subscript immediately following the main shell designation.

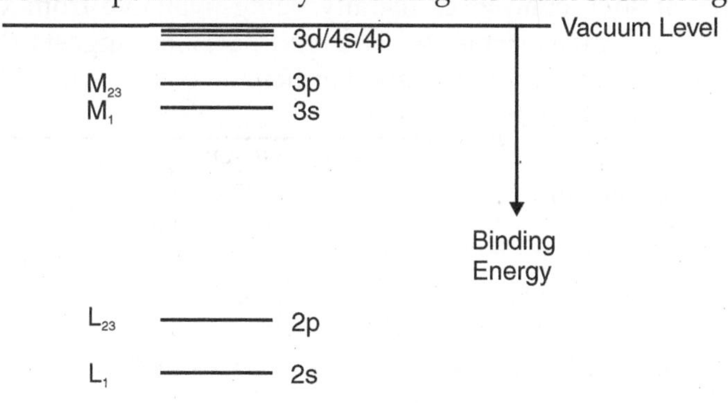

Levels with a non-zero value of the orbital angular momentum quantum number ($l > 0$), *i.e.* p,d,f,.. levels, show spin-orbit splitting. The magnitude of this splitting, however, is too small to be evident on this diagram - hence, the double subscript for these levels (*i.e.* $L_{2,3}$ represents both the L_2 and L_3 levels).

Electronic Structure - Solid State

In the solid state the core levels of atoms are little perturbed and essentially remain as discrete, localised (*i.e.* atomic-like) levels. The valence orbitals, however, overlap significantly with those of neighbouring atoms generating bands of spatially-delocalised energy levels. The energy level diagram for the solid is therefore closely resemblant of that of the corresponding isolated atom, except for the levels closest to the vacuum level.

The diagram below shows the electronic structure of Na metal:

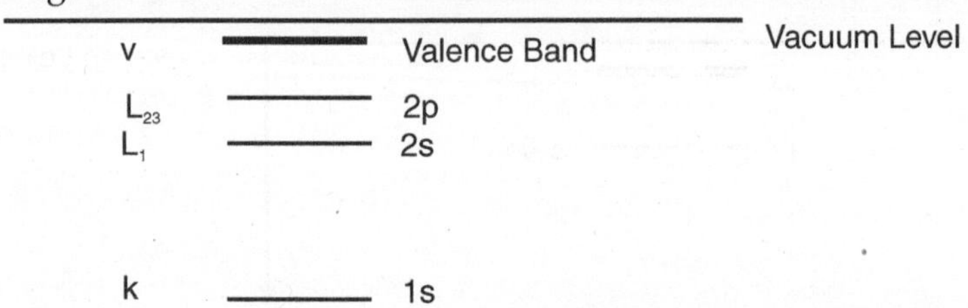

APPLICATIONS OF VIBRATIONAL SPECTROSCOPY

One of the classic examples of an area in which vibrational spectroscopy has contributed significantly to the understanding of the surface chemistry of an adsorbate is that of:

Molecular Adsorption of CO on Metallic Surfaces

Adsorbed carbon monoxide usually gives rise to strong absorptions in both the IR and EELS spectra at the (C-O) stretching frequency. The metal-carbon stretching mode (ca. 400 $cm^{-1)}$ is usually also accessible to EELS. The interpretation of spectra of CO as an adsorbed surface species draws heavily upon IR spectra from related inorganic cluster and coordination complexes - the structure of such complexes usually being available from x-ray single crystal diffraction measurements. This comparison suggests that the CO stretching frequency can provide a good indication of the surface coordination of the molecule: as a rough guideline,

	v(C-O)
CO (gas phase)	2143 cm^{-1}
Terminal CO	2100 – 1920 cm^{-1}
Bridging (2f site)	1920 – 1800 cm^{-1}
Bridging (3f / 4f site)	< 1800 cm^{-1}

RAIR spectrum for CO chemisorbed on a clean Pt surface

The reduction in the stretching frequency of terminally-bound CO from the value observed for the gas phase molecule (2143 $cm^{-1)}$ can be explained in terms of the Dewar-Chatt or *Blyholder model* for the bonding of CO to metals. This simple model considers the metal-CO bonding to consist of two main components:

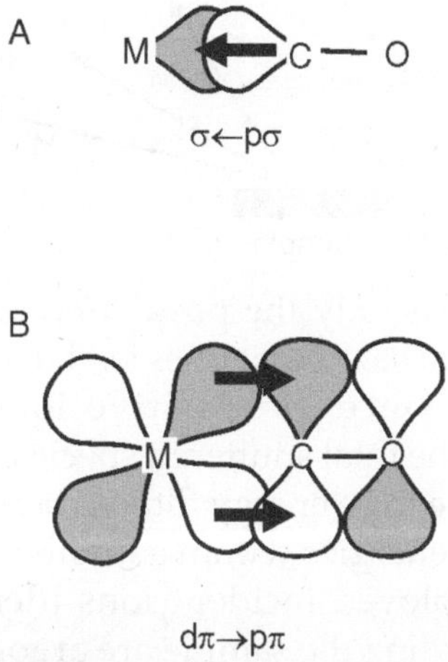

- A: This is a s bonding interaction due to overlap of a filled s -"lone pair" orbital on the carbon atom with empty metal orbitals of the correct symmetry - this leads to electron density transfer from the CO molecule to the metal centre.
- B: This is a p bonding interaction due to overlap of filled metal dp (and pp) orbitals with the p* antibonding molecular orbital of the CO molecule. Since this interaction leads to the introduction of electron density into the CO antibonding orbital there is a consequent reduction in the CO bond strength and its intrinsic vibrational frequency (relative to the isolated molecule).

SECONDARY ION MASS SPECTROMETRY

The technique of Secondary Ion Mass Spectrometry (SIMS) is the most sensitive of all the commonly-employed surface analytical techniques this is because of the inherent sensitivity associated with mass spectrometric-based techniques.

There are a number of different variants of the technique:

- Statics SIMS: used for sub-monolayer elemental analysis
- Dynamic SIMS: used for obtaining compositional information as a function of depth below the surface
- Imaging SIMS: used for spatially-resolved elemental analysis

All of these variations on the technique are based on the same basic physical process and it is this process, together with a brief introduction to the field of static SIMS. Further notes on dynamic and imaging SIMS can be obtained.

In SIMS the surface of the sample is subjected to bombardment by high energy ions - this leads to the ejection (or *sputtering*) of both neutral and charged (+/-) species from the surface. The ejected species may include atoms, clusters of atoms and molecular fragments.

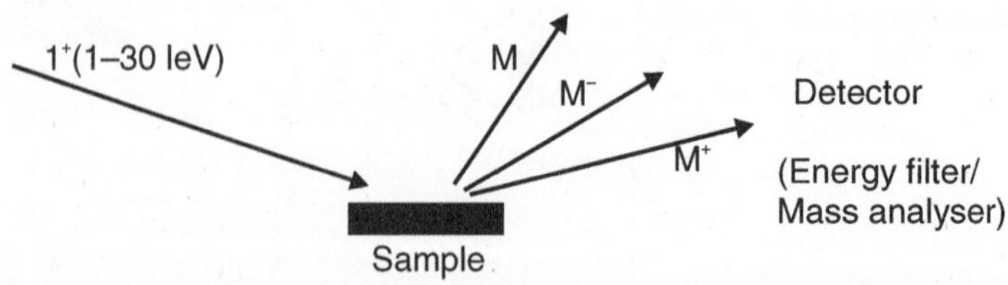

In traditional SIMS it is only the positive ions that are mass analysed - this is primarily for practical ease but it does lead to problems with quantifying the compositional data since the positive ions are but a small, non-representative fraction of the total sputtered species. It should be further noted that the displaced ions have to be energy filtered before they are mass analysed (*i.e.* only ions with kinetic energies within a limited range are mass analysed). The most commonly employed incident ions (denoted by I^+ in the above diagram) used for bombarding the sample are argon ions (Ar^+) but other ions (*e.g.* alkali metal ions, Ga^+) are preferred for some applications.

The mass analyser is typically a quadrupole MS analyser with unit mass resolution, but high specification time-of-flight (TOF) analysers are also used and provide substantially higher sensitivity and a much greater mass range (albeit at a higher cost).

Static SIMS

In static SIMS (SSIMS) the aim is to obtain sufficient signal to provide compositional analysis of the surface layer without actually removing a significant fraction of a monolayer - *i.e.* to be able to analyse less than ca. 10^{14} atoms/molecules (about 10 per cent of a monolayer for a $1cm^2$ sample). The technique is then capable of providing information about the topmost single atomic layer of the surface.

Example: static SIMS spectra from the surface of PTFE (polytetrafluoroethylene)

Note the substantial differences between the positive and negative ion spectra. In each case the most intense peaks correspond to relatively stable positive and negative ions respectively.

TEMPERATURE-PROGRAMMED TECHNIQUES

There are a range of techniques for studying surface reactions and molecular adsorption on surfaces which utilise temperature-programming to discriminate between processes with different activation parameters. Of these, the most useful for single crystal studies is:

Temperature Programmed Desorption (TPD)

When the technique is applied to a system in which the adsorption process is, at least in part, irreversible and *T*-programming leads to surface reactions, then this technique is often known as:

Temperature Programmed Reaction Spectroscopy (TPRS)

However, there is no substantive difference between TPRS and TPD.

The basic experiment is very simple, involving

1. Adsorption of one or more molecular species onto the sample surface at low temperature (frequently 300 K, but sometimes sub-ambient).
2. Heating of the sample in a controlled manner (preferably so as to give a linear temperature ramp) whilst monitoring the evolution of species from the surface back into the gas phase.

In modern implementations of the technique the detector of choice is a small, quadrupole mass spectrometer (QMS) and the whole process is carried out under computer control with quasi-simultaneous monitoring of a large number of possible products.

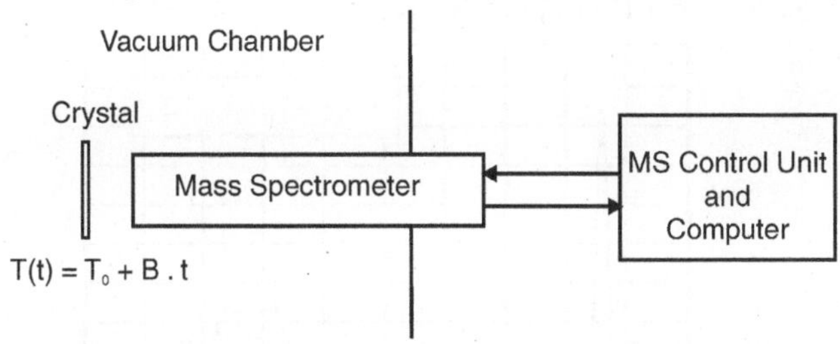

The data obtained from such an experiment consists of the intensity variation of each recorded mass fragment as a function of time / temperature. In the case of a simple reversible adsorption process it may only be necessary to record one signal - that attributable to the molecular ion of the adsorbate concerned.

The graph below shows data from a TPD experiment following adsorption of CO onto a Pd(111) crystal at 300 K.

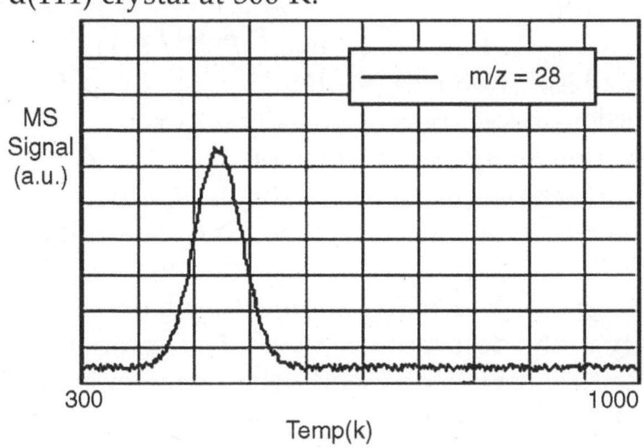

Since mass spectrometric detection is used the sensitivity of the technique is good with attainable detection limits below 0.1 per cent of a monolayer of adsorbate.

The following points are worth noting:

1. The area under a peak is proportional to the amount originally adsorbed, *i.e.* proportional to the surface coverage.

2. The kinetics of desorption (obtained from the peak profile and the coverage dependence of the desorption characteristics) give information on the state of aggregation of the adsorbed species *e.g.* molecular v's dissociative.
3. The position of the peak (the *peak temperature*) is related to the enthalpy of adsorption, *i.e.* to the strength of binding to the surface.

One implication of the last point, is that if there is more than one binding state for a molecule on a surface (and these have significantly different adsorption enthalpies) then this will give rise to multiple peaks in the TPD spectrum. The graph below shows data from a TPD experiment following adsorption of oxygen on Pt(111) at 80 K.

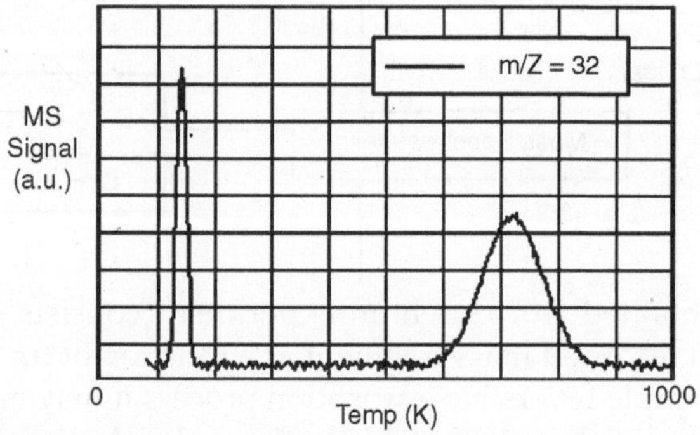

THEORY OF TPD

The rate of desorption of a surface species will in general be given by an expression of the form:

$$R_{des} = n.\, N^x \exp(-E_a^{des} / R\,T)$$

where R_{des} – desorption rate (= $-dN/dt$)

x– kinetic order of desorption (typically 0,1 or 2)

E_a^{des} – activation energy for desorption

In a temperature programmed desorption experiment in which the temperature is increased linearly with time from some initial temperature T_o, then:

$$T = T_o + \beta.t \text{ and } dT = \beta.dt$$

The intensity of the desorption signal, $I(T)$, is proportional to the rate at which the surface concentration of adsorbed species is decreasing. This is obtained by combining to give

$$I(T)\alpha - \frac{dN}{dT} = \frac{VN^x}{\beta}\exp\left(\frac{-E_a^{des}}{RT}\right)$$

This problem may also be considered in a rather simplistic graphical way -the key to this is to recognise that the expression for the desorption signal given in the above equation is basically a product of a coverage term (N^x - where N depends on T) and an exponential term (involving both E_a and T).

Initially, at low temperatures $E_a >> RT$ and the exponential term is vanishingly small. However, as the temperature is increased this term begins to increase very rapidly when the value of RT approaches that of the activation energy, E_a.

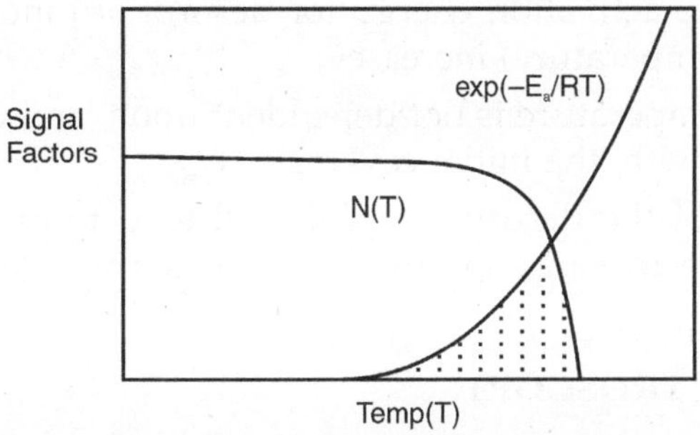

By contrast, the pre-exponential term is dependent upon the coverage, $N(T)$, at the temperature concerned - this term will remain at the initial value until the desorption rate becomes of significance as a result of the increasing exponential term. Thereafter, it will decrease ever more rapidly until the coverage is reduced to zero. The shaded area is an approximate representation of the product of these two functions, and hence also an approximate representation for the desorption signal itself - whilst this illustration may be overly simplistic it does clearly show why the desorption process gives rise to a well-defined desorption peak.

CASE I (Molecular adsorption) - in this case the desorption kinetics will usually be first order (*i.e.* $x = 1$). The maximum desorption signal in the $I(T)$ trace will occur when $(dI / dT) = 0$, *i.e.* when

$$\frac{d}{dT}\left[\frac{VN}{\beta}\exp\left(\frac{-E_a^{des}}{RT}\right)\right] = 0$$

Hence, remembering that the surface coverage changes with temperature *i.e.*

$$N = N(T), \frac{VN}{\beta}\cdot\frac{E_a}{RT^2}\exp\left(\frac{-E_a}{RT}\right) + \frac{v}{\beta}\exp\left(\frac{-E_a}{RT}\right)\cdot\frac{dN}{dT} = 0$$

where we have substituted E_a for E_a^{des} purely for clarity of presentation. Substituting for dN/dT from eqtn.then gives

$$\frac{vN}{\beta}\left[\cdot\frac{E_a}{RT^2} - \frac{v}{\beta}\exp\left(\frac{-E_a}{RT}\right)\right]\exp\left(\frac{-E_a}{RT}\right) = 0$$

The solution is given by setting the expression in square brackets to be equal to zero, *i.e.*

$$\frac{E_a}{RT_p^2} = \frac{v}{\beta}\exp\left(\frac{-E_a}{RT_p}\right)$$

where we have now defined the temperature at which the desorption maximum occurs to be $T = T_p$ (the *peak temperature*). Unfortunately, this equation cannot be re-arranged to make T_p the subject (*i.e.* to give a simple expression of the form T_p =.....), but we can note that:

1. As E_a^{des} (the activation energy for desorption) increases, then so T_p (the peak temperature) increases.
2. The peak temperature is not dependent upon, and consequently does not change with, the initial coverage, $N_{t=0}$.
3. The shape of the desorption peak will tend to be asymmetric, with the signal decreasing rapidly after the desorption maximum.

TEMPERATURE PROGRAMMED REACTION SPECTROSCOPY

In TPRS a number of desorption products will normally be detected in any one experiment - this is where mass spectrometric detection and multiple ion monitoring really becomes essential. A good example of TPRS results are those obtained from dosing formic acid (HCOOH) onto a copper surface.

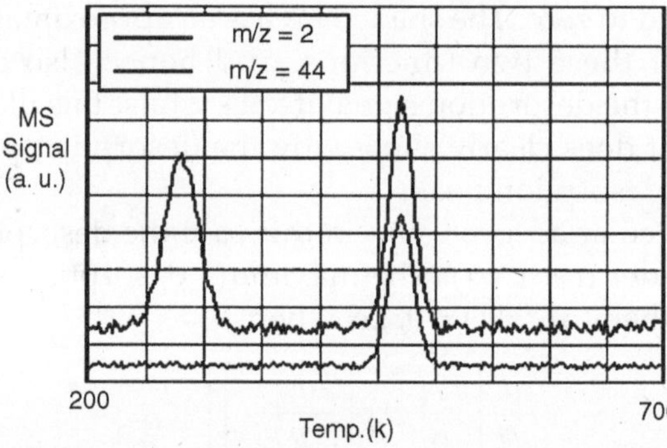

The spectra contain two regions of interest - the first at around room temperature where there is desorption of hydrogen (m/z = 2) alone, and the second between 400 & 500 K where there are coincident desorption peaks attributable to hydrogen and carbon dioxide (m/z = 44). The lower temperature hydrogen peak is at about the same temperature at which hydrogen atoms recombine and desorb as molecular hydrogen from a hydrogen-dosed Cu(110) surface *i.e.* the kinetics of its evolution into the gas phase are those of the normal recombinative desorption process. By contrast, the higher temperature hydrogen peak is observed well above the normal hydrogen desorption temperature-its appearance must be governed by the kinetics of decomposition of another surface species. The carbon dioxide peak must also be *decomposition limited* since on clean Cu(110) carbon dioxide itself is only weakly physisorbed at very low temperatures. More generally, you should note that coincident desorption of two species at temperatures well above their normal desorption temperatures is characteristic of the decomposition of a more complex surface species.

PHOTOELECTRON SPECTROSCOPY

Photoelectron spectroscopy utilizes photo-ionization and energy-dispersive analysis of the emitted photoelectrons to study the composition and electronic state of the surface region of a sample. Traditionally, when the technique has been used for surface studies it has been subdivided according to the source of exciting radiation into:

X-ray Photoelectron Spectroscopy (XPS)	using soft x-ray (200-2000 eV) radiation to examine core-levels.
Ultraviolet Photoelect ron Spectroscopy(UPS)	using vacuum UV (10-45 eV) radiation to examine valence levels.

The development of synchrotron radiation sources has enabled high resolution studies to be carried out with radiation spanning a much wider and more complete energy range (5 - 5000+ eV) but such work is, and will remain, a very small minority of all photoelectron studies due to the expense, complexity and limited availability of such sources.

Physical Basis

Photoelectron spectroscopy is based upon a single photon in/electron out process and from many viewpoints this underlying process is a much simpler phenomenon than the Auger process.

The energy of a photon is given by the Einstein relation:

$$E = h\,\mathrm{n}$$

where h - Planck constant (6.62×10^{-34} J s)

n - frequency (Hz) of the radiation

Photoelectron spectroscopy uses monochromatic sources of radiation (*i.e.* photons of fixed energy). In XPS the photon is absorbed by an atom in a molecule or solid, leading to ionization and the emission of a core (inner shell) electron. By contrast, in UPS the photon interacts with valence levels of the molecule or solid, leading to ionisation by removal of one of these valence electrons. The kinetic energy distribution of the emitted photoelectrons (*i.e.* the number of emitted photoelectrons as a function of their kinetic energy) can be measured using any appropriate electron energy analyser and a photoelectron spectrum can thus be recorded.

The process of photoionization can be considered in several ways: one way is to look at the overall process as follows:

$$\mathrm{A} + h\mathrm{n} \rightarrow \mathrm{A}^{+} + \mathrm{e}^{-}$$

Conservation of energy then requires that:

$$\mathrm{E(A)} + h\mathrm{n} = \mathrm{E(A^{+})} + \mathrm{E(e^{-})}$$

Since the electron's energy is present solely as kinetic energy (KE) this can be rearranged to give the following expression for the KE of the photoelectron.

$$KE = hn - (E(A^{+}) - E(A))$$

The final term in brackets, representing the difference in energy between the ionized and neutral atoms, is generally called the *binding energy* (BE) of the electron - this then leads to the following commonly quoted equation:

$$KE = hn - BE$$

An alternative approach is to consider a one-electron model along the lines of the following pictorial representation; this model of the process has the benefit of simplicity but it can be rather misleading.

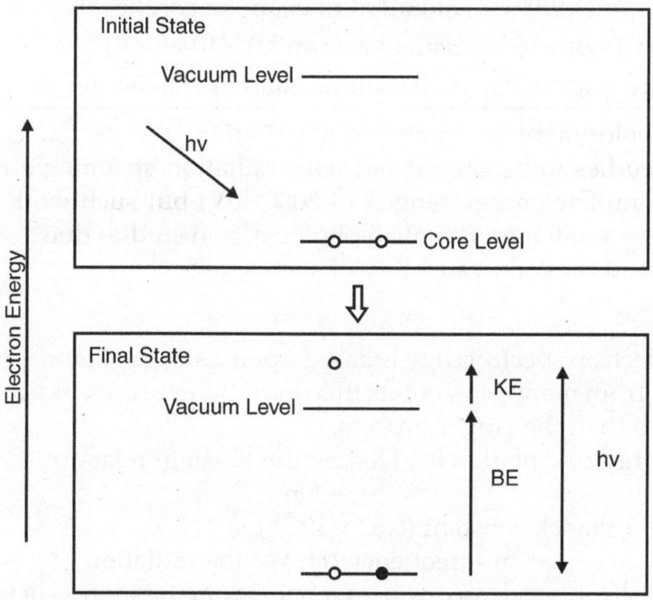

The BE is now taken to be a direct measure of the energy required to just remove the electron concerned from its initial level to the vacuum level and the KE of the photoelectron is again given by:

$$KE = hn - BE$$

The binding energies (BE) of energy levels in solids are conventionally measured with respect to the Fermi-level of the solid, rather than the vacuum level. This involves a small correction to the equation given above in order to account for the *work function* (Φ) of the solid.

Experimental Details

The basic requirements for a photoemission experiment (XPS or UPS) are:

1. A source of fixed-energy radiation (an x-ray source for XPS or, typically, a He discharge lamp for UPS)
2. An electron energy analyser (which can disperse the emitted electrons according to their kinetic energy, and thereby measure the flux of emitted electrons of a particular energy)

3. A high vacuum environment (to enable the emitted photoelectrons to be analysed without interference from gas phase collisions)

Such a system is illustrated schematically below:

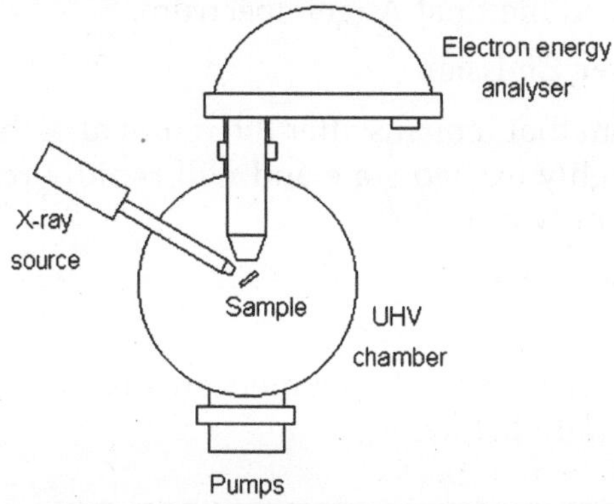

There are many different designs of electron energy analyser but the preferred option for photoemission experiments is a concentric hemispherical analyser (CHA) which uses an electric field between two hemispherical surfaces to disperse the electrons according to their kinetic energy.

THE AUGER PROCESS AND AUGER SPECTROSCOPY

Now let us return to the subject of Auger spectroscopy, the Auger process is illustrated using the K, L_1 & $L_{2,3}$ levels. These could be the inner core levels of an atom in either a molecular or solid-state environment.

IONIZATION

The Auger process is initiated by creation of a core hole - this is typically carried out by exposing the sample to a beam of high energy electrons (typically having a primary energy in the range 2 – 10 keV). Such electrons have sufficient energy to ionise all levels of the lighter elements, and higher core levels of the heavier elements.

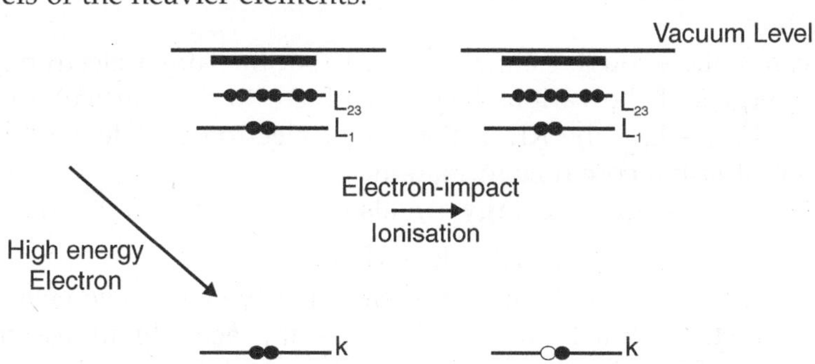

In the diagram above, ionisation is shown to occur by removal of a K-shell electron, but in practice such a crude method of ionisation will lead to ions with holes in a variety of inner shell levels.

In some studies, the initial ionisation process is instead carried out using soft x-rays (hn = 1000 - 2000 eV). In this case, the acronym XAES is sometimes used. As we shall see, however, this change in the method of ionisation has no significant effect on the final Auger spectrum.

Relaxation and Auger Emission

The ionized atom that remains after the removal of the core hole electron is, of course, in a highly excited state and will rapidly relax back to a lower energy state by one of two routes:

1. X-ray fluorescence, or
2. Auger emission

We will only consider the latter mechanism, an example of which is illustrated schematically below.

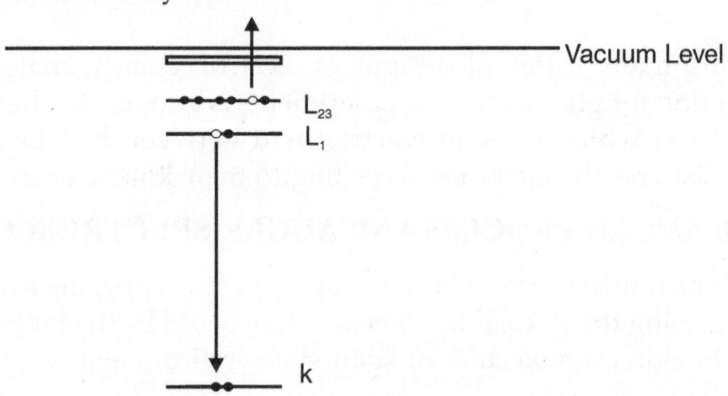

In this example, one electron falls from a higher level to fill an initial core hole in the K-shell and the energy liberated in this process is simultaneously transferred to a second electron; a fraction of this energy is required to overcome the binding energy of this second electron, the remainder is retained by this emitted *Auger electron* as kinetic energy. In the Auger process illustrated, the final state is a doubly-ionized atom with core holes in the L_1 and $L_{2,3}$ shells.

We can make a rough estimate of the KE of the Auger electron from the binding energies of the various levels involved. In this particular example, $KE = (E_K - E_{L1}) - E_{L23}$ the KE of the Auger electron is independent of the mechanism of initial core hole formation.

The expression for the energy can also be re-written in the form:

$$KE = E_K - (E_{L1} + E_{L23})$$

It should be clear from this expression that the latter two energy terms could be interchanged without any effect - *i.e.* it is actually impossible to say which electron fills the initial core hole and which is ejected as an Auger electron; they are indistinguishable. An Auger transition is therefore characterized primarily by.

1. The location of the initial hole
2. The location of the final two holes

Although the existence of different electronic states (terms) of the final doubly-ionized atom may lead to fine structure in high resolution spectra. When describing the transition, the initial hole location is given first, followed by the locations of the final two holes in order of decreasing binding energy. *i.e.* the transition illustrated is a $KL_1L_{2,3}$ transition. If we just consider these three electronic levels there are clearly several possible Auger transitions: specifically,

$K\,L_1\,L_1$ $\qquad$ $K\,L_1\,L_{2,3}\,K\,L_{2,3}\,L_{2,3}$

In general, since the initial ionisation is non-selective and the initial hole may therefore be in various shells, there will be many possible Auger transitions for a given element - some weak, some strong in intensity. Auger Spectroscopy is based upon the measurement of the kinetic energies of the emitted electrons.

Each element in a sample being studied will give rise to a characteristic spectrum of peaks at various kinetic energies. This is an Auger spectrum of Pd metal - generated using a 2.5 keV electron beam to produce the initial core vacancies and hence to stimulate the Auger emission process. The main peaks for palladium occur between 220 & 340 eV.

The peaks are situated on a high background which arises from the vast number of so-called *secondary electrons* generated by a multitude of inelastic scattering processes.

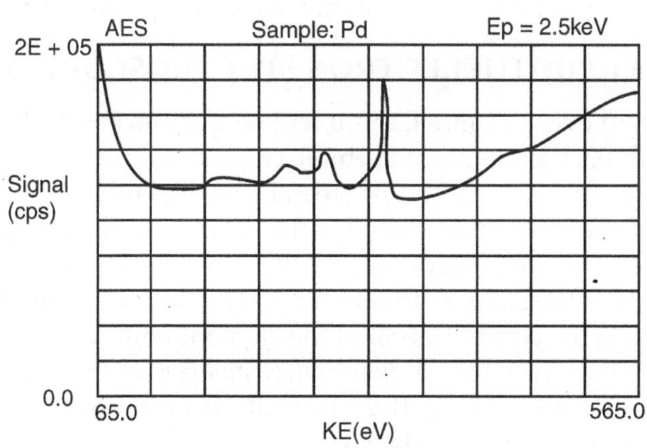

Auger spectra are also often shown in a differentiated form: the reasons for this are partly historical, partly because it is possible to actually measure spectra directly in this form and by doing so get a better sensitivity for detection. The plot below shows the same spectrum in such a differentiated form.

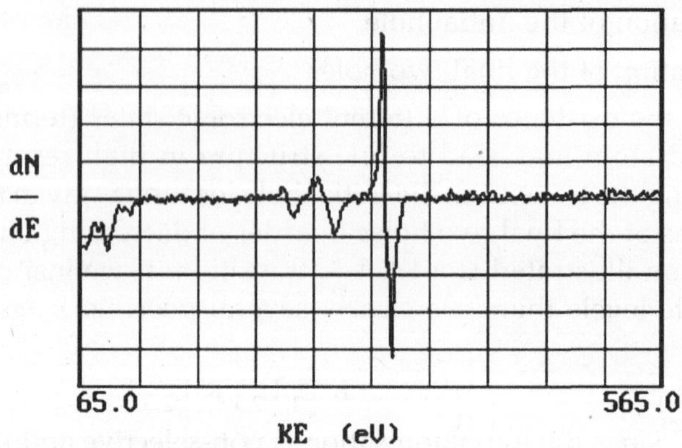

Auger Electron Spectroscopy (AES) is a surface-sensitive spectroscopic technique used for elemental analysis of surfaces; it offers

- High sensitivity (typically ca. 1 per cent monolayer) for all elements except H and He.
- A means of monitoring surface cleanliness of samples
- Quantitative compositional analysis of the surface region of specimens, by comparison with standard samples of known composition.

In addition, the basic technique has also been adapted for use in:

1. *Auger Depth Profiling*: providing quantitative compositional information as a function of depth below the surface.
2. *Scanning Auger Microscopy (SAM)*: providing spatially-resolved compositional information on heterogeneous samples.

X-RAY PHOTOELECTRON SPECTROSCOPY (XPS)

For each and every element, there will be a characteristic binding energy associated with each core atomic orbital *i.e.* each element will give rise to a characteristic set of peaks in the photoelectron spectrum at kinetic energies determined by the photon energy and the respective binding energies. The presence of peaks at particular energies therefore indicates the presence of a specific element in the sample under study - furthermore, the intensity of the peaks is related to the concentration of the element within the sampled region. Thus, the technique provides a *quantitative analysis of the surface composition* and is sometimes known by the alternative acronym, ESCA (Electron Spectroscopy for Chemical Analysis).

The most commonly employed x-ray sources are those giving rise to:

$$\text{Mg } K_\alpha \text{ radiation: } hn = 1253.6 \text{ eV}$$

$$\text{Al } K_\alpha \text{ radiation: } hn = 1486.6 \text{ eV}$$

The emitted photoelectrons will therefore have kinetic energies in the range of ca. 0 - 1250 eV or 0 - 1480 eV. Since such electrons have very short IMFPs in solids, the technique is necessarily surface sensitive.

The XPS spectrum of Pd Metal

The diagram below shows a real XPS spectrum obtained from a Pd metal sample using Mg K_a radiation

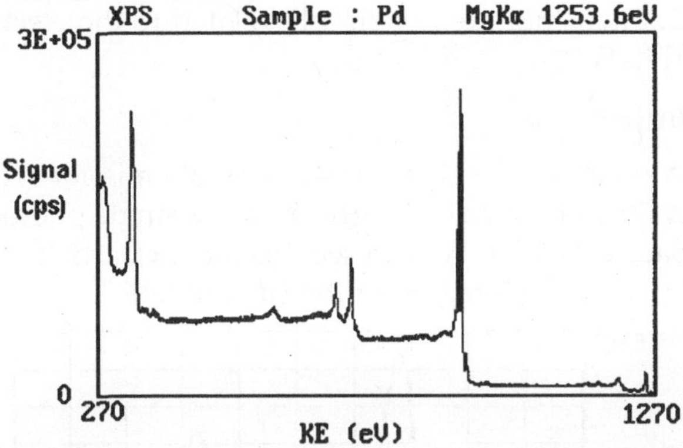

The main peaks occur at kinetic energies of ca. 330, 690, 720, 910 and 920 eV. Since the energy of the radiation is known it is a trivial matter to transform the spectrum so that it is plotted against BE as opposed to KE.

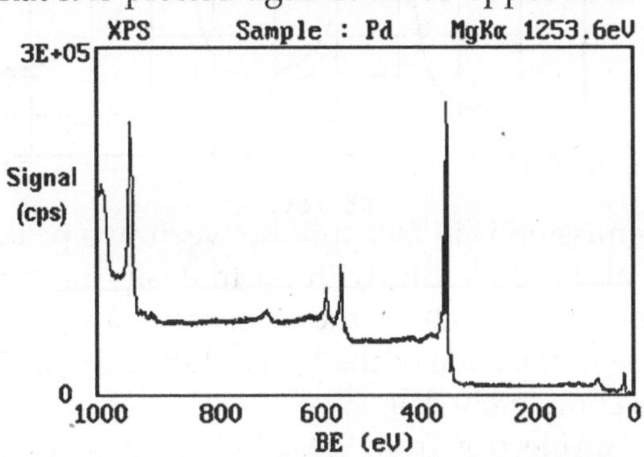

The most intense peak is now seen to occur at a binding energy of ca. 335 eV Working downwards from the highest energy levels.

1. The valence band (*4d*,*5s*) emission occurs at a binding energy of ca. 0 – 8 eV (measured with respect to the Fermi level, or alternatively at ca. 4 – 12 eV if measured with respect to the vacuum level).
2. The emission from the 4*p* and 4*s* levels gives rise to very weak peaks at 54 and 88 eV respectively
3. The most intense peak at ca. 335 eV is due to emission from the 3*d* levels of the Pd atoms, whilst the 3*p* and 3*s* levels give rise to the peaks at ca. 534/561 eV and 673 eV respectively.
4. The remaining peak is not an XPS peak at all ! - it is an Auger peak arising from x-ray induced Auger emission. It occurs at a kinetic energy of ca. 330 eV (in this case it is really meaningless to refer to an associated binding energy).

It may be further noted that

- There are significant differences in the natural widths of the various photoemission peaks
- The peak intensities are not simply related to the electron occupancy of the orbitals

Spin-Orbit Splitting

Closer inspection of the spectrum shows that emission from some levels (most obviously *3p* and *3d*) does not give rise to a single photoemission peak, but a closely spaced doublet. We can see this more clearly if, for example, we expand the spectrum in the region of the *3d* emission...

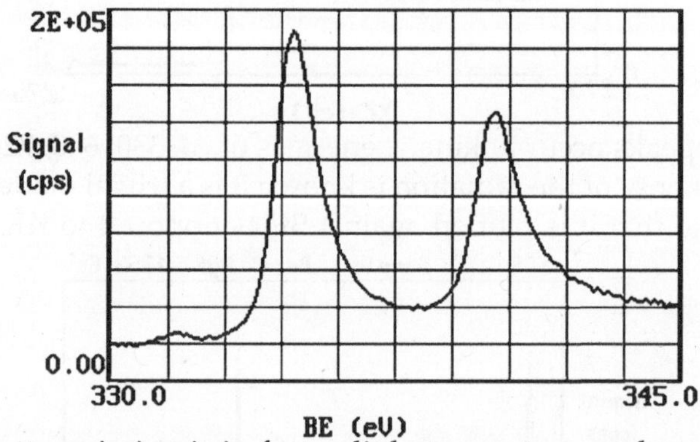

The *3d* photoemission is in fact split between two peaks, one at 334.9 eV BE and the other at 340.2 eV BE, with an intensity ratio of 3:2. This arises from spin-orbit coupling effects in the final state. The inner core electronic configuration of the initial state of the Pd is: $(1s)^2$ $(2s)^2$ $(2p)^6$ $(3s)^2$ $(3p)^6$ $(3d)^{10}$. with all sub-shells completely full.

The removal of an electron from the *3d* sub-shell by photo-ionization leads to a $(3d)^9$ configuration for the final state - since the *d*-orbitals (l = 2) have non-zero orbital angular momentum, there will be coupling between the unpaired spin and orbital angular momenta.

Spin-orbit coupling is generally treated using one of two models which correspond to the two limiting ways in which the coupling can occur - these being the *LS* (or Russell-Saunders) coupling approximation and the *j-j* coupling approximation.

If we consider the final ionised state of Pd within the Russell-Saunders coupling approximation, the $(3d)^9$ configuration gives rise to two states (ignoring any coupling with valence levels) which differ slightly in energy and in their degeneracy...

$$^2D_{5/2}\ g_J = 2 \times \{5/2\}+1 = 6$$

$$^2D_{3/2}\ g_J = 2 \times \{3/2\}+1 = 4$$

These two states arise from the coupling of the $L = 2$ and $S = 1/2$ vectors to give permitted J values of 3/2 and 5/2. The lowest energy final state is the one with maximum J (since the shell is more than half-full), *i.e.* $J = 5/2$, hence this gives rise to the "lower binding energy" peak. The relative intensities of the two peaks reflects the degeneracies of the final states ($g_J = 2J + 1$), which in turn determines the probability of transition to such a state during photoionization.

The Russell-Saunders coupling approximation is best applied only to light atoms and this splitting can alternatively be described using individual electron *l-s* coupling. In this case the resultant angular momenta arise from the single hole in the *d*-shell; a *d*-shell electron (or hole) has $l = 2$ and $s = 1/2$, which again gives permitted *j*-values of 3/2 and 5/2 with the latter being lower in energy. The peaks themselves are conventionally annotated as indicated - note the use of lower case lettering

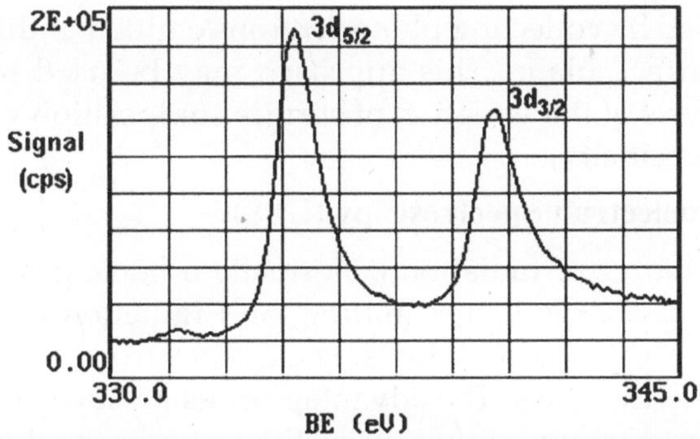

This spin-orbit splitting is of course not evident with *s*-levels ($l = 0$), but is seen with *p,d* & *f* core-levels which all show characteristic spin-orbit doublets.

Chemical Shifts

The exact binding energy of an electron depends not only upon the level from which photoemission is occurring, but also upon:

1. The formal oxidation state of the atom.
2. The local chemical and physical environment.

Changes in either (1) or (2) give rise to small shifts in the peak positions in the spectrum - so-called *chemical shifts*. Such shifts are readily observable and interpretable in XP spectra (unlike in Auger spectra) because the technique:

- Is of high intrinsic resolution (as core levels are discrete and generally of a well-defined energy).
- Is a one electron process (thus simplifying the interpretation).

Atoms of a higher positive oxidation state exhibit a higher binding energy due to the extra coulombic interaction between the photo-emitted electron and the ion core. This ability to discriminate between different oxidation states and chemical environments is one of the major strengths of the XPS technique. In practice, the ability to resolve between atoms exhibiting slightly different chemical shifts is limited by the peak widths which are governed by a combination of factors; especially

- The intrinsic width of the initial level and the lifetime of the final state.
- The line-width of the incident radiation - which for traditional x-ray sources can only be improved by using x-ray monochromators.
- The resolving power of the electron-energy analyser.

In most cases, the second factor is the major contribution to the overall line width.

Angle Dependent Studies

The degree of surface sensitivity of an electron-based technique such as XPS may be varied by collecting photoelectrons emitted at different emission angles to the surface plane. This approach may be used to perform non-destructive analysis of the variation of surface composition with depth (with chemical state specificity).

Ultraviolet Photoelectron Spectroscopy (UPS)

In UPS the source of radiation is normally a noble gas discharge lamp; frequently a He-discharge lamp emitting He I radiation of energy 21.2 eV. Such radiation is only capable of ionising electrons from the outermost levels of atoms - the valence levels. The advantage of using such UV radiation over x-rays is the very narrow line width of the radiation and the high flux of photons available from simple discharge sources. The main emphasis of work using UPS has been in studying:

1. The electronic structure of solids: detailed angle resolved studies permit the complete band structure to be mapped out in k-space.
2. The adsorption of relatively simple molecules on metals: by comparison of the molecular orbitals of the adsorbed species with those of both the isolated molecule and with calculations.

Vibrational Spectroscopy

Vibrational spectroscopy provides the most definitive means of identifying the surface species generated upon molecular adsorption and the species generated by surface reactions. In principle, any technique that can be used to obtain vibrational data from solid state or gas phase samples (IR, Raman etc.) can be applied to the study of surfaces - in addition there are a number of techniques which have been specifically developed to study the vibrations of molecules at interfaces (EELS, SFG etc.). There are, however, only two techniques that are routinely used for vibrational studies of molecules on surfaces - these are:

1. IR Spectroscopy (of various forms, *e.g*. RAIRS, MIR)
2. Electron Energy Loss Spectroscopy (EELS)

IR SPECTROSCOPY

There are a number of ways in which the IR technique may be implemented for the study of adsorbates on surfaces. For solid samples possessing a high surface area:

Transmission IR Spectroscopy: employing the same basic experimental geometry as that used for liquid samples and mulls. This is often used for studies on supported metal catalysts where the large metallic surface area permits a high concentration of adsorbed species to be sampled. The solid sample must, of course, be IR transparent over an appreciable wavelength range.

Diffuse Reflectance IR Spectroscopy (DRIFTS): in which the diffusely scattered IR radiation from a sample is collected, refocused and analysed. This modification of the IR technique can be employed with high surface area catalytic samples that are not sufficiently transparent to be studied in transmission.

For studies on low surface area samples (*e.g.* single crystals): Reflection-Absorption IR Spectroscopy (RAIRS): where the IR beam is specularly reflected from the front face of a highly-reflective sample, such as a metal single crystal surface. Multiple Internal Reflection Spectroscopy (MIR): in which the IR beam is passed through a thin, IR transmitting sample in a manner such that it alternately undergoes total internal reflection from the front and rear faces of the sample. At each reflection, some of the IR radiation may be absorbed by species adsorbed on the solid surface - hence the alternative name of Attenuated Total Reflection (ATR).

ADSORBATES ON METALLIC SURFACES BY REFLECTION IR SPECTROSCOPY

It can be shown theoretically that the best sensitivity for IR measurements on metallic surfaces is obtained using a grazing-incidence reflection of the IR light.

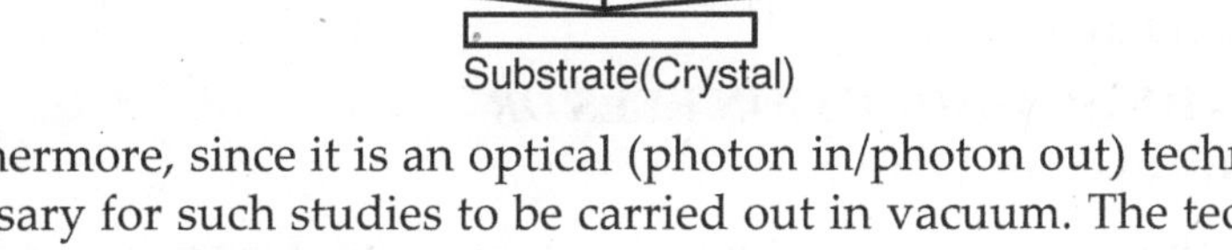

Furthermore, since it is an optical (photon in/photon out) technique it is not necessary for such studies to be carried out in vacuum. The technique is not inherently surface-specific, but

- There is no bulk signal to worry about.
- The surface signal is readily distinguishable from gas-phase absorptions using polarization effects.

One major problem, is that of sensitivity (*i.e.* the signal is usually very weak owing to the small number of adsorbing molecules). Typically, the sampled area is ca. 1 cm^2 with less than 10^{15} adsorbed molecules (*i.e.* about 1 nanomole). With modern FTIR spectrometers, however, such small signals (0.01 per cent – 2 per cent absorption) can still be recorded at relatively high resolution (ca. 1 cm^{-1}).

For a number of practical reasons, low frequency modes (< 600 cm^{-1}) are not generally observable - this means that it is not usually possible to see the vibration of the metal-adsorbate bond and attention is instead concentrated on the intrinsic vibrations of the adsorbate species in the range 600 - 3600 cm^{-1}. The observation of vibrational modes of adsorbates on metallic substrates is subject to the *surface dipole selection rule.*

This states that only those vibrational modes which give rise to an oscillating dipole perpendicular (normal) to the surface are IR active and give rise to an observable absorption band.

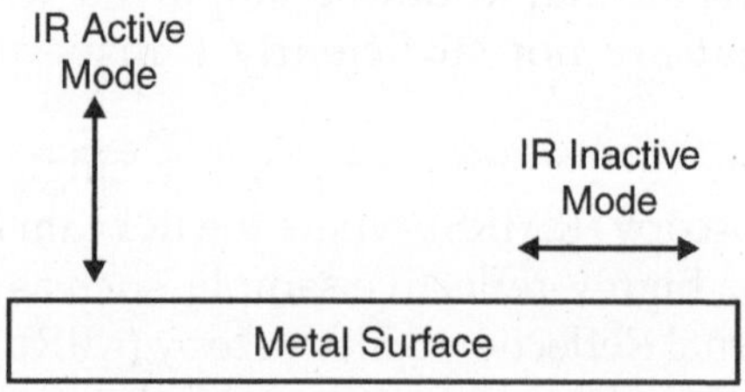

Further information on the selection rules for surface IR spectroscopy. It also needs to be remembered that even if a transition is allowed it may still be very weak if the transition moment is small.

REFLECTION IR SPECTRA

Eg.1 Nitric oxide (NO) adsorption on a Pt surface.

Eg.2 HCN adsorption on a Pt surface

Electron Energy Loss Spectroscopy (EELS)

This is a technique utilising the inelastic scattering of low energy electrons in order to measure vibrational spectra of surface species: superficially, it can be considered as the electron-analogue of Raman spectroscopy.

To avoid confusion with other electron energy loss techniques it is sometimes referred to as

HREELS - HIGH RESOLUTION EELS *OR*
VELS - VIBRATIONAL ELS

Since the technique employs low energy electrons, it is necessarily restricted to use in high vacuum (HV) and UHV environments - however, the use of such low energy electrons ensures that it is a surface specific technique and, arguably, it is the vibrational technique of choice for the study of most adsorbates on single crystal substrates. The basic experimental geometry is fairly simple as illustrated schematically below - it involves using an electron monochromator to give a well-defined beam of electrons of a fixed incident energy, and then analysing the scattered electrons using an appropriate electron energy analyser.

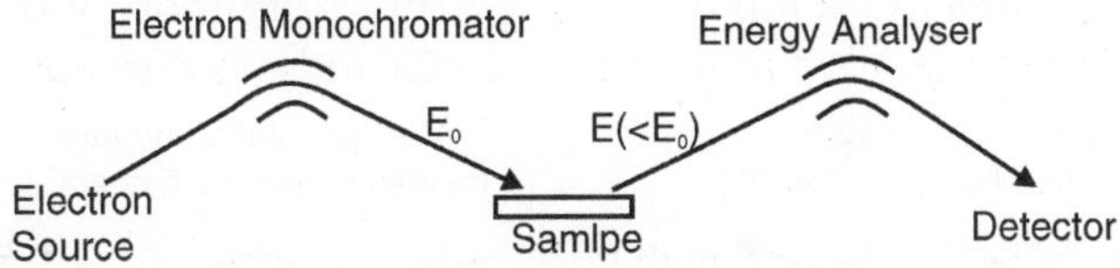

A substantial number of electrons are elastically scattered ($E = E_0$) - this gives rise to a strong *elastic peak* in the spectrum.

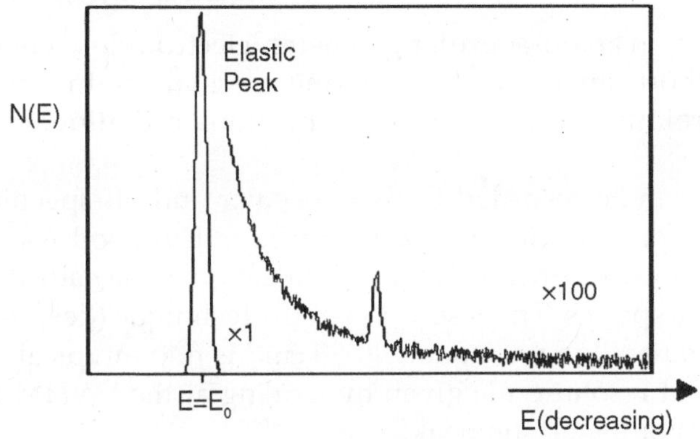

On the low kinetic energy side of this main peak ($E < E_0$), additional weak peaks are superimposed on a mildly sloping background. These peaks correspond to electrons which have undergone discrete energy losses during the scattering from the surface.

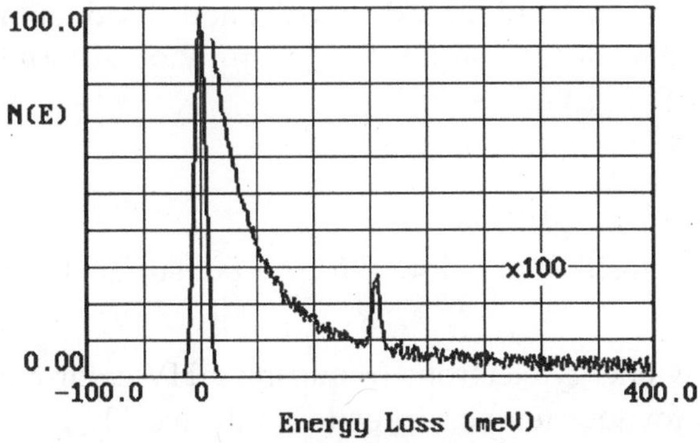

The magnitude of the energy loss, $\Delta E = (E_0 - E)$, is equal to the vibrational quantum (*i.e.* the energy) of the vibrational mode of the adsorbate excited in the inelastic scattering process. In practice, the incident energy (E_0) is usually in the range 5-10 eV (although occasionally up to 200 eV) and the data is normally plotted against the energy loss (frequently measured in meV).

The selection rules that determine whether a vibrational band may be observed depend upon the nature of the substrate and also the experimental geometry: specifically the angles of the incident and (analysed) scattered beams with respect to the surface. For metallic substrates and a specular geometry, scattering is principally by a long-range dipole mechanism. In this case the loss features are relatively intense, but only those vibrations giving rise to a dipole change normal to the surface can be observed.

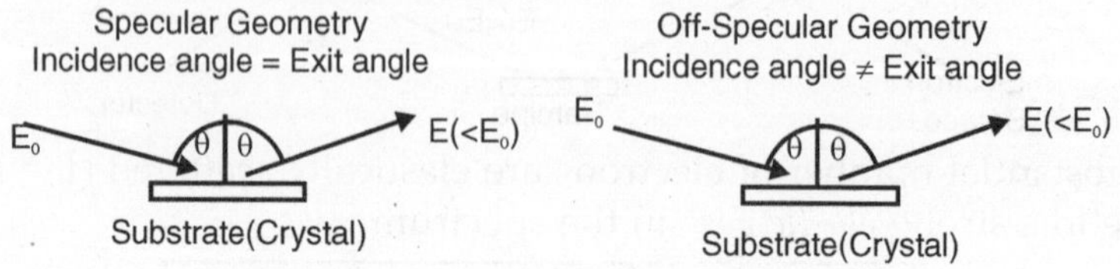

By contrast, in an off-specular geometry, electrons lose energy to surface species by a short-range impact scattering mechanism. In this case the loss features are relatively weak but all vibrations are allowed and may be observed.

If spectra can be recorded in *both* specular and off-specular modes the selection rules for metallic substrates can be put to good use - helping the investigator to obtain more definitive identification of the nature and geometry of the adsorbate species. The resolution of the technique (despite the HREELS acronym !) is generally rather poor; 40-80 cm^{-1} is not untypical. A measure of the instrumental resolution is given by looking at the FWHM (full-width at half maximum) of the elastic peak.

This poor resolution can cause problems in distinguishing between closely similar surface species - however, recent improvements in instrumentation have opened up the possibility of much better spectral resolution (< 10 cm^{-1}) and will undoubtedly enhance the utility of the technique.

In summary, there are both advantages and disadvantages in utilising EELS, as opposed to IR techniques, for the study of surface species It offers the advantages of.

- High sensitivity
- Variable selection rules
- Spectral acquisition to below 400 cm^{-1} but suffers from the limitations of.
- Use of low energy electrons (requiring a HV environment and hence the need for low temperatures to study weakly-bound species, and also the use of magnetic shielding to reduce the magnetic field in the region of the sample)
- Requirement for flat, preferably conducting, substrates
- Lower resolution.

8

Scanning Tunneling Spectroscopy

INTRODUCTION

Scanning tunneling spectroscopy (STS), an extension of scanning tunneling microscopy (STM), is used to provide information about the density of electrons in a sample as a function of their energy. In scanning tunneling microscopy, a metal tip is moved over a conducting sample without making mechanical contact. A bias voltage between the sample and tip allows a current to flow between the tip and the sample even though they are not in contact. This can occur because of quantum mechanical tunneling, hence the name of the instrument.

The scanning tunneling microscope is used to obtain "topographs" - topographic maps - of surfaces. The tip is rastered across a surface and (in constant current mode), a constant current is maintained between the tip and the sample by adjusting the height of the tip. A plot of the tip height at all measurement positions on the raster provides the topograph. These topographic images can obtain information that is atomically resolved, and images of metal and semiconductor surfaces can be obtained with atomic precision.

However, the scanning tunneling microscope does not measure the height of surface features. This can be shown when a molecule is adsorbed on a surface. The STM image may appear to have either increased or decreased height at that feature, although the geometry alone is certainly an increased height. A detailed analysis of the way in which an image is formed shows that the transmission of the electric current between the tip and the sample depends on two factors: (1) the geometry of the sample and (2) the arrangement of the electrons in the sample.

The arrangement of the electrons in the sample is described quantum mechanically by an "electron density". The electron density is a function of both position and energy, and is formally described as the local density of electron states, abbreviated as local density of states (LDOS), which is a function of energy. Spectroscopy, in its most general sense, refers to a measurement of the number of something as a function of energy. For scanning tunneling spectroscopy the scanning tunneling microscope is used to measure the number of electrons (the LDOS) as a function of the electron energy. The electron energy is set by the electrical potential difference (voltage) between the sample and the tip. The location is set by the position of the tip.

At its simplest, a "scanning tunneling spectrum" is obtained by placing a scanning tunneling microscope tip above a particular place on the sample. With the height of the tip fixed, the electron tunneling current is then measured as a function of electron energy by varying the voltage between the tip and the sample (the tip to sample voltage sets the electron energy). The change of the current with the energy of the electrons is the simplest spectrum that can be obtained, it is often referred to as an I-V curve. As is shown below, it is the slope of the I-V curve at each voltage (often called the dI/dV-curve) which is more fundamental because dI/dV corresponds tó the electron density of states at the local position of the tip, the LDOS.

Scanning tunneling spectroscopy (STS) studies the local electronic structure of a sample's surface. The electronic structure of an atom depends upon its atomic species (whether it is a gallium atom or an arsenic atom, for instance) and also upon its local chemical environment (how many neighbors it has, what kind of atoms they are, and the symmetry of their distribution). STS encompasses many methods: taking "topographic" (constant-current) images using different bias voltages and comparing them; taking current (constant-height) images at different heights; and ramping the bias voltage with the tip positioned over a feature of interest while recording the tunneling current. The last example results in current vs. voltage (I-V) curves characteristic of theelectronic structure at a specific x,y location on the sample surface. STMs can be set up to collect I-V curves at every point in a data set, providing a three-dimensional map of electronic structure. With a lock-in amplifier, dI/dV (conductivity) or dI/dz (work function) vs. V curves can be collected directly. All of these are ways of probing the local electronic structure of a surface using an STM.

FORCE VS. DISTANCE CURVES

Force vs. distance (F vs. d) curves are used to measure the vertical force that the tip applies to the surface while a contact-AFM image is being taken. This technique can also be used to analyse surface contaminants' viscosity, lubrication thickness, and local variations in the elastic properties of the surface. Strictly speaking, a force vs. distance curve is a plot of the deflection of the cantilever versus the extension of the piezoelectric scanner, measured using a position-sensitive photodetector. The van der Waals force curve represents just one contribution to the cantilever deflection. Local variations in the form of the F vs. d curve indicate variations in the local elastic properties. Contaminants and lubricants affect the measurement, as does the thin layer of water that is often present when operating an AFM in air.

In the laboratory, force vs. distance curves are quite complex and specific to the given system under study. The discussion here refers to Figure and represents a gross simplification, where shapes, sizes, and distances should not be taken literally. With this in mind, consider the simplest case of AFM in vacuum. At the left side of the curve, the scanner is fully retracted and the cantilever is undeflected since the tip is not touching the sample. As the scanner extends, the cantilever remains undeflected until it comes close enough to the sample surface for the tip to experience the attractive van der Waals force. The tip snaps into the surface. Equivalently, the cantilever suddenly bends slightly towards the surface.

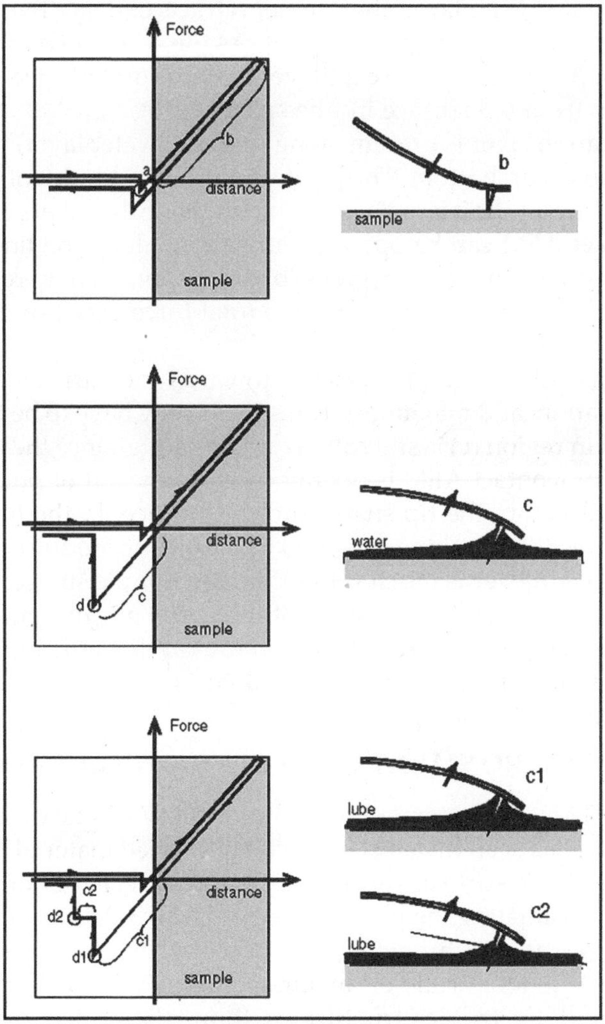

Fig. Forces vs. Distance Curves in Vacuum (top), Air (middle), and Air with a Contamination Layer (bottom)

As the scanner continues to extend, the cantilever deflects away from the surface, approximately linearly. After full extension, at the extreme right of the plot, the scanner begins to retract. The cantilever deflection retraces the same curve (in the absence of scanner hysteresis) as the scanner pulls the tip away from the surface.

In air, the retracting curve is often different because a monolayer or a few monolayers of water are present on many surfaces. This water layer exerts a capillary force that is very strong and attractive. As the scanner pulls away from the surface, the water holds the tip in contact with the surface, bending the cantilever strongly towards the surface. At some point, depending upon the thickness of the water layer, the scanner retracts enough that the tip springs free. This is known as the snap-back point. As the scanner continues to retract beyond the snap-back point, the cantilever remains undeflected as the scanner moves it away from the surface in free space.

If a lubrication layer is present along with the water layer, multiple snap-back points can occur, bottom. The positions and amplitudes of the snap-back points depend upon the viscosity and thickness of the layers present on the surface. Contact AFM can be operated anywhere along the linear portion of the force vs. distance curves, in regions (b) or (c). Operation in region (c) might be used for soft samples, to minimize the total force between the tip and the sample.

Operating with the cantilever bent towards the surface is inherently a less stable situation, and maximum scan speeds may have to be reduced. Note that operation in region (c) is still called contact mode, since the tip is touching the sample. Non-contact AFM is operated just to the left of point (a) on the F vs. d curve-just before the tip snaps into the surface. In the linear region of the F vs. d curves (b), the slope is related to the elastic modulus of the system.

When the cantilever is much softer than the sample surface, as is the case for non-destructive imaging, the slope of the curve mostly reflects the spring constant of the cantilever. However, when the cantilever is much stiffer than the sample surface, the slope of the F vs. d curve allows investigation of the elastic properties of the sample.

SPM ENVIRONMENTS AND ULTRA-HIGH VACUUM

The first STMs were operated primarily in UHV to study atomically clean surfaces. Silicon has been the most extensively studied material, and obtaining images of the 7 x 7 surface reconstruction of Si (111) is often used as a benchmark for evaluating the performance of UHV STM. A major application of UHV STM is scanning tunneling spectroscopy.

STS applied to atomically clean surfaces allows characterization of both topographic and electronic structure, without the added complication of contamination that is always present in air. Another application is the study of materials processes in situ-that is, in the same vacuum environment in which the materials are grown. In this way, processing may proceed without contaminating the sample. Although the first scanning probe microscope was a UHV STM, the development of UHV AFM came much later. Changing cantilevers and aligning the beam-bounce detection scheme - procedures that are not too difficult with an ambient AFM - were major obstacles to the development of a UHV AFM. However, the operation of both STM and AFM on the same vacuum flange is possible using piezoresistive cantilevers. A combined UHV STM/AFM enables scientists to work with conducting or non-conducting samples.

AMBIENT

The easiest, least expensive, and thus most popular environment for SPMs is ambient, or air. STM in air is difficult, since most surfaces develop a layer of oxides or other contaminants that interfere with the tunneling current.

One class of materials for which ambient STM works well is layered compounds. In graphite, MoS2, Nb3Se, and so forth, a clean, "fresh" surface can be prepared by peeling away older surfaces. AFM is indifferent to sample conductivity, and thus can image any solid surface in air, including the contaminants.

LIQUID

Liquid cells for the SPM allow operation with the tip and the sample fully submerged in liquid, providing the capability for imaging hydrated samples. Operation in liquid can also reduce the total force that the tip exerts on the sample, since the large capillary force is isotropic in liquid.

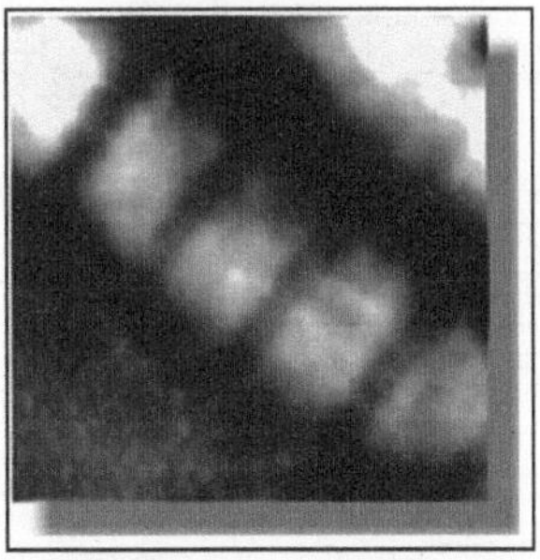

Fig. IC-AFM Image of Rabbit Muscle Myofibril taken in Liquid. Field of View 7 μm

A liquid environment is useful for a variety of SPM applications, including studies of biology, geologic systems, corrosion, or any surface study where a solid-liquid interface is involved. AFM, LFM, and IC-AFM, FMM, and PDM images can be taken in a liquid environment. A IC-AFM image of rabbit muscle myofibril taken in liquid.

ELECTROCHEMICAL

Like UHV, electrochemical cells provide a controlled environment for SPM operation. Usually provided as an option for ambient SPMs, EC-SPMs consist of a cell, a potentiostat, and software. Applications of EC-SPM include real-space imaging of electronic and structural properties of electrodes, including changes induced by chemical and electrochemical processes, phase formation, adsorption, and corrosion as well as deposition of organic and biological molecules in electrolytic solution.

THE SCANNING TUNNELING MICROSCOPE (STM)

The forefather of the STM was the topographiner. The topographiner was a device demonstrated in the early 1970's that utilizing field emission rather than tunneling as the surface sensing technique. Like the modern STM, the device consisted of a sharp metallic tip mounted on a piezoelectric element for positioning and an electronic feedback system that maintained the tip-sample spacing during a raster scan of the surface.

The resolution was limited to about 3 nm in the vertical direction and 400 nm in the horizontal due to the weak distance dependence of the field emission current. In the early 1980's, Gerd Binnig and Heinrich Rohrer began investigating the possibility of using tunneling electrons to probe the surface of conductors and semiconductors. In 1983, Binnig and Rohrer succeeded in producing an atomic resolution image of two unit cells of the 7 × 7 reconstruction surface of Silicon (111). It was this image that captured the attention of the surface science world and resulted in their receiving the Nobel Prize in 1986.

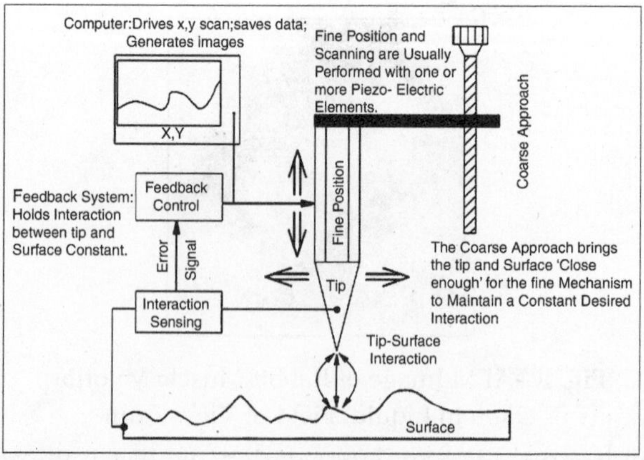

Fig. Schematic representation of a scanning probe microscope. The critical element in differentiating the different probe techniques and their relative resolutions is the tip-surface interaction and the method of sensing that interaction. In STM it is the tunneling current that is sensed.

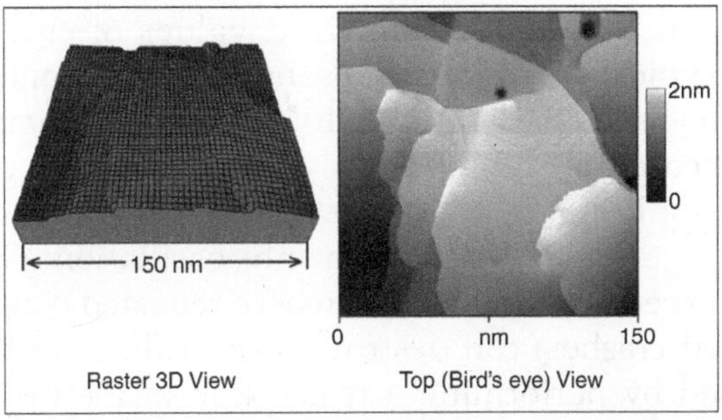

Fig. Comparison of different display methods for the same data set of constant current height versus position for the same large area of a Au(111) surface. The left hand image shows a portion of the 512 × 512 grid used for data collection with the heights mapped both as a 3D projection and by brightness. The right hand image is a standard top view where all the height information is contained in the color scale shown to the far right.

From the STM has evolved an entire menagerie of scanning probe microscopy (SPM) techniques. While this chapter is solely concerned with STM, it is useful to introduce it by way of the more general SPM approach. A schematic representation of the essential elements for any scanning probe microscope is presented in Figure. The tip-sample interaction defines the type of SPM and controls the spatial resolution possible with the particular technique. For example, if mechanical forces are measured by physical contact between tip and sample, then the radius of curvature of the tip and the elastic compliance of the substrate limit the possible spatial resolution in the x-y plane.

In STM it is the tunneling current that is probed. The exponential dependence of the current on tip-sample separation results in the possibility single atom resolution. This is generally explained by considering a tip formed from a single atom sitting in a 3 fold hollow site. For conditions normally used in high resolution STM (1nA set point and 300 mV bias), about 90 per cent of the current is carried by the apical atom because of the difference in distance between it and the atoms at its base. For the general SPM technique, the finer one can make the tip, the higher the x-y spatial resolution. Good STM tips are generally atomically sharp and methods for fabricating them will be discussed later. The z (or normal direction) resolution depends solely upon the z dependence of the interaction between tip and sample. Because of its exponential form, STM is the most sensitive SPM technique, easily achieving 0.005 nm sensitivity.

Another essential feature of all SPMs is the feedback control loop. This electronic system maintains the tip-sample interaction at a preset value by controlling the z position (or deflection) of the tip relative to the surface (or to an undeflected position). In the case of the STM, the absolute distance of the tip from the surface is difficult to impossible to determine, and the relative height is controlled through setting fixed values of current and voltage ($I \sim cV_{exp}(-dF^{1/2})$. Typical current values range from picoamps to nanoamps, while the bias voltage can vary from millivolts to volts. Commercial feedback loops generally incorporate both integral and proportional gain stages. As in all feedback loop applications, it is desirable to set the gain as high as possible but not so high as to drive the system into oscillation. Oscillation of the feedback loop is especially bad in STM where repeated contacts between tip and surface (called crashes) can destroy the usefulness of the tip. The SPM image is generated by performing a raster scan of the tip over the surface while recording the z deflection required to maintain the setpoint.

Almost all commercial instruments use one or more piezoelectric elements to perform the fine motion required to generate images with sub angstrom data intervals. Instruments designed for high resolution are usually limited to piezoelectric elements having 1 to 10 micron maximum scan widths, while 100 micron scan widths are available for lower resolution SPM studies. It is important to note that in the feedback controlled scan regime, one is almost never measuring a true height.

In STM, for example, the changes in tip height with position under feedback control reflect both the tip-sample separation and the spatial variation of the local density of surface states (LDOS) of the sample [$\Phi_{(x,y)}$ in the simple model]. Thus, the constant current image only reflects true height changes if the LDOS of the surface (the local work function) is constant across the surface. This would be the case for atomic steps on clean metal surfaces, but would NOT be the case for adsorbates on surfaces.

The range of z motion possible for the fine position control is generally only a few microns. Thus, some method of bringing the tip within a few microns of the surface without crashing it is required. This is identified as the coarse approach mechanism in the cartoon. Often, this course approach is accomplished by a fine pitch screw (as in the Veeco electrochemical STM) or a piezoelectric driven slip-stick (inertial approach) mechanism (as in the McAllister). Some designs incorporate x,y coarse motion into their coarse approach mechanism, as does the RHK design.

The computer control system does more than integrate the feedback loop with the scanning mechanism while saving piezo position coordinates. Another essential feature is in displaying the data in a manner easily interpreted by the human eye. The left hand image of figure shows the STM constant current height versus position of a large area of a Au(111) surface. While the data was actually acquired with a 512 × 512 grid, only a small portion of the grid is shown for clarity.

The change in tip height relative to the initial setpoint at each grid crossing is mapped both as a 3D projection and by brightness. The lighter the area, the higher the feature. While these 3D projection images are lovely, a real appreciation of the data requires the ability to rotate and tilt the image interactively relative to the observer. Almost all modern SPM software is capable of doing this real time, but this interactive image is not something that can be placed in a publication. Instead, a top view, sometimes called bird's eye view, is most frequently published. The right hand image is a standard top view of the same area as shown in the left image.

Here, all the height information is contained in the color scale which is displayed tothe far right. While virtually any commercial STM will do a good job of taking pictures of surfaces, spectroscopic applications place extra requirements on the instrumentation that must be met if even low quality data is desired.

1. The control and data acquisition system must be able to ramp the bias voltage or the tip sample separation while acquiring the tunneling current or some other voltage signal supplied by an external device. The rate at which data is acquired must be variable over a wide range in time (from about 5 microseconds to about 10 milliseconds per data point), in voltage (from about ±0.5 volts to ±10.0 volts, with the ability to set asymmetrical limits being highly desirable), or in z span (from about -2 nm to +5 nm), where negative distance moves the tip closer to the substrate than at the setpoint. The most commonly used auxiliary input is the output of a lock-in amplifier that detects a modulated signal in the tip current.
2. It must be possible to add modulation (typically sine or square wave) to the bias or tip position voltage. An old version of the Digital Instruments (now Veeco) software, version 3.2x, had built in square wave modulation and the software directly could display dI/dV determined from the resulting modulation in the current. This system worked very well, but was dropped from, or incorrectly implemented in, later versions of the software. To our knowledge, none of today's manufacturers offer such complete 'built in' spectroscopic capability.
3. The control system must be able to shut off the feedback loop during data acquisition, when desired.
4. The drift (both x,y and z) must be low. The extent of acceptable x,y drift is determined by the amount of spatial resolution desired for the spectroscopic data. If one is acquiring spatially averaged molecular spectra from a dense monolayer with a large curvature tip, drifts of the order of 1 nm/sec can be tolerated. On the other hand, if one is trying to locate particular atoms through their vibrational signature in STM based IETS, drifts of less than 0.1 nm/min are required. The amount of acceptable drift in the z direction is determined by the intensity of the spectral feature to be studied relative to the background tunneling current in the absence of that feature. Since this background increases (approximately) as I =

CVexp[-A(z + d)$\sqrt{\Phi}$], a small change in z, dz, results in an increase in relative current $(dI/I)_0$ of -A$\sqrt{\Phi}$dz. Or, using z in nm and Φ in volts, the background change in relative current with a small z drift of dz is given by, $(dI/I)_0$ ~ -0.1 dz. For true vibrational IETS (with no resonance enhancements), dI/I for the spectral transition, $(dI/I)_S$ ~ 0.002. Thus, less than 1 x 10^{-2} nm of z drift is allowed during the time required to scan a particular spectral band. For STM-OMTS, on the other hand, $(dI/I)_S$ ~ 0.1, and 100 times as much z drift is allowed. Moreover, because the OMTS signal is so much stronger than that of IETS, equal signal to noise can be obtained about $(0.1/0.002)^{1/2}$ ~ 7 times faster. Overall, STMOMTS is expected to be about 1000 times less z drift sensitive than STM-IETS of nonresonance enhanced transitions.

5. Sample and tip geometry and shielding. Because of the drift constraints discussed above, one needs to take spectra as fast as the electronic bandwidth allows. With the feedback loop turned off, the limiting term is the capacitance in the tip-sample assembly and the wires leading to the preamplifier circuit. Thus, the preamplifier needs to be close to the tip, and the tip and sample need to be electrically isolated and as small as possible.
6. An essential requirement for STM-IETS is that the working parts of the STM be at or below 10K. Otherwise, the thermal line width destroys the information inherent in identifying vibrational peak positions. It is not enough to cool the sample, the sample, tip, and all parts physically close must be cooled to this temperature. In the case of STM based OMTS or spectroscopic studies of density of states in metals or semiconductors, room temperature measurements are usually satisfactory, and many measurements can be made at significantly elevated temperatures.

COMMERCIAL INSTRUMENTS

There are currently available a wide range of SPM instruments which incorporate all of the features discussed above (except for tips which we will discuss later). Please note, however, that this is a rapidly evolving business and companies may broaden or narrow their offerings on short notice. This list is not exhaustive and the absence of a manufacturer from the list does not indicate any preference.

In the US, Veeco Metrology is the major supplier of ambient and solution phase SPM's, followed by Molecular Imaging Corporation. Asylum Research is an offshoot the original Digital Instruments (now owned by Veeco) and specializes in atomic force microscopy (AFM) and pico-force measurements. Novascan provides ambient scanning force microscopes, AFM tips, and chemically modified tips and samples. Quesant, in partnership with Novascan, provides a full range of ambient and liquid scanning probe microscopes. For

UHV systems made in the United States, one must turn either to McAllister Technical or RHK Technology. The McAllister system is very inexpensive and has provided very high resolution images in the hands of a number of research groups, but it is a room temperature only STM. RHK now offers both STM and AFM in UHV with variable temperature achieved by cooling the sample (only).

Note, ss indicated in a previous section, that for the purposes of measuring vibrational inelastic spectroscopy the entire microscope (sample and tip) must be cooled. To our knowledge there are currently no US suppliers of such microscopes. While commenting on spectroscopy, it is useful to note that STM heads where the sample and tip are electrically isolated from the rest of the microscope (including the sample holder) are especially desirable for spectroscopic purposes. These designs (like that of the RHK) minimize the capacitive coupling that can limit the sampling speed. The best known UHV STM and AFM system provider is the German company, Omicron. Omicron offers a range of UHV systems with cooled sample STM and AFM capability, and the multiprobe LT which is a dedicated low temperature STM wherein both the tip and sample are cooled. Witech is another German company that makes commercial AFM and SNOM equipment, but appears to have dropped their STM line.

Nanosurf is a Swiss company that manufactures extremely compact and inexpensive STM and AFM systems suitable for use in undergraduate laboratories. Both Witech and Nanosurf are respresented in the US by Nanoscience Instruments. Nanotech Electronica is a Spanish company specializing in scanning force microscopy and distributing a free SPM (including STM) data analysis program WS x M. In Russia, NT-MDT is a comprehensive supplier of STM, and AFM instruments and supplies, and also offers a small education oriented multi-purpose scanning probe microscope. Other SPM manufacturers include Danish Micro Engineering, JPK and attocube instruments specializing in AFM and SNOM, PSIA in Korea specializing in AFM, SIS ambient AFM systems, and Triple-O offering AFM and SNOM systems.

TIPS

Once an appropriate SPM has been purchased or built from scratch, one must obtain appropriate tips and samples. Since the emphasis here is on STM, suffice it to say that there are a number of commercial sources for silicon, silicon nitride, and variously coated tips appropriate for scanning force microscopy. While research grade AFM tips are commercially available at reasonable prices, this is not the situation for STM. While a few companies offer etched STM tips, they are expensive and generally unsuitable for spectroscopic applications because of their long exposure to various ambient environments. Thus, the STM practitioner almost always makes his own tips. These fall generally into two classes, mechanically formed and etched tips.

The first STM tips were mechanically formed by Binnig *et al.* These were formed by mechanical grinding (at 90°!) 1 mm diameter tungsten wire. To day, by far the most common mechanically formed tips are cut wire tips. While Au, $Pt_{0.9}Ir_{0.1}$, $Pt_{1-x}Rh_x$, and similar metal wires have been used, the most popular cut tip is made from 0.25 mm diameter $Pt_{0.8}Ir_{0.2}$ wire. Almost every laboratory has their own preferred cutting method. Some anneal the wire before cutting with dull scissors using a pulling motion. Others prefer to use very sharp dykes. Some suggest that a 60° angle cut is best, while others prefer larger or smaller angles.

Since the 'tip' is really an atomic asperity at the end of a rather rough mass of metal, it is not too surprising that the macroscopic and microscopic (as in microns) morphology have little relevance to the quality of the tip. In fact, the tips that give the best images do not always appear sharp under an optical microscope (200 to 400 power). In fact, there are almost always a number of 'tips' like fingers of a hand extending towards the surface.

Because of the strong exponential dependence of the tunneling current, only the longest 'finger' is important when measuring flat surfaces. When imaging rough surfaces, or intermediate to large features (a nm or more tall) on a flat surface, one may see 'ghost images' resulting from tunneling through secondary tips of nearly the same length as the longest. The offset between the ghost and the primary image is related to the offset between the longest and second longest asperity. The cure for ghost images is to re-cut the tip until they disappear. Cut $Pt_{0.8}Ir_{0.2}$ tips are inexpensive, quickly made, and have a fairly good yield (usually 1 in 3 will show atomic reconstruction lines on Au). Electrochemically etched tips offer a well defined geometry near the atomic asperity that functions as the tip.

Methods for preparing atomically sharp tips from a number of different metals were first developed primarily for field ion microscopy. Etched tips are commonly made from Pt, Ir, Au, W, Pd, Ni, and Ag. Prescriptions for a number of these metals were given by Nam *et al.* For solution, and especially electrochemical applications, etched $Pt_{0.8}Ir_{0.2}$ tips are preferred. Various methods for making these are described in the literature, including descriptions of how to limit Faradaic currents through tip coating. Au tips, both etched and coated have been reported in the literature, as have etched silver tips.

For spin polarized tunneling studies, Cavallini reports that etched nickel tips work well and have better oxidation resistance than tungsten tips. By far the most commonly used tip material for UHV studies is tungsten, and these tips are almost always etched tips. One has a great variety of methods to choose from when etching W tips. One may either follow the prescriptions and designs in the literature, or purchase commercial tip etching stands such as those offered by W-tech (through Omicron) or Shrodinger's Sharpener from Obbligato Objectives. Electrochemically etched W tips have an oxide layer on the surface. This oxide layer can be up to 20 nm thick. Dipping the W tips in 47 per cent HF prior to loading in the UHV chamber has been reported to improve tips, but an insulating layer is still apparently left on the tip surface. For simple imaging, the oxide layer on freshly prepared tips is usually not a problem. For spectroscopic studies, however, it is a major impediment since the spectrum can be dominated by the density of states of the oxide layer.

The stable form of the oxide is WO_3, and can be removed by heating above 800 °C. At this temperature WO_3 reacts with W to form WO_2 which then sublimes. W melts at 3410 °C, so a wide window exists to remove the oxide without deforming the tip. A convenient method of heating the tip to remove contaminants is to use electron bombardment. However care still needs to be taken since the local temperature at the tip can easily reached the melting point — even to the extent of forming an obvious round ball at the end of the tip wire. Some crude but effective in situ methods for tip cleaning have also been successful.

In the earliest days of STM, it was found that applying a 10 kHz 2 nm peak to peak (vertical) oscillation to a tip initially in contact with a platinum plate could produce clean sharp tips. Binnig thought that this procedure might clean the tip through some type of ultrasonic interaction. The application of a large voltage pulse (from 3 volts to hundreds of volts) has also been used for tip cleaning.

Field emission cleaning is proposed to account for why application of about 100 V between tip and sample at a distance suitable to produce nA to mA currents can result in clean tips. In the above two cases, one had best use either a clean portion of the sample (that is sacrificed), or another clean metal sample. Another common method for eliminating the oxide involves 'controlled' crashing of the tip on a clean gold surface. This method does work on occasion, but it can also lead to dull (poor resolution) or even bent (unrealistic surface images) tips. An extreme example of this is to simply let the tip scan over a large area of the surface for an extended time. For all the methods of this paragraph, it is not clear whether one cleans the W, or coats the tip with the metal counter surface.

SCANNING PROBE MICROSCOPES

In virtually all scanning probe microscopes, a piezoelectric scanner is used as an extremely fine positioning stage to move the probe over the sample (or the sample under the probe). The SPM electronics drive the scanner in a type of raster pattern. The scanner moves across the first line of the scan, and back. It then steps in the perpendicular direction to the second scan line, moves across it and back, then to the third line, and so forth. The path differs from a traditional raster pattern in that the alternating lines of data are not taken in opposite directions. SPM data are collected in only one direction - commonly called the fast-scan direction - to minimize line-to-line registration errors that result from scanner hysteresis. The perpendicular direction, in which the scanner steps from line to line, is called the slow-scan direction. While the scanner is moving across a scan line, the image data are sampled digitally at equally spaced intervals. The data are the height of the scanner in z for constant-force mode (AFM) or constant-current mode (STM). For constant-height mode, the data are the cantilever deflection (AFM) or tunneling current (STM).

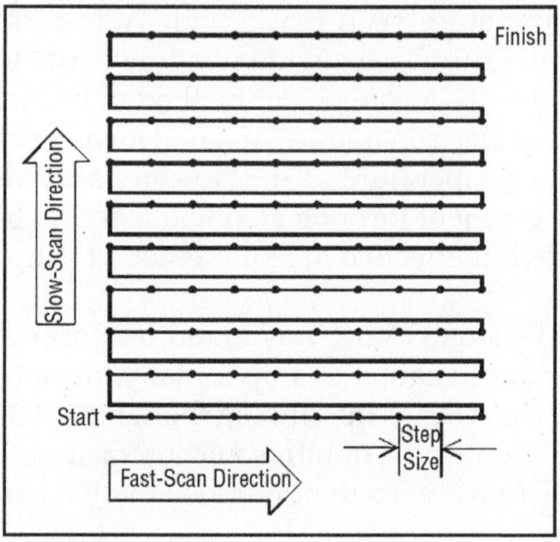

Fig. Scanner Motion During Data Acquisition

The spacing between the data points is called the step size. The step size is determined by the full scan size and the number of data points per line. In a typical SPM, scan sizes run from tens of angstroms to over 100 microns, and from 64 to 512 data points per line. (Some systems offer 1024 data points per line.) The number of lines in a data set usually equals the number of points per line. Thus, the ideal data set is comprised of a dense, square grid of measurements. The difficulties of achieving a perfectly square measurement grid in the plane of the sample surface are explored. And similar difficulties arise in the perpendicular (z) direction. To gain an understanding of the mechanisms underlying scanner operation, the fundamentals of piezoelectric scanners are discussed.

SCANNER DESIGN AND OPERATION

Piezoelectric materials are ceramics that change dimensions in response to an applied voltage. Conversely, they develop an electrical potential in response to mechanical pressure. Piezoelectric scanners can be designed to move in x, y, and z by expanding in some directions and contracting in others. Piezoelectric scanners for SPMs are usually fabricated from lead zirconium titanate, or PZT, with various dopants added to create specific materials properties. Scanners are made by pressing together a powder, then sintering the material. The result is a polycrystalline solid. Each of the crystals in a piezoelectric material has its own electric dipole moment. These dipole moments are the basis of the scanner's ability to move in response to an applied voltage.

After sintering, the dipole moments within the scanner are randomly aligned. If the dipole moments are not aligned, the scanner has almost no ability to move. A process called poling is used to align the dipole moments. During poling the scanners are heated to about 200°C to free the dipoles, and a DC voltage is applied to the scanner. Within a matter of hours most of the dipoles become aligned. At that point, the scanner is cooled to freeze the dipoles into their aligned state. The newly poled scanner can then respond to voltages by extending and contracting.

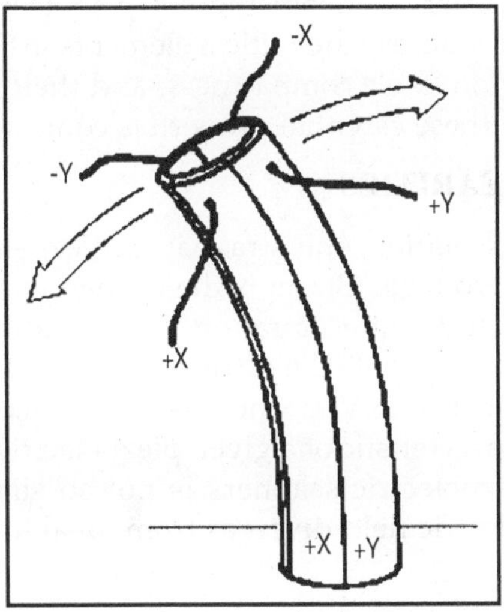

Fig. The Scanner Tube

Occasional use of the scanner will help maintain the scanner's polarization. The voltage applied to enact the scanning motion realigns the stray dipoles that relax into random orientation. If the scanner is not repoled by regular use, a significant fraction of the dipoles will begin to randomize (depolarize or depole) again over a period of weeks. Depoling is accelerated markedly if the scanner is subjected to temperatures above 150°C.

This means that if we want to add a heated stage to our SPM, we must isolate it thermally from the scanner. (The Curie temperature for PZT materials is about 150°C.) Many SPMs use variations of the simple tube design depicted in Figure. In this design, the scanner is a hollow tube. Electrodes are attached to the outside of the tube, segmenting it electrically into vertical quarters, for +x, +y, -x, and -y travel. An electrode is also attached to the centre of the tube to provide motion in the z direction. When alternating voltages are applied to the +x and -x electrodes, for example, the induced strain of the tube causes it to bend back and forth in the x direction. Voltages applied to the z electrode cause the scanner to extend or contract vertically. In most cases, the voltage applied to the z electrode of the scanner at each measurement point constitutes the AFM (constant-force) or the STM (constant-current) data set. In some cases, an external sensor is used to measure the height of the scanner directly). It is assumed that the data set consists of voltages applied to the z electrode. The maximum scan size that can be achieved with a particular piezoelectric scanner depends upon the length of the scanner tube, the diameter of the tube, its wall thickness, and the strain coefficients of the particular piezoelectric ceramic from which it is fabricated.

Typically, SPMs use scanners that can scan laterally from tens of angstroms to over 100 microns. In the vertical direction, SPM scanners can distinguish height variations from the sub-angstrom range to about 10 microns. Piezoelectric scanners are critical elements in SPMs, valued for their sub-angstrom resolution, their compactness, and their high-speed response. However, along with these essential properties come some challenges.

SCANNER NONLINEARITIES

As a first approximation, the strain in a piezoelectric scanner varies linearly with applied voltage. (Strain is the change in length divided by the original length) The following equation describes the ideal relationship between the strain and an applied electric field: S = d E where s is the strain in Å/m, E is the electric field in V/m, and d is the strain coefficient in Å/V. The strain coefficient is characteristic of a given piezoelectric material. In practice, the behaviour of piezoelectric scanners is not so simple-the relationship between strain and electric field diverges from ideal linear behaviour.

HYSTERESIS

To complicate matters, piezoelectric ceramics display hysteretic behaviour. Suppose we start at zero applied voltage, gradually increase the voltage to some finite value, and then decrease the voltage back to zero.

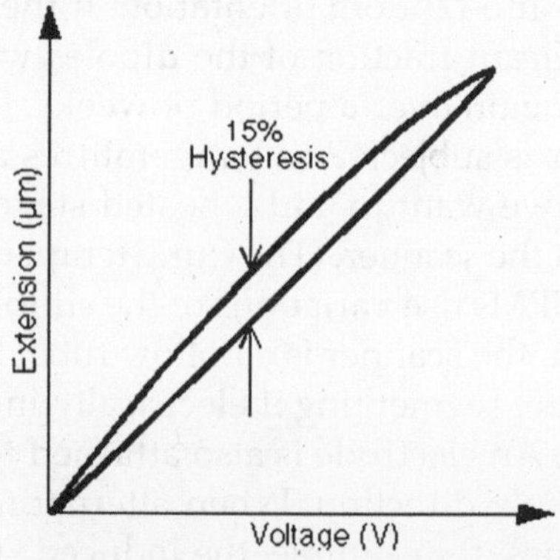

Fig. Hysteresis of a Scanner

If we plot the extension of the ceramic as a function of the applied voltage, the descending curve does not retrace the ascending curve - it follows a different path. It is assumed that the voltages change very slowly. The hysteresis of a piezoelectric scanner is the ratio of the maximum divergence between the two curves to the maximum extension that a voltage can create in the scanner (the change in Y/Ymax).

Hysteresis can be as high as 20 per cent in piezoelectric materials. As mentioned SPM data are usually collected in one direction to minimize registration errors caused by scanner hysteresis. Data collected in the return direction would be shifted slightly, distorting the measurement grid. Since most SPMs today have provisions for taking scans in any fast-scan direction, we can observe hysteresis in the plane of the sample by comparing data sets taken in opposite fast-scan directions. For example, we could compare a data set taken left to right with a data set taken right to left. Scanner hysteresis would cause a shift between the data sets.

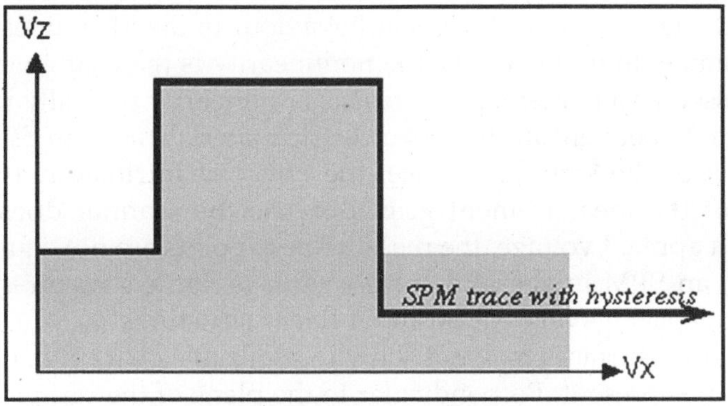

Fig. Effects of Hysteresis on a Step

Hysteresis in the direction perpendicular to the plane of the sample causes erroneous step-height profiles. Referring to the curve of Figure, we see that if the scanner is going up a step in the z direction, a certain voltage is required to allow the scanner to contract. But going down the same step the scanner extends, and extension takes more voltage than contraction for the same displacement. When the SPM image is represented by the voltage applied to the scanner. It is arbitrarily assumed that the strain coefficient in z was calibrated for a contracting scanner.

INTRINSIC NONLINEARITY

Suppose the applied voltages start from zero and gradually increase to some finite value. If the extension of the piezoelectric material is plotted as a function of the applied voltage, the plot is not a straight line, but an s-shaped curve, as seen in Figure. The straight line in the graph is a linear fit to the data.

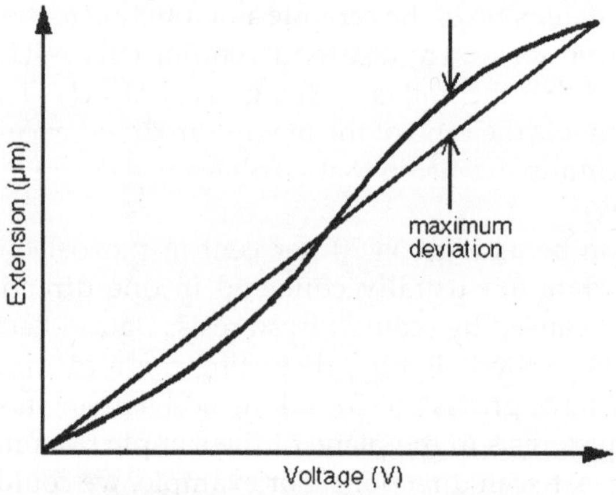

Fig. Intrinsic Nonlinearity of a Scanner

The intrinsic nonlinearity of a piezoelectric material is the ratio of the maximum deviation y from the linear behaviour to the ideal linear extension y at that voltage. In other words, the nonlinearity is the change in y divided by y, expressed as a percentage. Intrinsic nonlinearity typically ranges from 2 per cent to 25 per cent in the piezoelectric materials used in SPM systems. In the plane of the sample surface, the effect of intrinsic nonlinearity is distortion of the measurement grid. Because the scanner does not move linearly with applied voltage, the measurement points are not equally spaced. As a result, an SPM image of a surface with periodic structures will show non-uniform spacings and curvature of linear structures.

Less regular surfaces may not show recognizable distortion, even though the distortion is present. Perpendicular to the plane of the sample surface (in the z direction), intrinsic nonlinearity causes errors in height measurements. Height is usually calibrated for SPMs by scanning a sample with a known step height. By reading the voltage applied to the z electrode as it traverses the step, the z component of the strain coefficient can be calculated. However, if that strain coefficient is directly applied to measure a feature different in height from the calibrated height, the intrinsic nonlinearity of the scanner will generate errors.

CREEP

When an abrupt change in voltage is applied, the piezoelectric material does not change dimensions all at once. Instead, the dimensional change occurs in two steps: the first step takes place in less than a millisecond, the second on a much longer time scale. The second step, delta xc, is known as creep. Quantitatively, creep is the ratio of the second dimensional change to the first: the change in xc divided by the change in x. Creep is expressed as a percentage and is usually quoted with the characteristic time interval Tcr over which the creep occurs. Typical values of creep range from 1 per cent to 20 per cent, over times ranging from 10 to 100 seconds. The times involved during typical scans place the lateral motion of the scanner in the curved part of the response curve. As a result, two scans taken at different scan speeds show slightly different length scales (magnifications) when creep is present.

We can trust only the measurement made at the scan speed that we used to calibrate our SPM when creep is present. Another effect of creep in the plane of the sample is that it can slow down our work when we try to zoom in on a feature of interest. Suppose we want to characterize a defect on the surface of our sample.

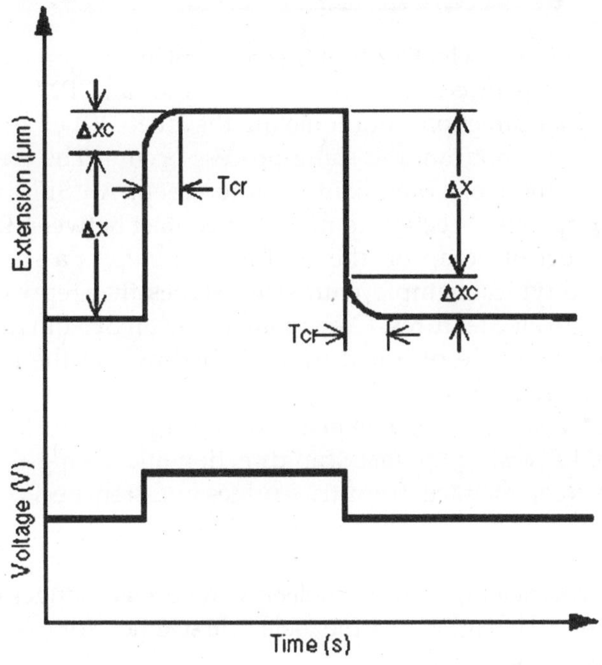

Fig. Plots Showing Creep of a Scanner.

Most likely, we 'll first take a large scan to find the defect. Suppose we see the defect, very tiny, in a corner of the large scan. What we would prefer to do is take a much smaller scan - a high-resolution scan-centered about the defect. An SPM applies an offset voltage to move the scanner to its new centre position to take the second scan. Because we have applied a sudden voltage step, in this case an offset voltage in the plane of the sample, creep will cause we to miss our target. If we continue to collect scans without changing the offset voltage, the scanner will continue to relax until the defect eventually drifts (creeps) into the centre of our scan. The process will likely take a matter of minutes. To examine the effects of creep in the z direction. As the tip traverses the step from bottom to top, the scanner contracts immediately with a voltage corresponding to the full step height. However, over the next few seconds the scanner continues to contract slowly as creep occurs.

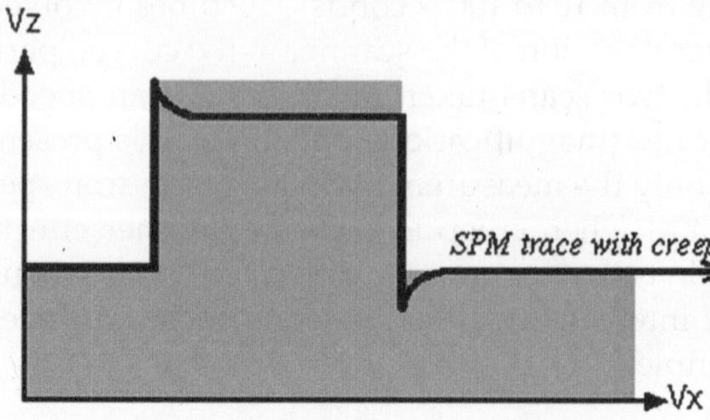

Fig. Effects of Creep on a Step.

To keep the tip in contact with the sample, an SPM has to apply a voltage in the other direction, counteracting the creep. When the tip traverses the step from top to bottom, the same process occurs. The scanner extends to accommodate the step, then continues to creep. An SPM must apply a voltage in the opposite direction to maintain contact between the tip and the sample. The effect of creep on the profile of a step is a simple example. Consider a more typical sample with steep slopes that represent defects or grains or other surface features. Creep may cause an SPM image to look like it has ridges on one side of a feature and shadows on the other side of a feature. If there are shadows where the tip has just traveled down a steep slope, or ridges where the tip has just traveled up a steep slope, creep may be responsible. Reversing the fast-scan direction and taking the same image helps separate creep artifacts from true ridges and trenches.

AGING

The strain coefficient, d, of piezoelectric materials changes exponentially with time and use. The aging of a piezoelectric scanner, for cases of both high and low voltage usage.

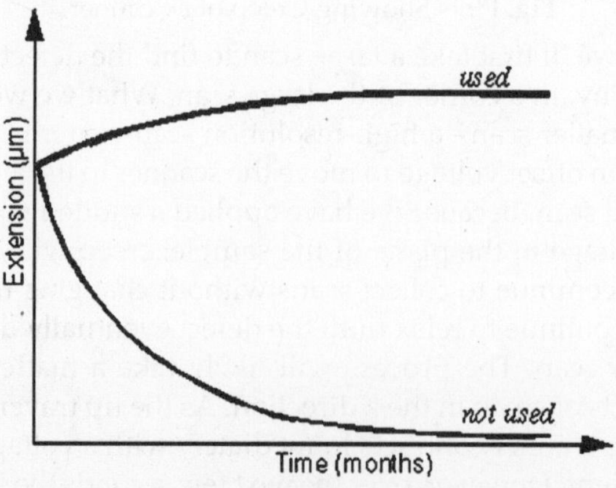

Fig. Aging of a Sscanner, with Use or Left Idle.

The aging rate is the change in strain coefficient per decade of time. When a scanner is not used, the deflection achieved for a given voltage gradually decreases. The aging rate of scanners for an SPM can produce a decrease in lateral strain coefficients (and therefore, an error in length measurements taken from SPM images) over time. When a scanner is used regularly, the deflection achieved for a given voltage increases slowly with use and time. These two phenomena are part of the same process. Each of the tiny crystals that compose the scanner has its own dipole moment. Repeated application of voltage in the same direction-such as the voltage applied during scanning-causes more and more of the dipoles to align themselves along the axis of the scanner. The amount of deflection achieved for a given voltage depends upon how many dipoles are aligned.

Thus, the more the scanner is used, the farther the scanner will travel. On the other hand, if the scanner is not used, the dipole moments of the crystals will gradually become randomly oriented again. As a result, fewer dipoles contribute to the deflection of the scanner. When we buy an SPM, the scanners have already been "poled," which means that they have already been exercised to the point where the deflection of the scanner is close to its maximum. The dependence of the scanner deflection on time and usage means that the scanner may not be extending the same distance for a given applied voltage as it did when it was first calibrated. As a result, when we measure a feature on an SPM image, the values of lateral and vertical dimensions may be in error.

CROSS COUPLING

The term cross coupling refers to the tendency of *x*-axis or *y*-axis scanner movement to have a spurious *z*-axis component. It arises from several sources and is fairly complex. For example, the electric field is not uniform across the scanner. The strain fields are not simple constants, but actually complex tensors. Some "cross talk" occurs between *x*, *y*, and *z* electrodes. But the largest effect is geometric. Geometric cross coupling has its basis in the way piezoelectric scanners are constructed, usually as segmented tubes or as tripods.

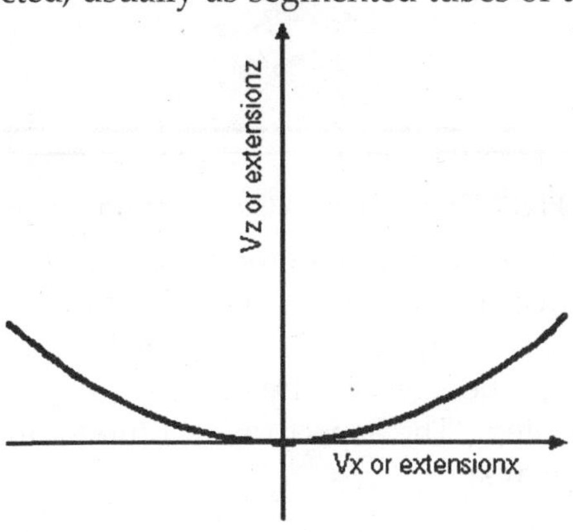

Fig. Cross Coupling Motion of the Scanner

The *x-y* motion of the scanner tube is produced when one side of the tube shrinks and the other side expands. As a result, a piezoelectric tube scans in an arc, not in a plane. A voltage applied to move the piezoelectric tube along the x or y axis (parallel to the surface of the sample) necessitates that the scanner extend and contract along the z axis (perpendicular to the surface of the sample) to keep the tip in contact with the sample.

A tripod scanner is designed with three mutually perpendicular bars or tubes glued together at one end. This design is also susceptible to cross coupling because the three bars of piezoelectric material are attached to one another. When the x bar extends or contracts, it causes rotation of the y and z bars. Cross coupling can cause an SPM to generate a bowl-shaped image of a flat sample.

A profile of such an image is shown in Figure with an example of a step. In interpreting, remember that the SPM image is based on the voltage required to compensate for the curvature generated by the arc of the scanner. The bowl shape may not always be evident in the final image because the curved background can be subtracted out using image-processing software. The best way to determine if the scanner is subject to cross coupling is to image a sample with a known radius of curvature, such as a lens. Software corrections can only flatten it or leave it alone. In the first case, we end up with a spuriously flat image, and in the second case the curvature of the scanner will be added to the curvature of the lens. The true curvature of a lens can be measured only when cross coupling is eliminated.

Fig. Effects of Cross Coupling on a Step

The example of a single step to demonstrate hysteresis, creep, and cross coupling in the vertical direction is used. To show a single image in the laboratory that illustrates each of these effects in isolation is virtually impossible. The sum of the effects of hysteresis, creep, and cross coupling in the image of a single step. (The aspect ratio of the tip may also contribute to the shape of the sidewalls)

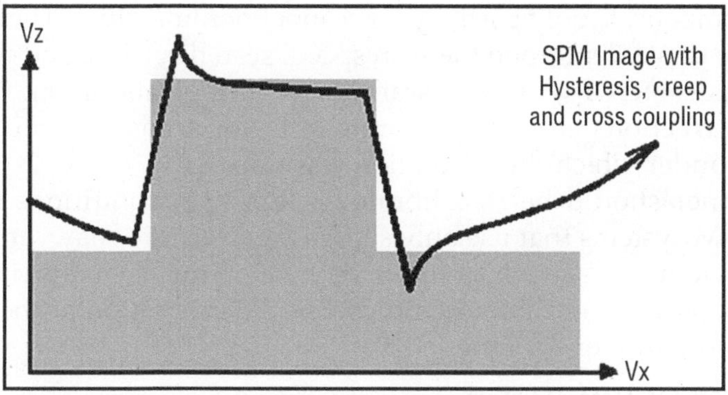

Fig. Effects of Hysteresis, Creep, and Cross Coupling on a Step

Traditionally, the nonlinear behaviour of piezoelectric scanners described above has been addressed imperfectly using software corrections. Some systems on the market use hardware solutions that eliminate most of the nonlinearities instead of correcting them. Hardware solutions are divided into optical, capacitive, and strain-gauge techniques. The best systems combine hardware and software corrections, using the strengths of each.

SOFTWARE CORRECTION

The first step in developing a software correction for scanner nonlinearities is to image a calibration grating. The system compares the measured data with the known characteristics of the grating. Then the system determines how to adjust the measured data to conform to the known characteristics. It stores that information in a file, or lookup table. Afterwards, the system can compensate for nonlinearity while collecting data by adjusting the voltage applied to the scanner in accordance with the file or the look-up table.

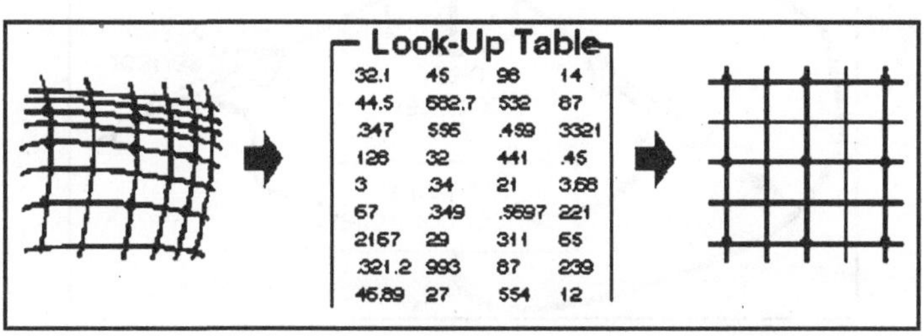

Fig. Software Correction, Shown Using a Look-up table

More sophisticated software correction algorithms include equations that model the scanner tube's non-linear response to applied voltages. The equations are used to calculate the voltages that are applied to the scanner to position it during a scan. Again, scanner calibration procedures are necessary to determine parameters for the algorithm. Software solutions in general are relatively simple and inexpensive to implement. Their main disadvantage is that they compensate only partially for scanner nonlinearities. The corrections are strongly dependent upon the scan speed, scan direction, and whether the scanner was centered within its scan range during calibration. As a result, software corrections are most accurate only for scans that reproduce the conditions under which the calibration was done.

A scanner should be re-calibrated when scan conditions change. In practice, SPM systems that use only software corrections may vary from true linear behaviour by as much as 10 per cent. Often this deviation is unnoticed by the SPM user who calibrates and checks the calibration of the SPM with the same procedure every time.

HARDWARE CORRECTION

Systems with hardware solutions sense the scanner's actual position with external sensors. The signal read from the sensor of each axis is compared to a signal that represents the intended scanner position along that axis. A feedback system applies voltage to the scanner to drive it to the desired position.

In this way, the scanner can be driven in a linear fashion. The external sensors themselves must be stable and immune to all sources of nonlinearity, since their purpose is to improve the linearity of the SPM.

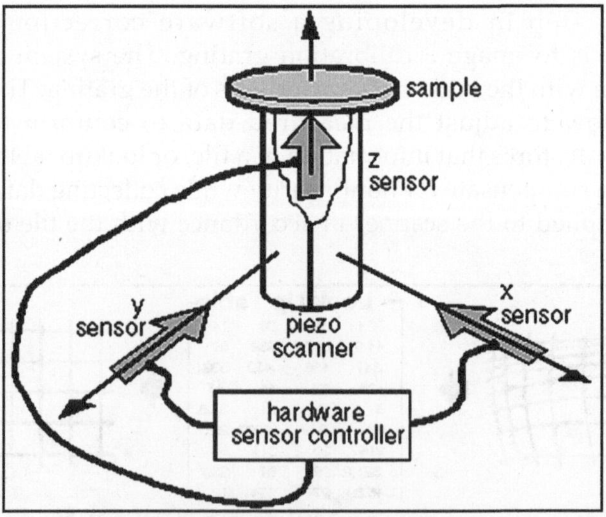

Fig. Schematic of a Scanner, Showing External Detectors

Because the scanner's actual position is measured, systems using hardware sensors compensate for intrinsic nonlinearity, hysteresis, creep, aging, and cross coupling. They can reduce the total nonlinearity of the system to less than 1 per cent. Figure illustrates how SPMs equipped with hardware correction enable we to zoom in on a feature of interest in an image without the effects of creep. The left image is a 40μm scan of a grating. The box in the image shows the intended location of the zoomed scan. The right image is the zoomed 2μm scan of that location. An image of the same area acquired at a later time would look identical to the right image, indicating that creep is not present in SPMs with hardware correction schemes.

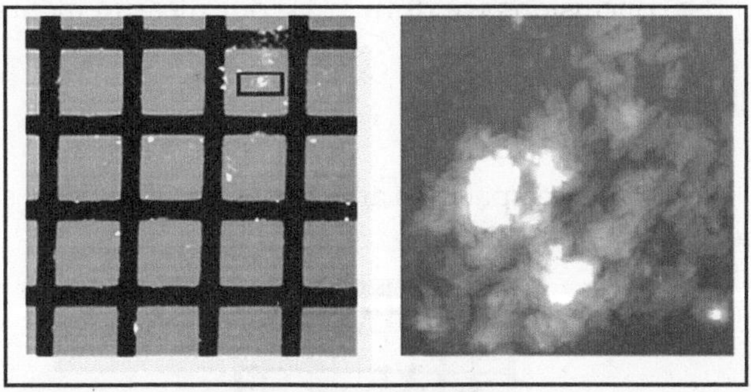

Fig. Images of a Grating, Demonstrating the Ability of Hardware Correction Systems to Zoom in on a Feature of Interest without Creep.

Optical Techniques

In optical solutions to scanner nonlinearity, a "flag", or reflector, is attached to the scanner. A light shines on the flag, and a detector responding to either the phase or the position of the light determines the actual position of the scanner. The system then uses feedback to adjust the voltage to the scanner to place it in the proper position to generate a linear scan.

Capacitive Techniques

In capacitive solutions, a metal electrode is placed on the scanner and another is mounted nearby. An electrical circuit then measures the capacitance between the electrodes, which varies with their separation. Having located the scanner in this way, the system uses feedback to adjust the scanner's position to compensate for nonlinearity.

Strain-Gauge Techniques

In strain-gauge solutions to scanner nonlinearities, a strain gauge is mounted on the scanner. The active element in a strain gauge is a piece of piezoresistive material-material whose resistance varies when it is under tensile or compressive stress. When the scanner moves, the resistance in the strain gauge changes. The magnitude of the resistance of the strain gauge reflects the amount of bending in the scanner, thereby measuring its position. An SPM uses feedback to adjust the scanner's position and to generate a linear scan.

Tests for Scanner Linearity

If we suspect that nonlinear behaviour of the piezoelectric scanner has created artifacts in an SPM image, we can perform various tests. The symptoms and gives a test procedure to identify each source of nonlinearity. When we are evaluating an SPM, make sure that it has adequate compensation for intrinsic nonlinearity, hysteresis, creep, aging, and cross coupling. Otherwise, our SPM images may be distorted, the measurements we make from them may be erroneous, and the conclusions we draw from our measurements may lead we astray.

Intrinsic Nonlinearity

X-Y Plane

Symptom: Non-uniform spacings and curvature appear in structures that we know or believe to be linear.

Fig. Image of a Grating, Showing non-uniform Spacing and Curvature

Test: Image a calibration grating. The image should show consistent spacing and straight lines.

Cantilevers

Cantilevers and their tips are critical components of an atomic force microscope system because they determine the force applied to the sample and the ultimate lateral resolution of the system. Figure depicts a scanning electron microscope (SEM) image of a cantilever.

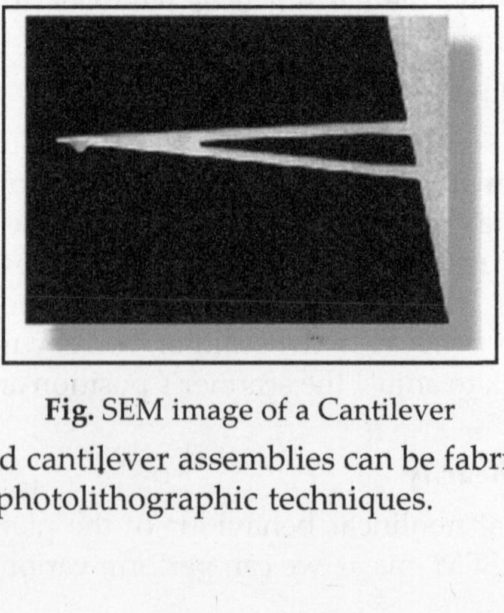

Fig. SEM image of a Cantilever

Integrated tip and cantilever assemblies can be fabricated from silicon or silicon nitride using photolithographic techniques.

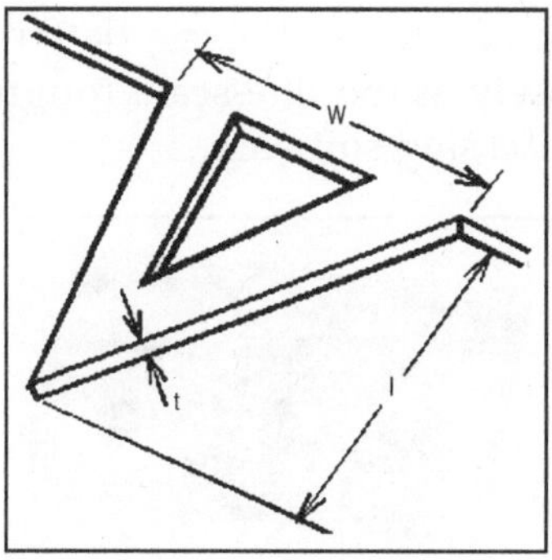

Fig. V-shaped cantilever, showing length (l), width (w), and thickness (t)

More than 1,000 tip and cantilever assemblies can be produced on a single silicon wafer. V-shaped cantilevers are the most popular, providing low mechanical resistance to vertical deflection, and high resistance to lateral torsion. Cantilevers typically range from 100 to 200μm in length, 10 to 40 μm in width, and 0.3 to 2 μm in thickness.

CONVOLUTION WITH OTHER PHYSICS

Another common category of SPM artifacts derives from the sensitivity of an SPM to physical properties other than those responsible for mapping the surface topography. For example, regions of variable conductivity are convolved with topographic features for STM, and soft or elastic surfaces can deform under the pressure of the AFM tip.

FEEDBACK ARTIFACTS

If the feedback loop of an SPM is not optimized, the SPM image can be affected. When feedback gains are too high the system can oscillate, generating high frequency periodic noise in the image. This may occur throughout the image or be localized to features with steep slopes. On the other hand, when feedback gains are too low, the tip cannot track the surface well. In the extreme case, the image loses detail, appearing smooth or "fuzzy". A less obvious effect is "ghosting". On sharp slopes, an overshoot can appear in the image as the tip travels up the slope, and an undershoot can appear as the tip travels down the slope. This feedback artifact commonly appears on steep features, represented as bright ridges on the uphill side and/or dark shadows on the downhill side of the feature.

IMAGE PROCESSING CAPABILITIES

Sophisticated image-processing software is available on all commercial SPMs. Beautiful images can be created by introducing artificial light sources, 3-dimensional rendering, colour tables, and curvature enhancement algorithms, by retouching areas of bad data, by filtering environmental noise, and by magnifying or reducing the vertical scale of the image to enhance or minimize surface features of interest. We can flatten an image of a surface of a tiny sphere to look closely at the fine-scale roughness. These are some of the benefits of image-processing software.

Fig. 3-D Rendered Image of a Semiconductor Surface

However, when image processing and image enhancement are used carelessly or irresponsibly, data can be misrepresented.

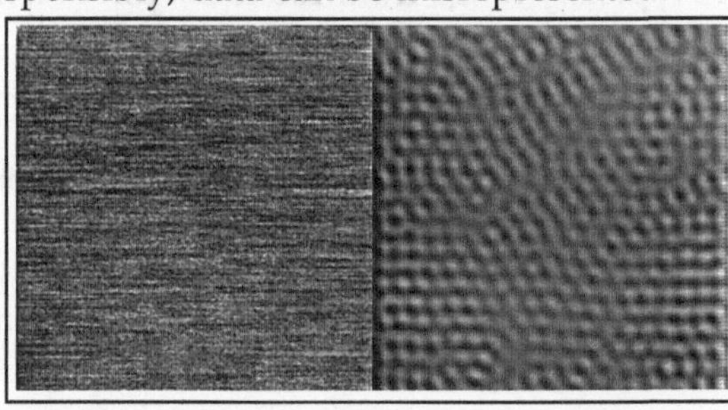

Fig. AFM Image of Random noise. Left Image is Raw Data

The best way to compare the quality of data sets from different SPMs is to compare unprocessed "raw" data in a gray-scale image. As an example of misuse of image processing, Figure shows a wholly specious image of a monolayer of an organic film, created entirely from random noise by a carefully crafted filter. Right image is a false image, produced by applying a narrow band pass filter to the raw data

KEY FEATURES OF SPMS

Considerations outside of the scientific realm are likely to be a part of our choice of an SPM. An instrument that is easy to use will have lower operating costs because a highly skilled operator is not required. Minimal training time is desirable in a multi-user facility. Finally, an instrument that is easy to run will be used more-and produce more results.

USER INTERFACE

Evaluate the complexity of the user interface. It should be laid out in a straightforward, manner with easy-to-use controls. For example, the user interface should have buttons with names that represent commonly performed functions such as "APPROACH" and "IMAGE". After we have seen someone else demonstrate the instrument, ask to operate it ourself. We are then in the best position to evaluate the simplicity of the user interface. Some SPMs are built upon operating systems that allow multi-tasking. If we can process data while collecting additional data, our throughput will be increased measurably.

OPTICAL MICROSCOPE

All commercial SPMs now include optical microscopes to help monitor the tip-to-sample approach and to select the areas of interest on the sample surface. A good optical microscope speeds up our work because it enables we to position the tip quickly and accurately, exactly where we want to take an SPM image.

If we intend to image very rough or oddly shaped samples (for example geological samples, or teeth), or cross sections of any kind that require landing the tip on a narrow edge, a good optical microscope is indispensable. Figure shows a real-time video image of a cantilever positioned over an integrated circuit.

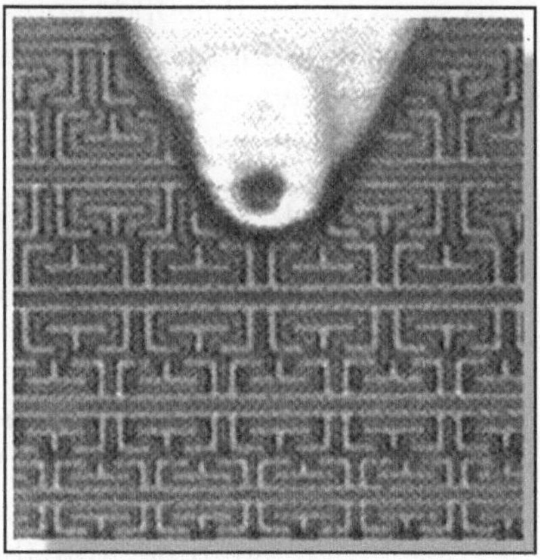

Fig. Real-time Video Image of a Cantilever Positioned Over an Integrated Circuit

When evaluating the optical microscope, look for the following features:

- Optical image clarity
- Useful magnification range
- Zoom lens
- Integrated video camera

Motorized focus and zoom controls are timesavers over manual models, but optical image quality is the most important. The magnification should enable we to navigate rough surfaces and cross sections, to a high magnification, where the field of view allows we to place the tip on micron-sized features. Note that a quoted magnification range is subject to "cheating". If we attach a larger monitor to the same video camera, we end up with higher magnification, but no better resolution. The best way to specify the optical microscope is by field of view or resolution.

PROBE HANDLING

Under normal operating conditions, an AFM tip should last for a couple of days, so changing the probe should be a straightforward process. However, the chips upon which cantilevers are mounted are very small and can be unwieldy to handle. New designs permit pre-aligned, pre-mounted probes to be changed very quickly, with minimal alignment of the beam-bounce detection system. As with the user interface, we should evaluate how difficult it is to change a probe. Change a probe ourself, instead of watching a highly skilled operator change it.

SYSTEM ACCESSIBILITY

If we intend to perform unique experiments with an SPM, we may be very interested in having access to the software, the electronics, and the mechanics of the system in order to make our own modifications. Investigate the design of the system to see whether it is open enough for our customized work. Consider also the technical strength and availability of the technical support staff at the factory. We may want their advice more often than a user who is happy to operate the machine off-the-shelf.

PROPERTIES OF CANTILEVERS

Atomic force microscopes require not only sharp tips, but also cantilevers with optimized spring constants - lower than the spring constants between atoms in a solid, which are on the order of 10 N/m. The spring constant of the cantilever depends on its shape, its dimensions, and the material from which it is fabricated. Thicker and shorter cantilevers tend to be stiffer and have higher resonant frequencies. The spring constants of commercially available cantilevers range over four orders of magnitude, from thousandths of a newton per meter to tens of newtons per meter. Resonant frequencies range from a few kilohertz to hundreds of kilohertz providing high-speed response and allowing for non-contact AFM operation.

HOW TO SELECT A CANTILEVER

The desirable properties for a cantilever depend on the imaging mode and the application. In contact mode, soft cantilevers are preferable because they deflect without deforming the surface of the sample. In non-contact mode, stiff cantilevers with high resonant frequencies give optimal results.

TIP SHAPE AND RESOLUTION

The lateral resolution of an AFM image is determined by two factors: the step size of the image, and the minimum radius of the tip. Consider an image taken with 512 by 512 data points. Such a scan 1 μm by 1μm would have a step size-and lateral resolution-of about 20Å (1 μm ÷ 512). The sharpest tips available commercially can have a radius as small as 50Å. Because the interaction area between the tip and the sample is a fraction of the tip radius, these tips typically provide a lateral resolution of 10 to 20Å. Thus, the resolution of AFM images larger than 1 μm by 1μm is usually determined not by the tip but by the image's step size. The quoted *best resolution* of an AFM, however, depends upon how resolution is defined.

In the microscopy community, two asperities (peaks) are considered resolved if the image satisfies Rayleigh's criterion. In this application, Rayleigh's criterion requires that the height of the image dip at least 19 per cent between the asperities.

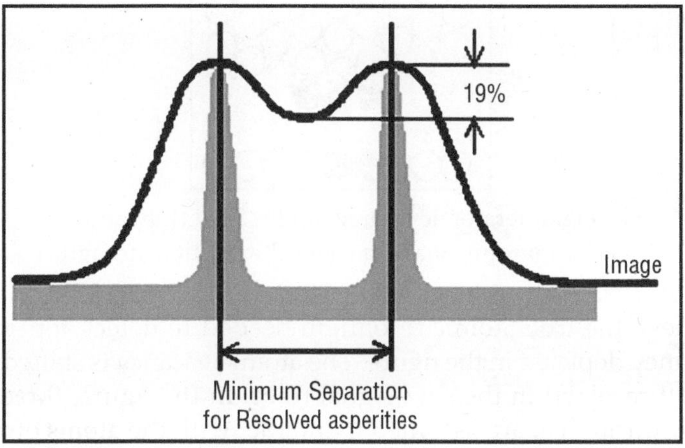

Fig. Definition of Lateral Resolution using Rayleigh's Criterion

To determine the lateral resolution of an SPM experimentally, the asperities are brought closer and closer together until the image no longer dips by 19 per cent between peaks. The minimum separation between resolved asperities determines the best lateral resolution of the system.

Using this definition, the lateral resolution of an AFM with the sharpest tips commercially available is 10 to 20Å. At first glance the figure of 10 to 20Å resolution seems to conflict with the ubiquitous images of atomic lattices in AFM brochures. The distinction between *imaging atomic-scale features* with accurate lattice spacing and symmetry, and *true atomic resolution* bears some comment.

An STM gives true atomic resolution. Because the dependence of the tunneling current on the tip-to-sample separation is exponential, only the closest atom on a good STM tip interacts with the closest atom on the sample, as shown in Figure, top. For AFMs, the dependence of cantilever deflection on tip-to-sample separation is weaker. The result is that several atoms on the tip interact simultaneously with several atoms on the sample.

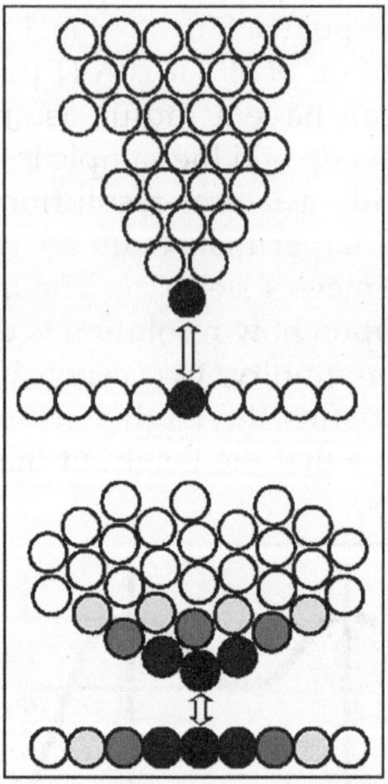

Fig. Interatomic Interaction for STM (top) and AFM (bottom). Shading shows Interaction Strength

Figure illustrates how this multiple interaction affects AFM images: AFMs cannot achieve the true atomic resolution needed to detect, for example, the atomic vacancy depicted in the figure. The atomic vacancy is shifted in position and not well resolved in the sum signal trace. In the figure, three tip atoms, numbered 1 through 3, are assumed to interact with the atoms of the surface. Three signal traces are also shown, representing the contribution from each atom. The traces show signals varying as a function of time, where the signal reflects the interatomic force between the given tip atom and the atom beneath it on the surface.

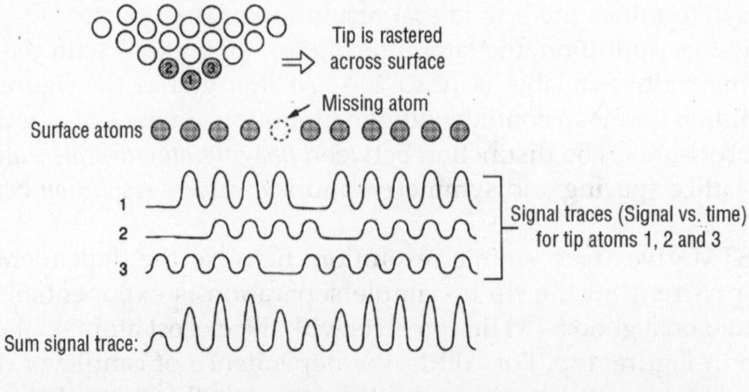

Fig. AFM Scan of Atomic-scale Periodic Lattice, Showing the Sum Signal Trace from Contributions of three Tip Atoms

If only atom number 1 on the very end of the tip interacts with the surface, then the signal trace would be straightforward: It would make sense to define time equals zero at the point when atom 1 is over the first atom encountered on the surface. The signal trace labeled 1 would show four bumps, then a hole (or no bump), then six more bumps. The atomic vacancy would be detected in its correct position, and true atomic resolution would be achieved. However, if three tip atoms are assumed to interact with the surface, then the image is not so straightforward. In this case, it makes sense to define time equals zero when atom 3 is over the first atom encountered on the surface. When atom 3 encounters the first surface atom, some signal is registered, apparent from the signal trace of atom 3.

At the next instant in time (assuming a step size on the order of one atomic spacing), atom 1 detects the first surface atom, atom 3 detects the second surface atom, and atom 2 still registers no signal. At the third instant in time, all three tip atoms register signals which vary in intensity according to their distance from the surface. This example shows that, since the tip atoms are displaced laterally, they produce signal traces that are offset from one and other. The vacancy appears in a different location for each signal trace. An image is the result of the sum of the contributions from all interacting tip atoms. Because the three interacting tip atoms are displaced from one and other laterally, there is no clear vacancy apparent in the sum signal. A minimum is apparent, but its location is shifted.

In this simple case, only three tip atoms were assumed to interact with the surface. In reality, the number of interacting tip atoms for contact AFM is large, and at each instant in time their interaction is with several surface atoms. For larger numbers of interacting atoms, features such as vacancies are even less well-defined and more shifted. Thus, while the periodicity of the lattice is reproduced, true atomic resolution is not achieved. For both contact and non-contact imaging, we should choose a tip that is sharper than the smallest features on our sample. If the tip is larger than the surface features, an artifact know as *tip imaging* occurs.

We may not always want to choose the sharpest tip available, since the sharpest tips are more expensive and can be less durable. We need the sharpest tips only when we require the best resolution. For LFM, blunter tips can be preferable precisely because they present a larger interaction area between tip and sample. A larger area of interaction can produce a stronger lateral deflection of the cantilever. The larger cantilever deflection must be balanced against the lower lateral resolution delivered by the blunter tip. AFM manufacturers offer microfabricated tips in three geometries: pyramidal, tetrahedral, and conical. Conical tips can be made sharp, with high aspect ratios (the ratio of tip length to tip width). Tip radii as small as 50Å have been observed. Pyramidal tips have lower aspect ratios and nominal tip radii of a few hundred angstroms, but they are more durable.

AFM tips are fabricated from silicon or silicon nitride. The fabrication process is different for the two materials. The features of tips made from each material are governed by the fabrication process as well as the material properties.

Silicon conical tips are made by etching into the silicon around a silicon dioxide cap. The high aspect ratio of conical tips makes them suitable for imaging deep, narrow features such as trenches, but they may break more easily than the pyramidal or tetrahedral geometries, as described above. Silicon also has the advantage that it can be doped, so that tips can be made electrically conducting. Conducting tips are useful for controlling the bias between the tip and the sample, or for preventing unwanted charge buildup on the tip. Silicon nitride tips are fabricated by depositing a layer of silicon nitride over an etched pit in a crystalline silicon surface. This method produces the pyramidal or tetrahedral tip geometry. The aspect ratio of a silicon nitride tip is thus limited by the crystallographic structure of the etch pit material, silicon. The tips are broader than conical silicon tips, making them sturdier but less suitable for imaging deep, narrow features. Silicon nitride is a harder material than silicon, which also makes silicon nitride tips more durable than silicon tips.

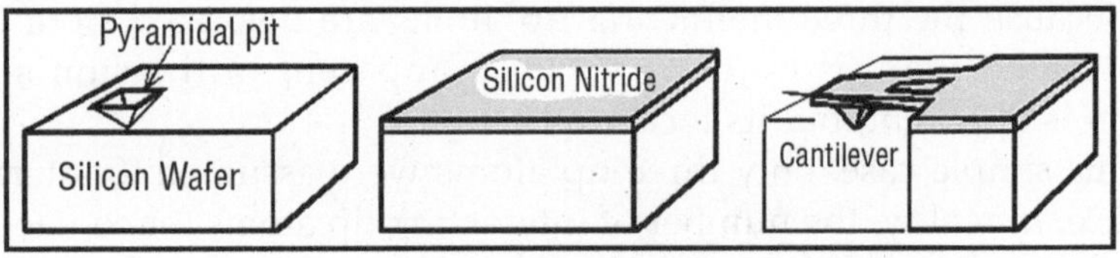

Fig. Fabrication of Silicon Nitride Tip

Silicon nitride films, however, contain residual stresses which make them deform as the film thickness increases. For this reason, applications that call for thick cantilevers - or cantilevers with high resonant frequencies - use silicon cantilevers. The thickness of silicon nitride cantilevers is usually less than one micron, whereas silicon cantilevers can be several microns thick.

A more specialized type of cantilever that has a tip grown in a scanning electron microscope is also available commercially. (A specialized tip can also be produced by "milling" an existing tip using a focused ion beam.) The SEM is used in an unusual way: to deposit contamination in a column. The tips are not microfabricated in bulk; they are grown one at a time on top of a bare cantilever or square-pyramidal tip. These tips have the advantage of almost unlimited aspect ratio. However, they are not normally sharp, and can bend or break easily.

IMAGE ARTIFACTS

SPM images are among the easiest to interpret of images generated by any microscopy technique. With an electron or optical microscope, contrast is based on complex electromagnetic diffraction effects. Determining whether a feature is protruding from the surface or recessed into it can be difficult with an image from an optical or electron microscope. SPMs, however, collect three-dimensional data. In an SPM image, a peak is unambiguously a peak, and a valley is clearly a valley. With an optical or electron microscope, artificial contrast can occur when a sample consists of reflecting material embedded in an absorbing matrix, for example. SPMs, on the other hand, are largely indifferent to variations in optical or electronic properties, and measure true surface topography. Despite the apparent simplicity of images from SPMs, SPM images are subject to some artifacts of their own. Fortunately, artifacts are relatively easy to identify.

TIP CONVOLUTION

Most imaging artifacts in an SPM image arise from a phenomenon known as tip convolution or tip imaging. Every data point in an image represents a spatial convolution (in the general sense, not in the sense of Fourier analysis) of the shape of the tip and the shape of the feature imaged. As long as the tip is much sharper than the feature, the true edge profile of the feature is represented. However, when the feature is sharper than the tip, the image will be dominated by the shape of the tip. Figure demonstrates the origin of tip convolution.

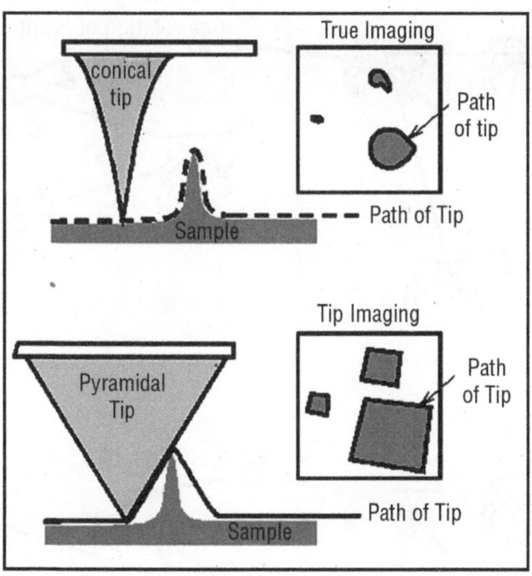

Fig. Comparison between True Imaging and Tip Imaging

The first commercial AFM tips were square pyramidal, formed by CVD deposition of Si3N4 on an etch pit in Si (100). The etch pit is bounded by (111) faces, which means that the resulting tip has a sidewall angle of about 62.5° along the flat side of the pyramid (45° along the corner edge of the pyramid). If such a tip is used, the edge profiles of SPM images of all features with sides steeper than 62.5° are dominated by the profile of the tip. Tips with higher aspect ratios have been marketed successfully since then, but microfabricated tips remain limited to a sidewall angle of about 80°. Further enhancement of the sidewall angle of the tip can result in degraded durability, or even bending of the tip during scanning. Because many samples have features with steep sides, tip imaging is a common occurrence in images. Sidewall angles on images should be measured routinely, to determine whether the slope is limited by that of the tip, or truly represents the topography of the sample. One consolation is that the height of the feature is reproduced accurately as long as the tip touches bottom between features. Thus, height measurements and roughness statistics remain fairly accurate. The lateral dimensions from a tip-imaged scan, on the other hand, can provide the user with only a maximum value. In other words, if we measure a tip-imaged feature as having a width of 200Å, we 'll know that the feature is at most 200Å wide.

To recognize tip imaging, look for a particular shape that is repeated throughout an image. The shape can be different sizes as the tip is convolved with features of different sizes, but it will always maintain the same orientation. If we suspect tip imaging is occurring, we can rotate our sample and image it again. If the tip is dominating the image, the orientation of the tip shape will be the same before and after rotation.

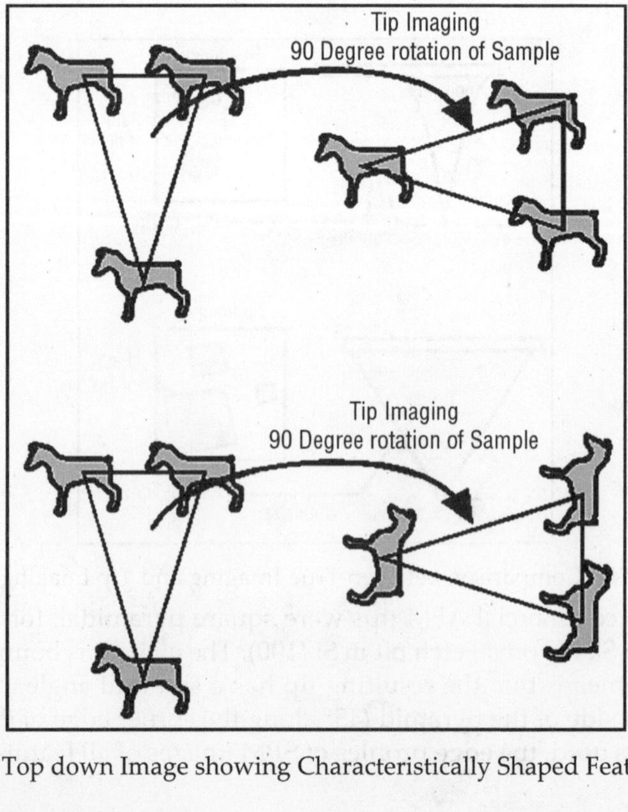

Fig. Top down Image showing Characteristically Shaped Features

If the image is a true representation of the surface, the shapes in the image will rotate along with the sample. This test is represented schematically in Figure. If our image is dominated by tip-convolution effects, change to another tip. For STM, the imaging portion of the tip is formed by an atom or cluster of atoms at the end of a long wire. Because the dependence of the tunneling current upon the tip-to-sample distance is exponential, the closest atom on the tip will image the closest atom on the sample.

If two atoms on the tip are equidistant from the surface, all of the features in the image will appear doubled. This is an example of multiple-tip imaging. The best way to alleviate this problem is to apply a voltage pulse to change the tip configuration by field emission. Alternatively, we can press the tip gently against the sample to form a new tip shape, and take another image.

9

Theory of Electromagnetic Radiation

Electromagnetic energy is a term used to describe all the different kinds of energies released into space by stars such as the Sun. These kinds of energies include some that you will recognize and some that will sound strange. They include:

- Radio Waves
- TV waves
- Radar waves
- Heat (infrared radiation)
- Light
- Ultraviolet Light (This is what causes Sunburns)
- X-rays (Just like the kind you get at the doctor's office)
- Short waves
- Microwaves, like in a microwave oven
- Gamma Rays

All these waves do different things (for example, light waves make things visible to the human eye, while heat waves make molecules move and warm up, and x rays can pass through a person and land on film, allowing us to take a picture inside someone's body) but they have some things in common. They all travel in waves, like the waves at a beach or like sound waves, and also are made of tiny particles. Scientists are unsure of exactly how the waves and the particles relate to each other. The fact that electromagnetic radiation travels in waves lets us measure the different kind by wavelength or how long the waves are. That is one way we can tell the kinds of radiation apart from each other.

Although all kinds of electromagnetic radiation are released from the Sun, our atmosphere stops some kinds from getting to us. For example, the ozone layer stops a lot of harmful ultraviolet radiation from getting to us, and that's why people are so concerned about the hole in it. We humans have learned uses for a lot of different kinds of electromagnetic radiation and have learned how to make it using other kinds of energy when we need to. DS1 would not be able to communicate with Earth, for example, if it could not produce radio waves.

HISTORY OF ELECTROMAGNETIC RADIATION

Electromagnetic radiation (EM radiation or EMR) is a form of energy emitted and absorbed by charged particles which exhibits wave-like behavior as it travels through space. EMR has bothelectric and magnetic field components, which stand in a fixed ratio of intensity to each other, and which oscillate in phase perpendicular to each other and perpendicular to the direction of energy and wave propagation. In a vacuum, electromagnetic radiation propagates at a characteristic speed, the speed of light.

Electromagnetic radiation is a particular form of the more general electromagnetic field (EM field), which is produced by moving charges. Electromagnetic radiation is associated with EM fields that are far enough away from the moving charges that produced them, that absorption of the EM radiation no longer affects the behavior of these moving charges. These two types or behaviors of EM field are sometimes referred to as the near and far field. In this language, EMR is merely another name for the far-field. Charges and currents directly produce the near-field. However, charges and currents produce EMR only indirectly—rather, in EMR, both the magnetic and electric fields are associated with changes in the other type of field, not directly by charges and currents. This close relationship assures that the electric and magnetic fields in EMR exist in a constant ratio of strengths to each other, and also to be found in phase, with maxima and nodes in each found at the same places in space.

EMR carries energy—sometimes called radiant energy—through space continuously away from the source (this is not true of the near-field part of the EM field). EMR also carries both momentum and angular momentum. These properties may all be imparted to matter with which it interacts. EMR is produced from other types of energy when created, and it is converted to other types of energy when it is destroyed. The photon is the quantum of the electromagnetic interaction, and is the basic "unit" or constituent of all forms of EMR. The quantum nature of light becomes more apparent at high frequencies (or high photon energy). Such photons behave more like particles than lower-frequency photons do.

In classical physics, EMR is considered to be produced when charged particles are accelerated by forces acting on them. Electrons are responsible for emission of most EMR because they have low mass, and therefore are easily accelerated by a variety of mechanisms. Rapidly moving electrons are most sharply accelerated when they encounter a region of force, so they are responsible for producing much of the highest frequency electromagnetic radiation observed in nature. Quantum processes can also produce EMR, such as whenatomic nuclei undergo gamma decay, and processes such as neutral pion decay. EMR is classified according to the frequency of its wave. The electromagnetic spectrum, in order of increasing frequency and decreasingwavelength, consists of radio waves, microwaves, infrared radiation, visible light, ultraviolet radiation, X-rays and gamma rays. The eyesof various organisms sense a somewhat variable but relatively small range of frequencies of EMR called the visible spectrum or light. The effects of EMR upon biological systems (and also to many other chemical systems, under standard conditions) depends both upon the radiation's power and frequency. For lower frequencies of EMR up to those of visible light (*i.e.*, radio, microwave, infrared), the damage done to cells and also to many ordinary materials under such conditions is determined mainly by heating effects, and thus by the radiation power. By contrast, for higher frequency radiations at ultraviolet frequencies and above (*i.e.*, X-rays and gamma rays) the damage to chemical materials and living cells by EMR is far larger than that done by simple heating, due to the ability of single photons in such high frequency EMR to damage individual molecules chemically.

INFRARED RADIATION

Infrared Radiation, emission of energy as electromagnetic waves in the portion of the spectrum just beyond the limit of the red portion of visible radiation. The wavelengths of infrared radiation are shorter than those of radio waves and longer than those of light waves. They range between approximately 10^{-6} and 10^{-3} (about 0.0004 and 0.04 in). Infrared radiation may be detected as heat, and instruments such as bolometers are used to detect it. Infrared radiation is used to obtain pictures of distant objects obscured by atmospheric haze, because visible light is scattered by haze but infrared radiation is not. The detection of infrared radiation is used by astronomers to observe stars and nebulas that are invisible in ordinary light or that emit radiation in the infrared portion of the spectrum.

An opaque filter that admits only infrared radiation is used for very precise infrared photographs, but an ordinary orange or light-red filter, which will absorb blue and violet light, is usually sufficient for most infrared pictures. Developed about 1880, infrared photography has today become an important diagnostic tool in medical science as well as in agriculture and industry. Use of infrared techniques reveals pathogenic conditions that are not visible to the eye or recorded on X-ray plates.

Remote sensing by means of aerial and orbital infrared photography has been used to monitor crop conditions and insect and disease damage to large agricultural areas, and to locate mineral deposits. In industry, infrared spectroscopy forms an increasingly important part of metal and alloy research, and infrared photography is used to monitor the quality of products. Infrared devices such as those used during World War II enable sharpshooters to see their targets in total visual darkness. These instruments consist essentially of an infrared lamp that sends out a beam of infrared radiation, often referred to as black light, and a telescope receiver that picks up returned radiation from the object and converts it to a visible image.

SPECTRUM

Spectrum, rainbowlike series of colors, in the order violet, blue, green, yellow, orange, and red, produced by splitting a composite light, such as white light, into its component colors. Indigo was formerly recognized as a distinct spectral color. The rainbow is a natural spectrum, produced by meteorological phenomena. A similar effect can be produced by passing sunlight through a glass prism. The first correct explanation of the phenomenon was advanced in 1666 by the English mathematician and physicist Sir Isaac Newton. When a ray of light passes from one transparent medium, such as air, into another, such as glass or water, it is bent; upon reemerging into the air, it is bent again. This bending is called refraction; the amount of refraction depends on the wavelength of the light. Violet light, for example, is bent more than red light in passing from air to glass or from glass to air. A mixture of red and violet light is thus dispersed into the two colors when it passes through a wedge-shaped glass prism.

A device for producing and observing a spectrum visually is called a spectroscope; a device for observing and recording a spectrum photographically is called a spectrograph; a device for measuring the brightness of the various portions of spectra is called a spectrophotometer; and the science of using spectroscopes, spectrographs, and spectrophotometers to study spectra is called spectroscopy. For extremely accurate spectroscopic measurements, an interferometer is used. During the 19th century, scientists discovered that beyond the violet end of the spectrum, radiations could be detected that were invisible to the human eye but that had marked photochemical action; these radiations were termed ultraviolet. Similarly, beyond the red end of the spectrum, infrared radiations were detected that, although invisible, transmitted energy, as shown by their ability to raise the temperature of a thermometer. The definition of spectrum was then revised to include these invisible radiations, and has since been extended to include radio waves beyond the infrared, and X rays and gamma rays beyond the ultraviolet.

The term *spectrum* is often loosely applied today to any orderly array produced by analysis of a complex phenomenon. A complex sound such as noise, for example, may be analyzed into an audio spectrum of pure tones of various pitches. Similarly, a complex mixture of elements or isotopes of different atomic weights can be separated into an orderly sequence called a mass spectrum in order of their atomic weights. Spectroscopy has not only provided an important and sensitive method of chemical analysis, but has also been the chief tool for discoveries in the apparently unrelated fields of astrophysics and atomic theory. In general, changes in motions of the outer electrons of atoms produce spectra in the visible, infrared, and ultraviolet regions. Changes in motions of the inner electrons of heavy atoms produce X-ray spectra. Changes in the configurations of the nucleus of an atom produce gamma-ray spectra. Changes in the configurations of molecules produce visible and infrared spectra. Different colors of light are similar in consisting of electromagnetic radiations that travel at a speed of approximately 300,000 km per sec (about 186,000 mi per sec). They differ in having varying frequencies and wavelengths, the frequency being equal to the speed of light divided by wavelength.

Two rays of light having the same wavelength also have the same frequency and the same color. The wavelength of light is so small that it is conveniently expressed in nanometers (nm), which are equal to one-billionth of a meter. The wavelength of violet light varies from about 400 to 450 nm, and of red light from about 620 to 760 nm, or from about 0.000016 to 0.000018 in for violet, and from 0.000025 to 0.000030 in for red.

Light

Light, form of energy visible to the human eye that is radiated by moving charged particles. Light from the Sun provides the energy needed for plant growth. Plants convert the energy in sunlight into storable chemical form through a process called photosynthesis. Petroleum, coal, and natural gas are the remains of plants that lived millions of years ago, and the energy these fuels release when they burn is the chemical energy converted from sunlight. When animals digest the plants and animals they eat, they also release energy stored by photosynthesis. Scientists have learned through experimentation that light behaves like a particle at times and like a wave at other times. The particle-like features are called photons. Photons are different from particles of matter in that they have no mass and always move at the constant speed of about 300,000 km/sec (186,000 mi/sec) when they are in a vacuum. When light diffracts, or bends slightly as it passes around a corner, it shows wavelike behavior. The waves associated with light are called electromagnetic waves because they consist of changing electric and magnetic fields.

II THE NATURE OF LIGHT

To understand the nature of light and how it is normally created, it is necessary to study matter at its atomic level. Atoms are the building blocks of matter, and the motion of one of their constituents, the electron, leads to the emission of light in most sources.

A Light Emission

Light can be emitted, or radiated, by electrons circling the nucleus of their atom. Electrons can circle atoms only in certain patterns called orbitals, and electrons have a specific amount of energy in each orbital. The amount of energy needed for each orbital is called an energy level of the atom. Electrons that circle close to the nucleus have less energy than electrons in orbitals farther from the nucleus. If the electron is in the lowest energy level, then no radiation occurs despite the motion of the electron. If an electron in a lower energy level gains some energy, it must jump to a higher level, and the atom is said to be excited. The motion of the excited electron causes it to lose energy, and it falls back to a lower level. The energy the electron releases is equal to the difference between the higher and lower energy levels. The electron may emit this quantum of energy in the form of a photon. Each atom has a unique set of energy levels, and the energies of the corresponding photons it can emit make up what is called the atom's spectrum. This spectrum is like a fingerprint by which the atom can be identified. The process of identifying a substance from its spectrum is called spectroscopy. The laws that describe the orbitals and energy levels of atoms are the laws of quantum theory. They were invented in the 1920s specifically to account for the radiation of light and the sizes of atoms.

B Electromagnetic Waves

The waves that accompany light are made up of oscillating, or vibrating, electric and magnetic fields, which are force fields that surround charged particles and influence other charged particles in their vicinity. These electric and magnetic fields change strength and direction at right angles, or perpendicularly, to each other in a plane (vertically and horizontally for instance). The electromagnetic wave formed by these fields travels in a direction perpendicular to the field's strength (coming out of the plane). The relationship between the fields and the wave formed can be understood by imagining a wave in a taut rope. Grasping the rope and moving it up and down simulates the action of a moving charge upon the electric field. It creates a wave that travels along the rope in a direction that is perpendicular to the initial up and down movement.

Because electromagnetic waves are transverse—that is, the vibration that creates them is perpendicular to the direction in which they travel, they are similar to waves on a rope or waves traveling on the surface of water. Unlike these waves, however, which require a rope or water, light does not need a medium, or substance, through which to travel. Light from the Sun and distant stars reaches Earth by traveling through the vacuum of space. The waves associated with natural sources of light are irregular, like the water waves in a busy harbor. Scientists think of such waves as being made up of many smooth waves, where the motion is regular and the wave stretches out indefinitely with regularly spaced peaks and valleys. Such regular waves are called monochromatic because they correspond to a single color of light.

B1 WAVELENGTH, FREQUENCY, AND AMPLITUDE

The wavelength of a monochromatic wave is the distance between two consecutive wave peaks. Wavelengths of visible light can be measured in meters or in nanometers (nm), which are one-billionth of a meter (or about

0.4 ten-millionths of an inch). Frequency corresponds to the number of wavelengths that pass by a certain point in space in a given amount of time. This value is usually measured in cycles per second, or hertz (Hz). All electromagnetic waves travel at the same speed, so in one second, more short waves will pass by a point in space than will long waves. This means that shorter waves have a higher frequency than longer waves. The relationship between wavelength, speed, and frequency is expressed by the equation: wave speed equals wavelength times frequency, or c = lf
Where *c* is the speed of a light wave in m/sec (3×10^8 m/sec in a vacuum), *l* is the wavelength in meters, and *f* is the wave's frequency in Hz. The amplitude of an electromagnetic wave is the height of the wave, measured from a point midway between a peak and a trough to the peak of the wave. This height corresponds to the maximum strength of the electric and magnetic fields and to the number of photons in the light.

Electromagnetic Spectrum

The electromagnetic spectrum refers to the entire range of frequencies or wavelengths of electromagnetic waves. Light traditionally refers to the range of frequencies that can be seen by humans. The frequencies of these waves are very high, about one-half to three-quarters of a million billion (5×10^{14} to 7.5×10^{14}) Hz. Their wavelengths range from 400 to 700 nm. X rays have wavelengths ranging from several thousandths of a nanometer to several nanometers, and radio waves have wavelengths ranging from several meters to several thousand meters. Waves with frequencies a little lower than the range of human vision (and with wavelengths correspondingly longer) are called infrared. Waves with frequencies a little higher and wavelengths shorter than human eyes can see are called ultraviolet. About half the energy of sunlight at Earth's surface is visible electromagnetic waves, about 3 percent is ultraviolet, and the rest is infrared.

Each different frequency or wavelength of visible light causes our eye to see a slightly different color. The longest wavelength we can see is deep red at about 700 nm. The shortest wavelength humans can detect is deep blue or violet at about 400 nm. Most light sources do not radiate monochromatic light. What we call white light, such as light from the Sun, is a mixture of all the colors in the visible spectrum, with some represented more strongly than others. Human eyes respond best to green light at 550 nm, which is also approximately the brightest color in sunlight at Earth's surface.

B3 Polarization

Polarization refers to the direction of the electric field in an electromagnetic wave. A wave whose electric field is oscillating in the vertical direction is said to be polarized in the vertical direction. The photons of such a wave would interact with matter differently than the photons of a wave polarized in the horizontal direction. The electric field in light waves from the Sun vibrates in all directions, so direct sunlight is called unpolarized. Sunlight reflected from a surface is partially polarized parallel to the surface. Polaroid sunglasses block light that is horizontally polarized and therefore reduce glare from sunlight reflecting off horizontal surfaces.

C Photons

Photons may be described as packets of light energy, and scientists use this concept to refer to the particle-like aspect of light. Photons are unlike conventional particles, such as specks of dust or marbles, however, in that they are not limited to a specific volume in space or time. Photons are always associated with an electromagnetic wave of a definite frequency. In 1900 the German physicist Max Planck discovered that light energy is carried by photons. He found that the energy of a photon is equal to the frequency of its electromagnetic wave multiplied by a constant called *h*, or Planck's constant. This constant is very small because one photon carries little energy. Using the watt-second, or joule, as the unit of energy, Planck's constant is 6.626×10^{-34} (a decimal point followed by 33 zeros and then the number 6626) joule-seconds in exponential notation. The energy consumed by a one-watt light bulb in one second, for example, is equivalent to two and a half million trillion photons of green light. Sunlight warms one square meter at the top of Earth's atmosphere at noon at the equator with the equivalent of about 14 100-watt light bulbs. Light waves from the Sun, therefore, produce a very large number of photons.

D Sources of Light

Sources of light differ in how they provide energy to the charged particles, such as electrons, whose motion creates the light. If the energy comes from heat, then the source is called incandescent. If the energy comes from another source, such as chemical or electric energy, the source is called luminescent

Photoelectric Effect

The photoelectric effect is a process in which an atom absorbs a photon that has so much energy that the photon sets one of the atom's electrons free to move outside the atom. Part of the photon's energy goes toward releasing the electron from the atom. This energy is called the activation energy of the electron. The rest of the photon's energy is transferred to the released electron in the form of motion, or kinetic energy. Since the photon energy is proportional to frequency, the released electron, or photoelectron, moves faster when it has absorbed high-frequency light. Metals with low activation energies are used to make photodetectors and photoelectric cells whose electrical properties change in the presence of light. Solar cells use the photoelectric effect to convert sunlight into electricity. Solar cells are used in place of electric batteries in remote applications like space satellites or roadside emergency telephones. Hand-held calculators and watches often use solar cells so that battery replacement is unnecessary.

E2 Photochemical Detection

The change induced in photographic film exposed to light is an example of photochemical detection of photons. Light induces a chemical change in photosensitive chemicals on film. The film is then processed to convert the chemical change into a permanent image and to remove the photosensitive chemicals from the film so it will not continue to change when it is viewed in full light. Human vision works on a similar principle. Light of different frequencies causes different chemical changes in the eye. The chemical action generates nerve impulses that our brains interpret as color, shape, and location of objects.

III BEHAVIOR OF LIGHT

Light behavior can be divided into two categories: how light interacts with matter and how light travels, or propagates through space or through transparent materials. The propagation of light has much in common with the propagation of other kinds of waves, including sound waves and water waves.

A Interaction with Material

When light strikes a material, it interacts with the atoms in the material, and the corresponding effects depend on the frequency of the light and the atomic structure of the material. In transparent materials, the electrons in the material oscillate, or vibrate, while the light is present. This oscillation momentarily takes energy away from the light and then puts it back again. The result is to slow down the light wave without leaving energy behind. Denser materials generally slow the light more than less dense materials, but the effect also depends on the frequency or wavelength of the light. Under certain laboratory conditions, scientists can slow light down. In 2001 scientists brought a beam of light to a halt by temporarily trapping it within an extremely cold cloud of sodium atoms.

Materials that are not completely transparent either absorb light or reflect it. In absorbing materials, such as dark colored cloth, the energy of the oscillating electrons does not go back to the light. The energy instead goes toward increasing the motion of the atoms, which causes the material to heat up. The atoms in reflective materials, such as metals, re-radiate light that cancels out the original wave. Only the light re-radiated back out of the material is observed. All materials exhibit some degree of absorption, refraction, and reflection of light. The study of the behavior of light in materials and how to use this behavior to control light is called optics.

A1 Refraction

Refraction is the bending of light when it passes from one kind of material into another. Because light travels at a different speed in different materials, it must change speeds at the boundary between two materials. If a beam of light hits this boundary at an angle, then light on the side of the beam that hits first will be forced to slow down or speed up before light on the other side hits the new material.

This makes the beam bend, or refract, at the boundary. Light bouncing off an object underwater, for instance, travels first through the water and then through the air to reach an observer's eye. From certain angles an object that is partially submerged appears bent where it enters the water because light from the part underwater is being refracted. The refractive index of a material is the ratio of the speed of light in a vacuum to the speed of light inside the material. Because light of different frequencies travels at different speeds in a material, the refractive index is different for different frequencies. This means that light of different colors is bent by different angles as it passes from one material into another.

This effect produces the familiar colorful spectrum seen when sunlight passes through a glass prism. The angle of bending at a boundary between two transparent materials is related to the refractive indexes of the materials through Snell's Law, a mathematical formula that is used to design lenses and other optical devices to control light.

Reflection

Reflection also occurs when light hits the boundary between two materials. Some of the light hitting the boundary will be reflected into the first material. If light strikes the boundary at an angle, the light is reflected at the same angle, similar to the way balls bounce when they hit the floor. Light that is reflected from a flat boundary, such as the boundary between air and a smooth lake, will form a mirror image. Light reflected from a curved surface may be focused into a point, a line, or onto an area, depending on the curvature of the surface.

A3 Scattering

Scattering occurs when the atoms of a transparent material are not smoothly distributed over distances greater than the length of a light wave, but are bunched up into lumps of molecules or particles. The sky is bright because molecules and particles in the air scatter sunlight. Light with higher frequencies and shorter wavelengths is scattered more than light with lower frequencies and longer wavelengths. The atmosphere scatters violet light the most, but human eyes do not see this color, or frequency, well. The eye responds well to blue, though, which is the next most scattered color. Sunsets look red because when the Sun is at the horizon, sunlight has to travel through a longer distance of atmosphere to reach the eye. The thick layer of air, dust and haze scatters away much of the blue. The spectrum of light scattered from small impurities within materials carries important information about the impurities. Scientists measure light scattered by the atmospheres of other planets in the solar system to learn about the chemical composition of the atmospheres.

B How Light Travels

The first successful theory of light wave motion in three dimensions was proposed by Dutch scientist Christiaan Huygens in 1678. Huygens suggested that light wave peaks form surfaces like the layers of an onion. In a vacuum, or a uniform material, the surfaces are spherical. These wave surfaces advance, or spread out, through space at the speed of light. Huygens also suggested that each point on a wave surface can act like a new source of smaller spherical waves, which may be called wavelets, that are in step with the wave at that point.

The envelope of all the wavelets is a wave surface. An envelope is a curve or surface that touches a whole family of other curves or surfaces like the wavelets. This construction explains how light seems to spread away from a pinhole rather than going in one straight line through the hole. The same effect blurs the edges of shadows. Huygens's principle, with minor modifications, accurately describes all forms of wave motion.

B1 Interference

Interference in waves occurs when two waves overlap. If a peak of one wave is aligned with the peak of the second wave, then the two waves will produce a larger wave with a peak that is the sum of the two overlapping peaks. This is called constructive interference. If a peak of one wave is aligned with a trough of the other, then the waves will tend to cancel each other out and they will produce a smaller wave or no wave at all. This is called destructive interference. In 1803 English scientist Thomas Young studied interference of light waves by letting light pass through a screen with two slits. In this configuration, the light from each slit spreads out according to Huygens's principle and eventually overlaps with light from the other slit. If a screen is set up in the region where the two waves overlap, a point on the screen will be light or dark depending on whether the two waves interfere constructively or destructively.

If the difference between the distance from one slit to a point on the screen and the other slit to the same point on the screen is an exact number of wavelengths, then light waves arriving at that point will be in step and constructively interfere, making the point bright. If the difference is an exact odd number of half wavelengths, then light waves will arrive out of step, with one wave's trough arriving at the same time as another wave's peak.

The waves will destructively interfere, making the point dark. The resulting pattern is a series of parallel bright and dark lines on the screen. Instruments called interferometers use various arrangements of reflectors to produce two beams of light, which are allowed to interfere. These instruments can be used to measure tiny differences in distance or in the speed of light in one of the beams by observing the interference pattern produced by the two beams.

Holography is another application of interference. A hologram is made by splitting a light wave in two with a partially reflecting mirror. One part of the light wave travels through the mirror and is sent directly to a photographic plate. The other part of the wave is reflected first toward a subject, a face for example, and then toward the plate. The resulting photograph is a hologram. Instead of being an image of the face, it is an image of the interference pattern between the two beams. A normal photograph records only the light and dark features of the face and ignores the positions of peaks and troughs of the light wave that form the interference pattern. Since the full light wave is restored when a hologram is illuminated, the viewer can see whatever the original wave contained, including the three-dimensional quality of the original face.

B2 Diffraction

Diffraction is the spreading of light waves as they pass through a small opening or around a boundary. Young's principle of interference can be applied to Huygens's explanation of diffraction to explain fringe patterns in diffracted light. As a beam of light emerges from a slit in an illuminated screen, the light some distance away from the screen will consist of overlapping wavelets from different points of the light wave in the opening of the slit. When the light strikes a spot on a display screen across from the slit, these points are at different distances from the spot, so their wavelets can interfere and lead to a pattern of light and dark regions.

The pattern produced by light from a single slit will not be as pronounced as a pattern from two slits. This is because there are an infinite number of interfering waves, one from each point emerging from the slit, and their interference patterns overlap one another.

ULTRAVIOLET RADIATION

Ultraviolet Radiation, electromagnetic radiation that has wavelengths in the range between 4000 angstrom units (Å), the wavelength of violet light, and 150 Å, the length of X rays. Natural ultraviolet radiation is produced principally by the sun. Ultraviolet radiation is produced artificially by electric-arc lamps. Ultraviolet radiation is often divided into three categories based on wavelength, UV-A, UV-B, and UV-C. In general shorter wavelengths of ultraviolet radiation are more dangerous to living organisms. UV-A has a wavelength from 4000 Å to about 3150 Å. UV-B occurs at wavelengths from about 3150 Å to about 2800 Å and causes sunburn; prolonged exposure to

UV-B over many years can cause skin cancer. UV-C has wavelengths of about 2800 Å to 150 Å and is used to sterilize surfaces because it kills bacteria and viruses. The earth's atmosphere protects living organisms from the sun's ultraviolet radiation. If all the ultraviolet radiation produced by the sun were allowed to reach the surface of the earth, most life on earth would probably be destroyed. Fortunately, the ozone layer of the atmosphere absorbs almost all of the short-wavelength ultraviolet radiation, and much of the long-wavelength ultraviolet radiation. However, ultraviolet radiation is not entirely harmful; a large portion of the vitamin D that humans and animals need for good health is produced when the human's or animal's skin is irradiated by ultraviolet rays.

When exposed to ultraviolet light, many substances behave differently than when exposed to visible light. For example, when exposed to ultraviolet radiation, certain minerals, dyes, vitamins, natural oils, and other products become *fluorescent*—that is, they appear to glow. Molecules in the substances absorb the invisible ultraviolet light, become energetic, then shed their excess energy by emitting visible light. As another example, ordinary window glass, transparent to visible light, is opaque to a large portion of ultraviolet rays, particularly ultraviolet rays with short wavelengths. Special-formula glass is transparent to the longer ultraviolet wavelengths, and quartz is transparent to the entire naturally occurring range.

SUN

Sun, closest star to Earth. The Sun is a huge mass of hot, glowing gas. The strong gravitational pull of the Sun holds Earth and the other planets in the solar system in orbit. The Sun's light and heat influence all of the objects in the solar system and allow life to exist on Earth. The Sun is an average star—its size, age, and temperature fall in about the middle of the ranges of these properties for all stars. Astronomers believe that the Sun is about 4.6 billion years old and will keep shining for about another 7 billion years. For humans, the Sun is beautiful and useful, but also powerful and dangerous. As Earth turns, the Sun rises over the eastern horizon in the morning, passes across the sky during the day, and sets in the west in the evening. This movement of the Sun across the sky marks the passage of time during the day. The Sun's movement can produce spectacular sunrises and sunsets under the right atmospheric conditions.

At night, reflected sunlight makes the Moon and planets bright in the night sky. The Sun provides Earth with vast amounts of energy every day. The oceans and seas store this energy and help keep the temperature of Earth at a level that allows a wide variety of life to exist. Plants use the Sun's energy to make food, and plants provide food for other organisms. The Sun's energy also creates wind in Earth's atmosphere. This wind can be harnessed and used to produce power. While it lights our day and provides energy for life, sunlight can also be harmful to people. Human skin is sensitive to ultraviolet light emitted from the Sun. Earth's atmosphere blocks much of the harmful light, but sunlight is still strong enough to burn skin under some conditions. Sunburn is one of the most important risk factors in the development of skin cancers, which can be fatal. Sunlight is also very harmful to human eyes. A person should never look directly at the Sun, even with sunglasses or during an eclipse. The Sun influences Earth with more than just light. Particles flowing from the Sun can disrupt Earth's magnetic field, and these disruptions can interfere with electronic communications.

Physical Characteristics

The Sun is large and massive compared to the other objects in the solar system. The Sun's radius (the distance from its center to its surface) is 695,508 km (432,169 mi), 109 times as large as Earth's radius. If the Sun were hollow, a million Earths could fit inside it. The Sun has a mass of 1.989×10^{27} metric tons. This number is very large. Written out, it would be the digits 1989 followed by 24 zeroes. The Sun is 333,000 times as massive as Earth is. Despite its large mass, the Sun has a lower density, or mass per unit volume, than Earth. The Sun's average density is only 1.409 g/cu cm (1.188 oz/cu in), which is a quarter of the average density of Earth. The Sun produces an enormous amount of light. It generates 3.83×10^{26} watts of power in the form of light. In comparison, an incandescent lamp emits 60 to 100 watts of power. The temperature of the outer, visible part of the Sun is 5500°C (9900°F).

From Earth the Sun looks small, because it is far away. Its average distance from Earth is 150 million km (93 million mi). Light from the Sun takes about eight minutes to reach Earth. This light is still strong enough when it reaches Earth, however, to damage human eyes when viewed directly. The Sun is much closer to Earth than any other star is. The Sun's nearest stellar neighbor, Proxima Centauri (part of the triple star Alpha Centauri), is 4.3 light-years from our solar system, meaning light from Proxima Centauri takes 4.3 years to reach the Sun. The Sun is so much closer to Earth than all other stars are that the intense light of the Sun keeps us from seeing any other stars during the day.

Importance to Earth

Earth would not have any life on it without the Sun's energy, which reaches Earth in the form of heat and light. This energy warms our days and illuminates our world. Green plants absorb sunlight and convert it to food, which these plants then use to live and grow. In this process, the plants give off the oxygen that animals breathe. Animals eat these plants for nourishment. All plant and animal life relies on the Sun's presence. The Sun also provides—directly or indirectly—much of the energy on Earth that people use for fuel. Devices called solar cells turn sunlight into electricity. Sunlight can heat a gas or liquid, which can then be circulated through a building to heat the building. The energy stored in fossil fuels originally came from the Sun. Ancient plants used sunlight as fuel to grow. Animals consumed these plants. The plants and animals stored the energy of sunlight in the organic material that composed them. When the ancient plants and animals died and decayed, this organic material was buried and gradually turned into the petroleum, coal, and natural gas people use today.

The Sun's energy produces the winds and the movements of water that people harness to produce electricity. The Sun heats Earth's oceans and land, which in turn heat the air and make it circulate in the atmosphere as wind. The Sun fuels Earth's water cycle, evaporating water from the oceans, seas, and lakes. This water returns to the ground in the form of precipitation, flowing back to the oceans through the ground and in rivers. The energy of water's motion in rivers can be harnessed with dams.

Role in the Solar System

The Sun's gravitational pull holds the solar system together. The planets, asteroids, comets, and dust that make up our solar system are strongly attracted to the Sun's huge mass. This gravitational attraction keeps these bodies in orbit around the Sun. The Sun also influences the solar system with its diffuse outer atmosphere, which expands outward in all directions. This expanding atmosphere fills the solar system with a constant flow of tiny, fast, electrically charged particles. This flow is called the solar wind. The region through which the solar wind blows is called the heliosphere. Estimates vary about the extent of the heliosphere, ranging from about 86 to about 100 times the distance between Earth and the Sun. Interstellar winds may give the heliosphere an egg shape. The solar wind spreads out as it leaves the Sun. The point at which the solar wind is so diffuse that it stops having an effect on its surroundings is called the heliopause. The heliopause marks the outer edge of the solar system. Within the heliosphere, the Sun provides most of the heat and light that are present, and the particles in the solar wind interact with the planets and satellites in the solar system. The solar wind causes auroras—displays of colored light—in the atmosphere of Earth's polar regions. The solar wind also carries remnants of the Sun's magnetic field, which affect the magnetic fields of the planets and larger satellites. The solar wind pushes the planets' magnetic fields away from the Sun, turning them into elongated, windsock shapes. For more information.

THE SUN'S ENERGY

The Sun produces an amazing amount of light and heat through nuclear reactions (Nuclear Energy). The process that produces the Sun's energy is called nuclear fusion. In nuclear fusion, two atoms come together to produce a heavier atom. Fusion reactions release energy and tiny elementary particles.

Scale of the Sun's Energy

In just one second the Sun emits more energy than humans have used in the last 10,000 years. The Sun has been shining relatively steadily for 4.6 billion years. Until the early 20th century, humans did not know of any process that could explain the energy production of the Sun. Even if a fire, such as those that occur on Earth, were as large as the Sun, the fire would consume the mass of the Sun in a few thousand years. Scientists now know that the Sun is mainly composed of hydrogen, the lightest and most abundant element in the universe. The Sun contains an enormous amount of hydrogen, however, which makes the Sun very massive. All matter inside the Sun is gravitationally attracted to all the other matter in the Sun, and this attraction tends to pull the Sun's mass together. This inward pull creates high pressures and temperatures inside the Sun.

The center is so violent and hot that collisions between atoms break the hydrogen atoms apart into their subatomic ingredients. A hydrogen atom is made up of a nucleus that contains a positively charged proton, and a negatively charged electron that orbits the nucleus. In the Sun, collisions separate the electron from the nucleus, freeing each to move about the solar interior. The positively charged nuclei, or protons, are called ions. A gas in which particles are ionized, or have electric charges, is called plasma. Scientists often consider plasma, such as the material inside the Sun, to be a fourth state of matter—the three more familiar states of matter are gas, liquid, and solid.

Nuclear Fusion in the Core

The separation of hydrogen nuclei from their electrons makes nuclear fusion possible at the Sun's core, producing the Sun's light and heat. With their electrons gone, hydrogen nuclei (protons) can be packed much more tightly than complete atoms. At great depths inside the Sun, the pressure of overlying material is enormous, the protons are squeezed tightly together, and the material is very hot and densely concentrated. At the Sun's center, the temperature is 15.6 million degrees C (28.1 million degrees F), and the density is more than 13 times that of solid lead. This is hot and dense enough to make the nuclei fuse together. Outside the solar core, where the overlying weight and compression are less, the gas is cooler and thinner, and nuclear fusion cannot occur.

The nuclear fusion reaction that powers the Sun involves four protons that fuse together to make one nucleus of helium. Two of the original protons become neutrons (electrically neutral particles about the same size as protons). The result is a helium nucleus, containing two protons and two neutrons. The helium nucleus is slightly less massive (by a mere 0.7 percent) than the four protons that combine to make it. The fusion reaction turns the missing mass into energy, and this energy powers the Sun. The relationship between energy and the missing matter was explained in 1905 by German-born American physicist Albert Einstein. The mass loss, *m,* during the transformation of four protons into one helium nucleus, supplies an energy, *E,* according to the relation $E = mc^2$, where *c* is the speed of light. The speed of light is a constant number equal to 3×10^8 m/s (1×10^9 ft/s). Every second, fusion reactions convert about 700 million metric tons of hydrogen into helium within the Sun's energy-generating core. In doing so, about 5 million metric tons of this matter become energy. This energy leaves the Sun as radiation, and the part of this radiation that constitutes visible light is what makes the Sun shine.

The rate of nuclear reactions in the Sun is relatively low, because protons repel each other. This repulsion often prevents them from getting close enough to each other to fuse. Protons push each other away because they have the same electrical charge. The particles must overcome this repulsion in order to fuse together. Only a tiny fraction of the protons inside the Sun are moving fast enough to overpower this repulsive electrical force. The nuclei that are moving fast enough can get very close together, and a force called the strong nuclear force takes over. The strong nuclear force is, as its name implies, very powerful, but only over very short distances. It pulls the nuclei together and holds them together. In this way, nuclear reactions proceed at a relatively slow pace inside the Sun. If the pace were much quicker, the Sun would explode like a giant hydrogen bomb.

The Proton-Proton Chain

Four protons do not combine directly to form a helium nucleus, since the protons are constantly moving and are almost never in the same place at the same time. Moreover, the electrical repulsion between four protons is too great to overcome, even if the four protons happen to come together at an appropriate speed at the same time. Instead, the protons come together in a series of steps to form a helium nucleus, and these steps are called the proton-proton chain. In the first step of the proton-proton chain, two exceptionally fast protons meet head on and merge into each other, tunneling through the electrical barrier between them. The two protons combine, with most of their mass forming a deuteron, the nucleus of a heavy form of hydrogen known as deuterium. A deuteron contains one proton and one neutron, so one of the protons must become a neutron in this step. The conversion of a proton to a neutron releases a much smaller particle called a neutrino. There are several types of neutrinos—the type that the proton-proton chain produces is called an electron neutrino. The reaction also creates a positron, a positively charged particle the size of an electron. The symbolic representation of the first step of the proton-proton chain is

$$p + p \rightarrow {}^2D + e^+ + \upsilon_e$$

where p represents the protons, ^2D represents deuterium, e^+ represents the positron, and υ_e represents the electron neutrino. In the second step of the chain, the deuteron collides with another proton to form a nucleus of light helium, which has two protons and one neutron. Less energy is needed to maintain a light helium nucleus than is needed to maintain a deuteron and a proton separately. The extra energy is released as a photon, or a packet of light energy. In symbolic terms, the second step is

$$^{2}D + p \rightarrow {}^{3}He + \gamma$$

where ^{3}He is light helium and g represents a photon. In the final step of the proton-proton chain, two light helium nuclei meet and fuse together to form a nucleus of normal heavy helium, which has two protons and two neutrons. This reaction also releases two unattached hydrogen nuclei that return to the solar gas. In symbolic terms, the third step is

$$^{3}He + {}^{3}He \rightarrow {}^{4}He + 2p$$

where ^{4}He represents a normal helium nucleus with two protons and two neutrons.

The positron created in the first step of the chain eventually collides with a free electron. The positron and the electron are opposite particles—the positron is the antimatter equivalent of the electron. When the positron and the electron collide, they annihilate each other, releasing energy. The electron and the positron disappear, their mass transformed into two photons:

$$e^{+} + e^{-} \rightarrow 2\gamma$$

where e^{-} represents the electron. The net result of the proton-proton chain is the transformation of four hydrogen nuclei into a helium nucleus (with two protons and two neutrons), two neutrinos, and six photons:

$$4p \rightarrow {}^{4}He + 2\upsilon_{e} + 6\gamma.$$

Solar Neutrinos

The conversion of two protons into two neutrons in the proton-proton chain produces two tiny, elusive, fast-moving neutral particles called neutrinos. Nuclear reactions in the Sun's central furnace create prodigious quantities of neutrinos. Every second the Sun releases 2×10^{38} neutrinos, and every second an estimated 70 billion of these solar neutrinos pass through every square centimeter of Earth that is facing the Sun. Neutrinos move at the velocity of light, have no electrical charge, and have so little mass that scientists are not sure that neutrinos have any mass at all.

The ghostlike neutrinos therefore travel almost unimpeded through the Sun, Earth, and nearly any amount of matter. Scientists can snag small numbers of neutrinos in massive underground detectors called neutrino telescopes. These telescopes are placed so deep underground that only neutrinos can reach them. Scientists using these telescopes have detected solar neutrinos, confirming that the Sun is indeed powered by nuclear fusion. The number of neutrinos detected by these telescopes, however, is only one-third to one-half of the total number of neutrinos predicted to exist by the theory of solar neutrino production. This discrepancy between the number of detected neutrinos and the number predicted is known as the solar neutrino problem. There are two possible explanations—scientists might not understand exactly how the Sun produces its energy, or they could have an incomplete knowledge of neutrinos.

Astronomers are convinced that their models of the Sun are correct and that their predictions for the expected amount of solar neutrinos are therefore correct. Studies of the interior of the Sun substantiate the current models of how the Sun produces its energy, so most scientists agree that the problem lies in their understanding of neutrinos. The theory scientists favor to explain the problem is that neutrinos from the Sun change on their way to Earth. Scientists know of at least three types of neutrinos. Nuclear fusion reactions in the Sun produce a type of neutrino called an electron neutrino.

The other two proven types of neutrinos are called muon neutrinos and tau neutrinos. Most neutrino telescopes, especially those devoted to solar research, can only detect electron neutrinos. In the 1990s studies of muon neutrinos (produced by reactions between particles called cosmic rays and Earth's atmosphere) showed that muon neutrinos might change into tau neutrinos. Research conducted since the late 1990s indicates that electron neutrinos from the Sun may also change into another type of neutrino. This change would mean the electron neutrino detectors miss many of the Sun's neutrinos.

The Sun's Spectrum

Sunlight appears yellowish, but it is actually a combination of a rainbow of colors. Scientists use special instruments called spectrographs to separate sunlight out into its different colors. These instruments do the same thing that water molecules in the atmosphere do when the molecules produce a rainbow. Each color corresponds to a different wavelength of light. Red has the longest wavelength of visible light, and violet has the shortest. The range of wavelengths of sunlight and the intensity at each wavelength are called the Sun's spectrum. The study of the spectra of the Sun and other objects or materials is called spectroscopy. When sunlight is spread out like a rainbow in the Sun's spectrum, many dark gaps separate one color from another in the row of colors. These gaps are called absorption lines. Each absorption line is created when sunlight passes through the gases in the Sun's photosphere. Atoms and ions of each element in the gas absorb light at certain wavelengths, creating dark gaps in the Sun's spectrum.

The dark absorption lines in the spectra of the Sun and other stars fingerprint the ingredients of these stars. Each chemical element produces a unique set of lines, and the presence of these lines shows that a particular element is present in the stellar photosphere. Darker absorption lines indicate greater absorption and therefore larger amounts of the element.

Absorption lines in the Sun's spectrum show that hydrogen is by far the most abundant element in the Sun. Other prominent absorption lines are produced by helium, sodium, calcium, and iron. Altogether, 92.1 percent of the atoms in the Sun are hydrogen atoms, 7.8 percent are helium atoms, and the other, heavier elements—sodium, calcium, iron, and other elements—make up only 0.1 percent of the atoms in the Sun. The Sun's absorption lines are called Fraunhofer lines, named after German physicist Joseph von Fraunhofer, who cataloged them in the 1800s. The most common Fraunhofer lines are listed below, by the letter Fraunhofer gave them, the color that they block, and the element that causes them.

A	(Extreme Red)	Made by Terrestrial Oxygen
B	(Red)	Made by Terrestrial Oxygen
C	(Red)	Made by Solar Hydrogen
D_1	(Yellow)	Made by Solar Sodium
D_2	(Yellow)	Made by Solar Sodium
E	(Green)	Made by Solar Iron
F	(Blue)	Made by Solar Hydrogen
G	(Violet)	Made by Solar Iron and Calcium Group
H	(Extreme Violet)	Made by Solar Calcium

The Fraunhofer lines designated A and B actually have nothing to do with the composition of the Sun. They only appear on spectra gathered within Earth's atmosphere. Earth's atmosphere absorbs sunlight at the wavelengths of the A and B Fraunhofer lines, creating dark lines on the Sun's spectrum. A spectrum gathered above Earth's atmosphere would not have these lines.

10

X-ray Emission Spectroscopy

INTRODUCTION

William Lawrence Bragg and William Henry Bragg, who were the 1915 Nobel Prize Winners, were the original pioneers in developing X-ray emission spectroscopy (also called X-ray fluorescence spectroscopy). He measured the X-ray wavelengths of many elements to high precision, using high-energy electrons as excitation source. He also painstakingly produced numerous diamond-ruled glass diffraction gratings for his spectrometers. Intense and wavelength-tunable X-rays are now typically generated with synchrotrons. In a material, the X-rays may suffer an energy loss compared to the incoming beam.

This energy loss of the re-emerging beam reflects an internal excitation of the atomic system, an X-ray analogue to the well-known Raman spectroscopy that is widely used in the optical region. In the X-ray region there is sufficient energy to probe changes in the electronic state (transitions between orbitals; this is in contrast with the optical region, where the energy loss is often due to changes in the state of the rotational or vibrational degrees of freedom).

For instance, in the ultra soft X-ray region (below about 1 keV), crystal field excitations give rise to the energy loss. The photon-in-photon-out process may be thought of as a scattering event. When the x-ray energy corresponds to the binding energy of a core-level electron, this scattering process is resonantly enhanced by many orders of magnitude. This type of X-ray emission spectroscopy is often referred to as resonant inelastic X-ray scattering (RIXS).

Due to the wide separation of orbital energies of the core levels, it is possible to select a certain atom of interest. The small spatial extent of core level orbitals forces the RIXS process to reflect the electronic structure in close vicinity of the chosen atom. Thus RIXS experiments give valuable information about the local electronic structure of complex systems, and theoretical calculations are relatively simple to perform.

X-RAY EMISSION SPECTRUM

INSTRUMENTATION

There exist several efficient designs for analyzing an X-ray emission spectrum in the ultra soft X-ray region. The figure of merit for such instruments is the spectral throughput, *i.e.* the product of detected intensity and spectral resolving power. Usually, it is possible to change the parameters within a certain range while keeping their product constant.

Grating Spectrometers

Typically, the X-rays emerging from a sample must pass a source-defining slit, then optical elements (mirrors and/or gratings) disperse them by diffraction according to their wavelength and, finally, a detector is placed at their focal points.

Spherical Grating Mounts

Henry Augustus Rowland (1848–1901) devised an instrument that allowed the use of a single optical element that combines diffraction and focusing: a spherical grating. Reflectivity of X-rays is low regardless of the used material and therefore grazing incidence upon the grating is necessary. X-ray beams impinging on a smooth surface at a few degrees glancing angle of incidence undergo external total reflection which is taken advantage of to enhance the instrumental efficiency substantially. Denote by R the radius of a spherical grating.

Imagine a circle with half the radius R tangent to the center of the grating surface. This small circle is called the *Rowland circle*. If the entrance slit is anywhere on this circle, then a beam passing the slit and striking the grating will be split into a specularly reflected beam, and beams of all diffraction orders, that come into focus at certain points on the same circle.

Plane Grating Mounts

Similar to optical spectrometers, a plane grating spectrometer first needs optics that turns the divergent rays emitted by the x-ray source into a parallel beam. This may be achieved by using a parabolic mirror. The parallel rays emerging from this mirror strike a plane grating (with constant groove distance) at the same angle and are diffracted according to their wavelength. A second parabolic mirror then collects the diffracted rays at a certain angle and creates an image on a detector.

A spectrum within a certain wavelength range can be recorded simultaneously by using a two-dimensional position-sensitive detector such as a microchannel photomultiplier plate or an X-ray sensitive CCD chip (film plates are also possible to use).

Interferometers

Instead of using the concept of multiple beam interference that gratings produce, the two rays may simply interfere. By recording the intensity of two such co-linearly at some fixed point and changing their relative phase one obtains an intensity spectrum as a function of path length difference. One can show that this is equivalent to a Fourier transformed spectrum as a function of frequency. The highest recordable frequency of such a spectrum is dependent on the minimum step size chosen in the scan and the frequency resolution (*i.e.* how well a certain wave can be defined in terms of its frequency) depends on the maximum path length difference achieved. The latter feature allows a much more compact design for achieving high resolution than for a grating spectrometer because x-ray wavelengths are small compared to attainable path length differences.

Early History of X-Ray Spectroscopy in the U.S.

Philips Gloelampen Fabrik, in Holland, was and still is a major manufacturer of light bulbs. It also has had one of the world's largest R&D labs. In 1941 Holland was overrun by Hitler's Germany. Somehow the company was able to transfer a substantial sum of money to a company that it set up as an R&D laboratory in an estate in Irvington on the Hudson in NY. As an extension to their work on light bulbs the Dutch company had developed a line of X-ray tubes. Put together with transformers, they became power supplies used for medical applications. They could also be used to power scientific X-ray instrumentations but there was very little commercial demand for the latter. So, the management decided to try to develop this market and set up development groups in their research labs in both Holland and the U.S.

They hired Dr. Ira Duffendack, a professor at University of Michigan and a world expert on infrared research to head the lab and to hire a staff. In 1951 he hired Dr. David Miller as Assistant Director of Research. Dr. Miller had done research on X-ray instrumentation at Washington University in St. Louis. Dr. Duffendack also hired Dr Bill Parish, a well known researcher in X-ray diffraction, to head up the section of the lab on X-ray instrumental development. X-ray diffraction units were widely used in academic research departments to do crystal analysis. An essential component of a diffraction unit was a very accurate angle measuring device known as a goniometer. Such units were not commercially available, so each investigator had do try to make their own. Dr Parrish decided this would be a good device to use to generate an instrumental market, so his group designed and learned how to manufacture a goniometer. This market developed quickly and, with the readily available tubes and power supplies, a complete diffraction unit was made available and was successfully marketed. The U.S. management did not want the laboratory to be converted to a manufacturing unit so it decided to set up a commercial unit to further develop the X-ray instrumentation market. In 1953 Norelco Electronics was established in Mount Vernon, NY, dedicated to the sale and support of X-ray instrumentation. It included a sales staff, a manufacturing group, an engineering department and an applications lab. .Dr. Miller was transferred from the lab to head up the engineering department. The sales staff sponsored three schools a year, one in Mount Vernon, one in Denver, and one in San Francisco. Weeklong, the school curriculums reviewed the basics of X-ray instrumentation and specific application of Noreclo products. The faculty was members of the engineering department and academic consultants. The schools were well attended by academic and industrial R&D scientists. The engineering department was also a new product development group. It added an X-ray spectrograph to the product line very quickly and contributed other related products for the next 8 years.

The applications lab was an essential sales tool. When the spectrograph was introduced as a quick and accurate analytical chemistry device, it met with widespread skepticism. All research facilities had a chemistry department and analytical analysis was done by "wet chemistry" methods. The idea of doing it by instruments was considered unreliable. To overcome this bias, the salesman would ask a prospective customer for a task the customer was doing by "wet methods". The task would be given to the applications lab and they would demonstrate how accurately and quickly it could be done using the X-ray units. This proved to be a very strong sales tool, particularly when the results were published in the Norelco Reporter, a technical journal issued monthly by the company with wide distribution in commercial and academic units.

X-RAY SPECTROGRAPH

An X-ray spectrograph consists of a high voltage power supply (50KV or 100KV), a broad band X-ray tube, usually with a tungsten anode and a beryllium window, a specimen holder, an analyzing crystal, a goniometer, and an X-ray detector device. These are arranged as shown in figure.

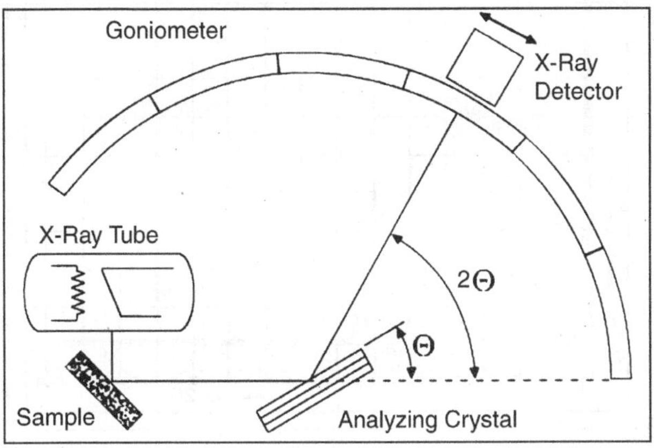

The continuous X-spectrum emitted from the tube radiates the specimen and excites the characteristic spectoral X-ray lines in the specimen. Each 92 elements can be excited to emit a characteristic spectrum. Unlike the optical spectrum the X-ray spectrum is quite simple. The strongest line, the alpha line, is all that is needed to identify the element and the intensity of this line can be used to determine the existence of and the amount of the element present in the specimen. Using a single crystal as a reflecting analyzer these characteristic lines emit X-ray quanta at specific angles which are captured at known specific angles in the detector as it rotates around the goniometer. The detector is a device that can be used to measure intensity by generating an electronic pulse for each quanta it receives. By setting the detector at a predetermined angular location for a known time a counting rate of pulses can be related to the line intensity at that particular angular position. These counts are collected by electronics and may be supplied by number or plotted on a curve by an appropriate display unit. The angular position for every X-ray spectral line is known and recorded and this information is readily available. A chart for a scan of a Molybdenum specimen is shown in Figure. The tall peak on the left side is the characteristic alpha line at a two theta of 12 degrees. Second and third order lines also appear.

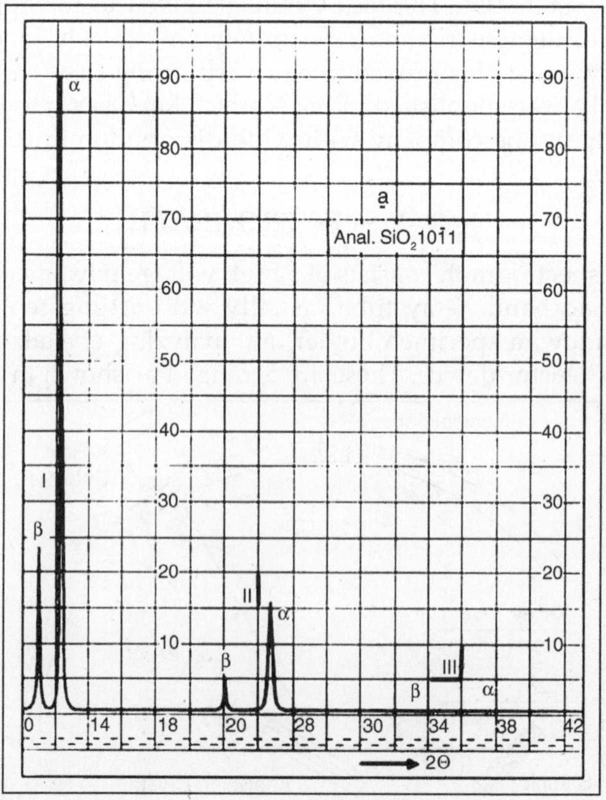

Since the alpha line is often the only line of interest in many industrial applications, the final device in the Norelco X- ray spectrographic instrument line was the Autrometer. This device could be programmed to automatically read at any desired two theta angle for any desired time interval. Soon after the Autrometer was introduced, Philips decided to stop trying to market X-ray instruments developed in both U.S. and Europe and settled on offering only the Eindhoven line of instruments.

In 1961, during the development of the Autrometer, Norelco was given a sub-contract from the Jet Propulsion Lab .The Lab was working on the instrument package for the Surveyor spaceship. The composition of the moon's surface was of major interest and the use of an X-ray detection instrument was a possible solution. Working with a power limit of 30 watts was very challenging but a device was delivered but it wasn't used. Later NASA developments did lead to an X-ray spectrographic unit that did make the desired moon soil analysis.

The Norelco efforts faded but the use of X-ray spectroscopy in units known as XRF instruments continued to grow. With a boost from NASA units were finally reduced to handheld size and are seeing widespread use. More information is available on-line at XRF instruments. Units are available from Bruker, Thermo Scientific and SPECRTO. The miniaturization that has been achieved is astounding. When a primary x-ray excitation source from an x-ray tube or a radioactive source strikes a sample, the x-ray can either be absorbed by the atom or scattered through the material. The process in which an x-ray is absorbed by the atom by transferring all of its energy to an innermost electron is called the "photoelectric effect."

During this process, if the primary x-ray had sufficient energy, electrons are ejected from the inner shells, creating vacancies. These vacancies present an unstable condition for the atom. As the atom returns to its stable condition, electrons from the outer shells are transferred to the inner shells and in the process give off a characteristic x-ray whose energy is the difference between the two binding energies of the corresponding shells. Because each element has a unique set of energy levels, each element produces x-rays at a unique set of energies, allowing one to non-destructively measure the elemental composition of a sample.

The process of emissions of characteristic x-rays is called "X-ray Fluorescence," or XRF. Analysis using x-ray fluorescence is called "X-ray Fluorescence Spectroscopy." In most cases the innermost K and L shells are involved in XRF detection. A typical x-ray spectrum from an irradiated sample will display multiple peaks of different intensities. Spectrum taken using Amptek XR-100CR 25mm^{2}X500μm X-Ray Detector (20μs shaping time) and Amptek MCA8000A Multichannel Analyzer. The characteristic x-rays are labeled as K, L, M or N to denote the shells they originated from. Another designation alpha (a), beta (b) or gamma (g) is made to mark the x-rays that originated from the transitions of electrons from higher shells. Hence, a Ka x-ray is produced from a transition of an electron from the L to the K shell, and a Kb x-ray is produced from a transition of an electron from the M to a K shell, etc. Since within the shells there are multiple orbits of higher and lower binding energy electrons, a further designation is made as a_1, a_2 or b_1, b_2, etc. to denote transitions of electrons from these orbits into the same lower shell.

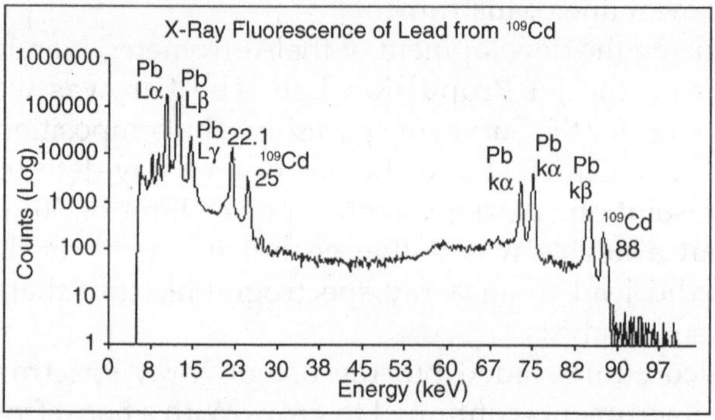

The XRF method is widely used to measure the elemental composition of materials. Since this method is fast and non-destructive to the sample, it is the method of choice for field applications and industrial production for control of materials. Depending on the application, XRF can be produced by using not only x-rays but also other primary excitation sources like alpha particles, protons or high energy electron beams.

Sometimes, as the atom returns to its stable condition, instead of emitting a characteristic x-ray it transfers the excitation energy directly to one of the outer electrons, causing it to be ejected from the atom. The ejected electron is called an "Auger" electron. This process is a competing process to XRF. Auger electrons are more probable in the low Z elements than in the high Z elements.

TOTAL REFLECTION X-RAY FLUORESCENCE

Total Reflection x-ray fluorescence (TXRF) and the fundamentally related Grazing emission x-ray fluorescence (GEXRF) rely on scatter properties near and below the Bragg angle to reduce background intensities and improve detections limits an order of magnitude or more over more traditional XRF instruments. If light is directed at a smooth surface at a very small angle (typically less than 0.5 degree for x-rays) virtually 100 per cent of the light will be reflected at an equally small angle.

This is the same principle relied on for polycapillary optics. A few x-rays will excite atoms immediately at the surface, and those atoms emit their characteristic radiation in all directions. Because there is virtually no backscatter into the detector, extraordinary detections limits can be achieve. GEXRF turns the theory around and takes advantage of the fact that when x-rays are directed at a surface they will not be scattered at an angle below the

Bragg angle. A detector that only detects x-rays coming off a surface at an angle less than the Bragg angle, will only detect fluorescence x-rays and not background scatter.

TXRF INSTRUMENTS

Hardware

TXRF instruments are usually very sophisticated and expensive pieces of equipment with finely tuned optics. The x-ray tubes are usually very high in power, several kilowatts, and must have a small spot size on the anode. A long collimator or wave-guide is needed to restrict the angle to less than the Bragg angle.

Using multilayers in the wave-guide can improve the efficiency. The sample needs to be finely and reproducibly polished and positioned precisely with respect to angle and height. A detector is positioned above the surface. Given the sophistication of these systems, Si(Li) or other high resolution detectors are used in most systems.

Some people prefer the GEXRF variation. The x-ray tube can be directed at the sample with little regard to spot size or angle. This saves on a lot of hardware expense. A detector and collimator assembly is positioned so that only x-rays coming from less than the Bragg angle are counted.

Advantages and Disadvantages

While these techniques can achieve amazing performance, they are seldom used. The principle problems are that only a few products are suitable for TXRF analysis without a substantial amount of sample preparation. The other problem is that the optical alignment is so critical that minor vibrations and temperature changes make it necessary to re-align the optics, and/or calibrate the instrument. These problems, in addition to the high cost of most existing systems, have limited the use of these techniques to date.

EDXRF VS WDXRF

In an effort to save money, space, sample preperation time, or simply to add an analytical instrument to their process many companies will decide to evaluate energy dispersive x-ray fluorescence (EDXRF) analyzers as a substitute for their standard wavelength dispersive x-ray fluorescence (WDXRF) analysis.

This is very common with geological applications where WDX is the benchmark, but it occurs with many other applications as well. What all these companies eventually discover is that EDXRF is not the low cost drop in replacement that they thought it would be but has significant differences, some positive and some negative, that must be considered in the evaluation process or else dealt with later when it may be less convenient.

More than just the Backwards Spectra

As most scientifically minded persons know, the energy of the light photon increases as the wavelength decreases, so in an EDX spectra the low atomic number elements are on the left while they are to the right of a WDX spectra. But the difference goes far beyond that.

WDXRF

The WDXRF analyzer uses a x-ray source to excite a sample. X-rays that have wavelengths that are characteristic to the elements within the sample are emmitted and they along with scattered source x-rays go in all directions. A crystal or other diffraction device is placed in the way of the x-rays coming off the sample.

A x-ray detector is position where it can detector the x-rays that are diffracted and scattered off the crystal. Depending on the spacing between the atoms of the crystal lattice (diffractive device) and its angle in relation to the sample and detector, specific wavelengths directed at the detector can be controlled. The angle can be changed in order to measure elements sequentially, or multiple crystals and detectors may be arrayed around a sample for simultaneous analysis.

EDXRF

The EDXRF analyzer also uses an x-ray source to excite the sample but it may be configured in one of two ways. The first way is direct excitation where the x-ray beam is pointed directly at the sample. Filter made of various elements may be placed between the source and sample to increase the excitation of the element of interest or reduce the background in the region of interest.

The second way uses a secondary target, where the source points at the target, the target element is excited and fluoresces, and then the target fluorescence is used to excite the sample. A detector is positioned to measure the fluorescent and scattered x-rays from the sample and a multichannel analyzer and software assigns each detector pulse an energy value thus producing a spectrum. Note that there is absolutely no reason why the spectra cannot be displayed in a wavelength dependant graph format.

Points of Comparison

- *Resolution*: It describes the width of the spectra peaks. The lower the resolution number the more easily an elemental line is distinguished from other nearby x-ray line intensities.
  - The resolution of the WDX system is dependant on the crystal and optics design, particularly collimation, spacing and positonal reproducibilty. The effective resolution of a WDX system may vary from 20 eV in an inexpensive benchtop to 5 eV or less in a

laboratory instrument. The resolution is not detector dependant.

- The resolution of the EDX system is dependent on the resolution of the detector. This can vary from 150 eV or less for a liquid nitrogen cooled Si(Li) detector, 150-220 eV for various solid state detectors, or 600 eV or more for gas filled proportional counter.
  a. *Advantage WDXRF*: High resolution means fewer spectral overlaps and lower background intensities.
  b. *Advantage EDXRF*: WDX crystal and optics are expensive, and are one more failure mode.

• *Spectral Overlaps*: Spectral deconvolutions are necessary for determining net intensities when two spectral lines overlap because the resolution is too high for them to be measured indepedantly.

- With a WDX instrument with very high resolution (low number of eV) spectral overlap corrections are not required for a vast majority of elements and applications. The gross intensities for each element can be determined in a single acquisition.
- The EDXRF analyzer is designed to detect a group of eleemnts all at once. The some type of deconvolution method must be used to correct for spectral overlaps. Overlaps are less of a problem with 150+ eV resolution systems, but are significant when compared to WDXRF. Spectral overlaps become more problematic at lower resolutions.
  a. Advantage WDXRF: Spectral deconvolution routines introduce error due to counting statistics for every overlap correction onto every other element being corrected for. This can double or triple the error.

• *Background*: The background radiation is one limiting factor for determining detection limits, repeatability, and reproducibilty.

- Since a WDX instrument usually uses direct radiation flux the background in the region of interest is directly related to the amount of continuum radiation within the region of interest the width of which is determined by the resolution.
- The EDXRF instrument uses filters and/or targets to reduce the amount of continuum radiation in the region of interest which is also resolution dependant, while producing a higher intensity x-ray peak to excite the element of interest.
  a. *Even*: WDX has an advantage due to resolution. If a peak is one tenth as wide it has one tenth the background.
  b. EDX counters with filters and targets that can reduce the background intensities by a factor of ten or more.

• *Source Efficiency*: How efficiently the source x-rays are utilized determines how much power is needed to make the system work optimally. Higher power costs much more money.

- Every time an x-ray beam is scattered off a surface the intensity is reduced by a factor of 100 or so. For any XRF system intensity is lost in the process of exciting the sample, but a WDX analyzer also looses a factor of 100 in intensity at the diffraction device, although some modern multilayers are more efficient. The sample to detector path length is often 10 cm or more introducing huge geometrical losses.
- With direct excitation the EDX system avoids wasting x-ray intensity. When filters are used the 3 to 10 times more energy is required, and when secondary targets are used 100 times more energy is required making the total energy budget simlar between Seconday target EDX and WDX systems before the path length is considered. An EDX system typically has sample to detector path lengths less than 1 cm.
  a. *Advantage EDXRF*: In order to achieve similar counts at the detector a WDX system needs 100-1000 times the flux of a direct excitation EDX system and 10-100 times the flux of a secondary target system. This one proinciple reason WDX systems cost more.

- *Excitation Efficiency*: Usually expressed in PPM per count-per-second (cps) or similar units, this is the other main factor for determining detection limits, repeatability, and reproducibility. The relative excitation efficiency is improved by having more source x-rays closer to but above the absorption edge energy for the element of interest.
  a. *WDXRF:* Generally uses direct unaltered x-ray excitation, which contains a continuum of energies with most of them not optimal for exciting the element of interest.
  b. *EDXRF*: Analyzers may use filter to reduce the continuum energies at the elemental lines, and effectively increaseing the percentage of x-rays above the element absorption edge. Filters may also be used to give a filter fluorescence line immediately above the absorption edge, to further improve excitation efficiency. Secondary targets provide an almost monochromatic line source that can be optimized for the element of interest to achieve optimal excitation efficiency.

ELECTRON MICROPROBE ANALYSIS

Micro x-ray fluorescence (μXRF) gets its name because these instruments are designed to analyze very small spot sizes. They are fundamentally like EDXRF systems, and have seen a similar development path through the years. There are a number of differences between μXRF and EDXRF to make it worthwhile to classify them separately.

MICRO X-RAY FLUORESCENCE

Hardware

The typical μXRF system has the usual EDX hardware, but has several key differences, most notably, downward facing optics. Special x-ray tubes are used with a smaller spot size on the target, so that there is less beam spread after collimation. Usually direct incident radiation is used, because filters tend to make the x-ray pattern broader. Systems usually have several collimators ranging from 12 microns to a millimeter or more, and can automatically switch between them.

Polycapillary optics that focus the x-rays are becoming increasing common in high end systems, and they are capable of producing spot sizes in the 50-200 micron range at much higher x-ray flux rates than with an ordinary through hole collimator. Samples are normally presented on a movable XYZ stage that can be manually operated in some low cost systems, or automatic and programmable in higher end system. The most sophisticated μXRF analyzers can even raster the sample producing a dot map of the surface. Cameras and lasers are typical devices used to aid in sample positioning. The detectors can vary from proportional counters in the low end systems to PIN diodes, Si(Li), or SDD detectors in higher end systems. There is even a μXRF system that uses WDX optics.

Applications

The biggest application by far is plating analysis, particularly for analyzing printed circuit boards. Literally thousands of instruments have been sold into this application, primarily for analyzing copper, gold, and tin-lead solder. But there are numerous other applications, and more sophisticated plating analysis systems can measure several layers at once. High-end μXRF systems are normally sold into forensic, fine art, and archeological applications. These systems can do an excellent job of elemental fingerprinting to either match evidence taken from a crime scene or match materials from a painting or other work of art with known authentic materials.

NONDISPERSIVE X-RAY FLUORESCENCE (NDXRF)

Non-dispersive x-ray fluorescence (NDXRF) got its start in the 1920's when Ross and other experimenters discovered that they could isolate an x-ray line for an element by using two filters made of different elements over two detectors. One filter absorbs the elements x-rays, while the other transmits them.

The difference in counts between the two matched detectors with balanced filters is the net intensity and is related to that elements concentration. When combined with earlier work that demonstrated that elements could be measured by measuring total x-ray intensities from some simple samples, a new and powerful method was born. Unfortunately it was almost 50 years later when small microprocessor based analyzers were built in the 1970's that NDXRF started to make a commercial impact.

Hardware

NDXRF has the least expensive hardware of any of the XRF methods, because it only requires a few low costs components. It needs an x-ray source, usually either a radioisotope such as Fe-55, Cd-109, Cm-244, Am-241 of Co-57, or a small x-ray tube. And it requires a detector such as an ionization chamber or Geiger-Mueller Counter, which does not need to be energy dispersive. While Ross used two detectors, the more common approach is to use a single detector and use a filter wheel or tray to position the filters over the detector in sequence. In addition to the Ross method a single filter (Hull Method) or no filter at all may be used to measure some elements.

In commercial devices it is most common to see a proportional counter used as the detector since it is a low resolution EDXRF detector. The advantage of a proportional counter is that it can be configured to not count the backscattered source x-rays making the overall background counts substantially lower. At the same time a proportional counter instrument may be used for EDXRF analysis, making it a hybrid EDX/NDX instrument. With x-ray tube source devices, x-ray tube filters may be used in combination with specially selected target anodes to produce optimal sources for exciting the elements in a sample.

The non-dispersive XRF method is very powerful, and cases where an appropriate filter pair exists and can successful isolate an elements wavelength it is often possible to match the performance of a WDX analyzer at a tenth the cost using 100 times less source intensity.

Example Application

One of the most common applications is measuring phosphorus, sulfur and chlorine in oil. Generally either a Fe-55 radioisotope, or an x-ray tube with either a Pd, Ag, or Ti target is used to excite those elements. By looking at the absorption edges for various materials it is easily seen that chlorine has an absorption edge above chlorine in energy, sulfur has and absorption edge between chlorine and sulfur, and phosphorus has an absorption edge between sulfur and phosphorus. X-rays do not readily excite hydrogen and carbon in the base matrix and the detector windows readily absorb their x-rays, and so they aren't measured.

In this case chlorine can be measured by using chlorine as a transmitting filter and sulfur as an absorbing filter. The difference between the counts of x-rays off the oil sample and through the filter correlates to the chlorine concentration. The filers are electronically balanced by measuring the intensity on a blank and introducing a coefficient that when multiplied by one intensity yields zero net count. Similarly a sulfur and phosphorus pair of filters can be used to measure sulfur. A single filter can be used to measure phosphorus since there are usually no measurable elements below phosphorus that would produce counts.

The matter is confused somewhat because it is difficult to produce good sulfur and phosphorus filters, so usually an element with an L absorption edge at the appropriate energy is used instead, Mo or Nb for S, and Zr for P. Since the heavy metals are denser the filters are usually much thinner than their K absorption edge counterparts.

BASIC THEORY OF X-RAY FLUORESCENCE

X-RAY FLUORESCENCE INTRODUCTION

Although X-ray fluorescence spectroscopy is no longer regarded as a new instrumental technique for elemental analysis, ongoing evolutionary developments continue to redefine the role of this important analytical tool. From the demonstration of the first principles in the 1960's to the development of the first commercial instruments in the 1970's, the increasing availability of affordable computational power has a least been as important to the desirability and acceptance of the technology as innovative hardware design. With the widespread availability and use of a 32-bit microprocessor personal computer as the industry standard platform, X-ray fluorescence spectroscopy has become a useful and complimentary laboratory tool to other techniques.

X-Ray Fluorescence Theory

An electron can be ejected from its atomic orbital by the absorption of a light wave (photon) of sufficient energy. The energy of the photon (hv) must be greater than the energy with which the electron is bound to the nucleus of the atom. When an inner orbital electron is ejected from an atom, an electron from a higher energy level orbital will be transferred to the lower energy level orbital. During this transition a photon maybe emitted from the atom. This fluorescent light is called the characteristic X-ray of the element.

The energy of the emitted photon will be equal to the difference in energies between the two orbitals occupied by the electron making the transition. Because the energy difference between two specific orbital shells, in a given element, is always the same (*i.e.* characteristic of a particular element), the photon emitted when an electron moves between these two levels, will always have the same energy.

Therefore, by determining the energy (wavelength) of the X-ray light (photon) emitted by a particular element, it is possible to determine the identity of that element. For a particular energy (wavelength) of fluorescent light emitted by an element, the number of photons per unit time (generally referred to as peak intensity or count rate) is related to the amount of that analyte in the sample. The counting rates for all detectable elements within a sample are usually calculated by counting, for a set amount of time, the number of photons that are detected for the various analytes' characteristic X-ray energy lines.

It is important to note that these fluorescent lines are actually observed as peaks with a semi-Gaussian distribution because of the imperfect resolution of modern detector technology. Therefore, by determining the energy of the X-ray peaks in a sample's spectrum, and by calculating the count rate of the various elemental peaks, it is possible to qualitatively establish the elemental composition of the samples and to quantitatively measure the concentration of these elements.

WAVELENGTH DISPERSIVE X-RAY FLUORESCENCE

Wavelength dispersive x-ray fluorescence (WDXRF) is the old timer among commercial x-ray spectrometers, since the method works without high-resolution solid-state detectors. Instead, WDXRF instruments rely on diffractive optics to give them high spectral resolution. WDX spectrometers with simple electronic counting circuits were around well before the computer age, and are still the workhouse and leading performer for routine XRF analysis.

Hardware

WDXRF can be relatively simple and inexpensive, or complex and very expensive depending on the number of optical components. WDX instruments use a x-ray tube source to directly excite the sample. Because the overall efficiency of the WDXRF system is low, x-ray tubes in larger systems are normally rated at 1-4 kilowatts. There are some specialized low power systems that operate at 50 to 200 watts. A diffraction device, usually a crystal or multilayer, is positioned to diffract x-rays from the sample toward the detector.

Diffracted wavelengths are those that satisfy the 2dnsin ? relationship, where d is the atomic spacing within the crystal, n is an integer, and theta is the angle between the sample and detector. Other wavelengths are scattered very inefficiently. Collimators are normally used to limit the angular spread of x-rays, to further improve the effective resolution of the WDX system. Because the detector is not relied on for the systems resolution it can be a proportional counter or other low-resolution counter capable of detecting a million or more counts per second.

All the components can be fixed to form a fixed single WDX channel that is ideal for analyzing a single element. A simultaneous WDX analyzer will have a number of fixed single channels usually formed in a circle around the sample with the x-ray tube facing upward in the middle. Other WDX analyzers use a goniometer to allow the angle (?) to be changed, so that one element after another may be measured in sequence. This type of instrument is a sequential WDX analyzer. There are also combined sequential/simultaneous instruments as well.

Applications

WDXRF can be used for a tremendous variety of elemental analysis applications. It can be used to measure virtually every element form Na to Pu in the periodic table, and some instruments can be used for quantitative or semi-quantitative work for even lighter elements. It can measure elemental concentrations ranging from a few ppm to nearly 100 percent. It can be used for monitoring major components in a product or process or the addition of minor additives. WDXRF is extremely popular in the geological field and is often used for measuring raw minerals, and finished products composed of minerals.

DEVELOPMENT OF THE COMMERCIAL X-RAY SPECTROMETERS

Four types of spectrometers are available to the analyst. From the 1950's to 1960's nearly all the x-ray spectrometers were wavelength dispersive spectrometers. In a wavelength dispersive spectrometer, a crystal separates the wavelengths of the fluorescence from the sample, similar to grating spectrometers for visible light. The other x-ray spectrometer available at that time was the electron microprobe, which uses a focused electron beam to excite x-rays in a solid sample as small as 10^{-12} cm^3.

The first microprobe was built by R. Castaing in 1951 and became commercially available in 1958. In the early 1970's, energy dispersive spectrometers became available, which use Li-drifted silicon or germanium detectors. The advantage these instruments brought was the ability to measure the entire spectrum simultaneously. With the help of computers, deconvolution methods can be performed to extract the net intensities of individual x-rays.

SAMPLE PREPARATION PAGE

Sample preparation is by far the most important step in any analytical technique. How well the sample is prepared and presented will affect the the ability to yield accurate data from the instrument. X-ray fluorescence is not immune to this crucial step, despite the ability to correct for standard counting errors, instrument variation errors, operatoration errors, and matrix effects. Choosing the right sample preparation technique will depend on the goals of accuracy and precision the analyst needs or wishes to attain.

Specimen preparation is crucial to the relationship between spectral line intensity and the element concentration. Factors such as surface roughness, particle shape, particle size, homogeneity, particle distribution, and mineralization can affect this relationship.

PLASTIC SAMPLES

Plastic samples present some special challenges since they come in numerous forms, powders, pellets, fibers, sheets, and variously shaped solids, and cannot be prepared by most of the usual methods.

Plastic Powders

Plastic powders like other powders can be presented in a sample cup either loose or manually pressed. Plastics are resistant to hydraulic pressing, unless heated to 100 to 150 °C before pressing or pressed using heated platens. Manually pressed samples are recommended for most applications.

Plastic Pellets, Beads, or Regrind

Plastic Pellets can also be analyzed loose in a sample cup as long as the pellets are fairly uniform in shape. Using a sample spinner helps with the analysis. It is also possible to heat pellets to 100 to 150 °C and press them into a pellet cap, or press them by normal methods. For most applications loose pellets presented in a spinning sample cup will give good results. Coarse regrind material may also be analyzed in this manner.

Plastic Fibers

Fibers are particularly tricky to work with. They can be coiled or balled up in a cup, but the reproducibility is generally not very good. It helps somewhat to insert a weight in the cup on top of the sample. An alternative approach that often works better is to wind the fibers on a spool. If the wound spool is thick and fairly uniform the results may be satisfactory. Ultimately most fibers must be ground to a powder using a freezer mill or other suitable plastic grinding apparatus and presented in powder form.

Plastic Films, Sheets and Fabric

Plastic films, sheets and fabric, can often be analyzed simply by cutting them to size and placing them in the analysis position. Several layers of thin films and sheets can measured to increase the net counts for elements with higher energy x-rays. It is also useful to weight some types of films or fabrics with a flat piece of solid metal or a metal ring, provided a metal is selected that doesn't interfere with the analysis. It is important that the part of the sample in the analysis region be uniformly flat and smooth, or else there will be reproducibility errors.

Solid Plastic Samples

Solid plastic parts usually make excellent samples. It is only important to remember that the surface must be uniformly flat and smooth in the analysis area. Many plastics shrink during curing plastic pieces giving them a concave surface. As long as all the parts have the same shape and can be position reproducibly, it may still be possible to analyze them without flattening the surface in some way. It is important to note that the composition of the surface skin is sometimes slightly different than in the interior of the part, since the molding process forces some additives to the surface.

To measure the average interior composition it may be necessary to grind or sand off the surface layer. It is then important to make sure that the surface finish is repeatable

SAMPLE PREPARATION OF LOOSE POWDERS PAGE

Loose Powders

The simplest approach to analyzing a loose powder sample is to simply fill a sample cup approximately ¾ full and analyze it without any additional preparation, or simply tapping the cup on a clean surface to pack it to a more consistent density.

This method is satisfactory in same cases where the reproducibility requirements are not very strict. The greatest advantage with this method is that it is easy. Spinning the sample may improve the consistency of the measurement by averaging over a larger area.

There are many shortcomings with this technique; the bulk density may not be consistent, grain size variation cause the readings to vary and finer grains can be forced to the surface during tap packing. Because of these problems this method works best with homogenous material that has been dried and ground to a uniform grain size. Of course, these recommendations apply to every sample presentation method.

Manually Pressed Powders

Several venders offer a sample press, a modified arbor press, that is designed to press a sample within a sample cup. The sample cup is first filled ¾ full or more and then positioned under the press and compacted to 15 to 30 Newton-meters. Higher pressures can be achieved depending on the press. The cup must be held down firmly to prevent the film from bulging, and it is usually a good idea to place a Kimwipe or other clean disposable material under the cup.

Place another Kimwipe over the cup before pressing if the sample material is likely to stick to the press. If the material compresses a lot it may be necessary to refill and press 2 or 3 times to achieve a reasonable sample depth. Manually pressed samples can still be prepared quickly since pressing generally takes less than a minute.

Because tap packing is unnecessary this method eliminates the problem with finer particles settling. Ultimately, by giving the samples more uniform bulk density it is possible to achieve analytical reproducibility that is often no more than 10 per cent worse than with a hydraulically pressed sample.

Hydraulically Pressed Powders

The preferred method for analyzing powders or samples that are usually ground to a powder to make them more homogeneous is the hydraulic pressed pellet. Several manufacturers make hydraulic presses that are capable of pressures ranging from 10 to 50 tons or more. The press uses a die set to contain and form the sample during pressing. There are a few types of in die sets and a lot of variations to the methods incorporating a lot of individual creativity, but one basic outline follows. The sample is first dried and ground to a fine consistency, 400 mesh or better is recommended. Remember that the x-ray wavelengths are still substantially smaller than the particles. Next the sample is usually blended with a binder that helps hold the pellet together, although some materials don't require it. A ring and puck mill or mixer mill is very useful for both the grinding and blending steps.

To prepare the press, the die set must first be cleaned with methanol or other solvent. The backing, usually an aluminum cap, is inserted into the die. A specific weight of sample is then poured into the cap, usually 5-10 grams. It is important to keep the mass constant because the sample may not be infinitely thick at high x-ray energies. Next a polished pellet is placed over the sample to produce a smooth finish. The plunger goes in after that, and then the die set is positioned in the press.

Follow the press instructions for pressing the sample to 10-20 tons, and holding it for a period of time usually from 10-100 seconds. Different materials produce better pellets at different pressures, so finding the best pressure may require some experimentation. Once the best pressure is determined however all samples of that type must be compacted to the same pressure and hold time to achieve optimal analytical results. The pellet is then removed from the die set, taking care not to crack it in the process. The finished pellet is uniform in composition, density, and mass per unit area, and has a smooth finish. These are all highly desirable traits for an XRF sample, so many operators will spend the needed five to ten minutes it takes to prepare the sample. The only remaining disadvantage with peletized samples is that there are still matrix affects due to the grain size being larger than the x-rays.

SAMPLE PREPARATION OF LIQUIDS PAGE

Liquid samples can be the easiest or most difficult samples to work with depending on the composition and stability. A sample cup is filled about ¾ full and then presented to the analyzer. The problems with liquid samples are that they evaporate, stratify, and precipitate. The liquid may attack or be absorbed by the window film, wick up and out of the cup.

Because of these issues liquid samples should be freshly prepared, preferably immediately prior to analysis, although some liquids are stable for a day or more. Solutions should be well mixed prior to pipetting them into a sample cup. The sample should be taken from the center of the container since some components may concentrate on the walls. There are some techniques that can be used to overcome most of the other problems.

Issues Associated with Use of Liquid Samples

Evaporation

All liquids have some vapor pressure at standard pressure and temperature. Some low vapor materials such as water or mineral oil may be stable for a day or more, while gasoline and other volatile liquids may only be stable for a few seconds. Highly volatile samples should be prepared and analyzed one at a time and the time from pouring to starting the analysis should be consistent within a few seconds.

A cover or cap may be placed over the cup to reduce evaporation, but the film will bulge causing poor reproducibility. Single end cups and caps for double-ended cups can be punctured with a pin, and some have a snap off device that leaves a hole. Window film covers can also be punctured to relieve pressure. Most cup manufactures also make baffled cups that are designed to minimize evaporation and in theory the sample.

Stratification

Liquids stratify when they either contain two or more immiscible liquids like oil and water, or if they contain several molecules that have different density such as a crude oil. Various organo-metalics are also higher in density and tend to settle out over a long period of time. Most liquids need only be well mixed prior to pipetting it into a cup, and they will be stable during analysis. Others will stratify during analysis and must be analyzed quickly.

Immiscible liquids are a special challenge. Some times it is possible to produce a meta-stable emulsion. Others times it may be necessary to analyze the two components separately. Lastly solidification techniques can be attempted.

Precipitates

As with stratification, liquid samples that precipitate may be well mixed and analyzed one at a time if the particles stay suspended long enough. Normally the supernatant liquid is pipetted off the top and measured separately from the precipitate, which should be dried, weighed, and measured as a powder. Solidification techniques have also been used successfully with rapidly precipitating samples.

Wicking

Wicking happens when the liquid is hydrophilic with respect to the window film. The liquid is drawn along the film, out, up, and over the bottom retaining ring. This can also happen even with a hydrophobic liquid because of gravity, when the liquid level is above the height of the bottom retaining ring. One cup vender attacks this problem by having a taller bottom ring, while others use collars that slide up the outside of the cup holding the window film above the sample height. Another technique that works in a pinch is to use tape around the outside of the cup to hold the window film up.

Alternative Liquid Sample Preparation Methods

Since many liquid samples are inherently unstable there are a number of alternative methods for stabilizing the samples. There are also a number of pre-concentration techniques that are available that have been used with some success.

Liquid Sample Solidification

Solidification is one useful method when the sample contains immiscible liquids such as oil and water, or the it precipitates rapidly. Several materials including, cellulose, carbon, gelatin, and alumina, have been used successfully as solidifying agents. Lower atomic number solidification agents are preferred when analyzing low Z elements. The ideal mixing ratio must be determined experimentally.

Thin Film Preparations

Another popular method to attempt, is to deposit a sample on a thin sample support and then measure it either wet or after drying. By reducing the sample thickness it is possible to reduce the background while keeping the elemental net intensities high, thus improving the detection limits. Drying the sample helps even more since much of the background due to backscatter is due to the hydrogen and other low atomic elements in the base matrix.

Common support materials include filter paper and IR cards. A few, typically 5-50, microliters of sample are be pipetted onto the support, and allowed to soak in and distribute. The method has also been done using an atomizer to spray the sample. It can be presented wet provided the moisture content and distribution is consistent from sample to sample or else it can be dried. The primary problem with these techniques is that they are often not reproducible enough for routine laboratory work.

Sample Concentration

There are several concentration/thin films techniques that have been used for XRF analysis. The simplest involves filling up a sample cup and dying it. This works best when the matrix is a highly volatile solvent. If the elements of interest are particulates in suspension then a filter can be used to filter a large volume of fluid. Then the filter can be analyzed wet or dried. One instrument vender uses this method to analyze trace metals such as iron in nuclear reactor coolant.

Another method involves the use of ion exchange filters or resins. A large amount of fluid can be moved through an ion exchange medium removing the ions of interest. The medium can be measured to determine concentrations of the elements of interests. Filters can be analyzed wet or dry. It is important to select an ion exchange medium that does not interfere with the XRF analysis.

11

Speed of Light Propagation

INTRODUCTION TO PROPAGATION OF LIGHT

Optics deals with the propagation of light through transparent media, and its interaction with mirrors, lenses, slits, *etc*. Optical effects can be divided into two broad classes. Firstly, those which can be explained without reference to the fact that light is fundamentally a wave phenomenon, and, secondly, those which can only be explained on the basis that light is a wave phenomenon.

Let us, for the moment, consider the former class of effects. It might seem somewhat surprising that any optical effects at all can be accounted for without reference to waves. Light really is a wave phenomenon. It turns out, however, that wave effects are only crucially important when the wavelength of the wave is either comparable to, or much larger than, the size of the objects with which it interacts.

When the wavelength of the wave becomes much smaller than the size of the objects with which it interacts then the interactions can be accounted for in a very simple geometric manner. Since the wavelength of visible light is only of order a micron, it is very easy to find situations in which its wavelength is very much smaller than the size of the objects with which it interacts.

Thus, "wave-less" optics, which is usually called geometric optics, has a very wide range of applications. In geometric optics, light is treated as a set of *rays,* emanating from a source, which propagate through transparent media according to a set of *three* simple laws. The first law is the *law of rectilinear propagation,* which states that light rays propagating through a homogeneous transparent medium do so in straight-lines.

The second law is the *law of reflection,* which governs the interaction of light rays with conducting surfaces (*e.g.,* metallic mirrors). The third law is the *law of refraction,* which governs the behaviour of light rays as they traverse a sharp boundary between two different transparent media (*e.g.,* air and glass).

GEOMETRIC OPTICS

Let us first consider the law of *rectilinear propagation*. The earliest surviving optical treatise, Euclid's *Catoptrics*, recognized that light travels in straight-lines in homogeneous media. However, following the teachings of Plato, Euclid thought that light rays emanate from the eye, and intercept external objects, which are thereby "seen" by the observer. The ancient Greeks also thought that the speed with which light rays emerge from the eye is very high, if not infinite. After all, they argued, an observer with his eyes closed can open them and immediately see the distant stars.

Hero of Alexandria, in his *Catoptrics* (first century BC), also maintained that light travels with infinite speed. His argument was by analogy with the free fall of objects. If we throw an object horizontally with a relatively small velocity then it manifestly does not move in a straight-line. However, if we throw an object horizontally with a relatively large velocity then it appears to move in a straight-line to begin with, but eventually deviates from this path. The larger the velocity with which the object is thrown, the longer the initial period of apparent rectilinear motion.

Hero reasoned that if an object were thrown with an infinite velocity then it would move in a straight-line forever. Thus, light, which travels in a straight-line, must move with an infinite velocity. The erroneous idea that light travels with an *infinite* velocity persisted until 1676, when the Danish astronomer Olaf Römer demonstrated that light must have a *finite* velocity, using his timings of the successive eclipses of the satellites of Jupiter, as they passed into the shadow of the planet. Ptolemy formulated a very inaccurate version of the law of refraction, which only works when the light rays are almost normally incident on the interface in question. Despite its obvious inaccuracy, Ptolemy's theory of refraction persisted for nearly 1500 years. The true law of refraction was discovered empirically by the Dutch mathematician Willebrord Snell in 1621. However, the French philosopher René Descartes was the first to publish, in his *La Dioptrique* the now familiar formulation of the law of refraction in terms of sines. Although there was much controversy at the time regarding plagiarism, Descartes was apparently unaware of Snell's work. Thus, in English speaking countries the law of refraction is called "Snell's law", but in French speaking countries it is called "Descartes' law".

In 1658, the French mathematician Pierre de Fermat demonstrated that all three of the laws of geometric optics can be accounted for on the assumption that light always travels between two points on the path which takes the *least time* (or, more rigorously, the extremal time). Fermat's ideas were an extension of those of Hero of Alexandria. Fermat's (correct) derivation of the law of refraction depended crucially on his (correct) assumption that light travels *more slowly* in dense media than it does in air. Unfortunately, many famous scientists, including Newton, maintained that light travels *faster* in dense media than it does in air.

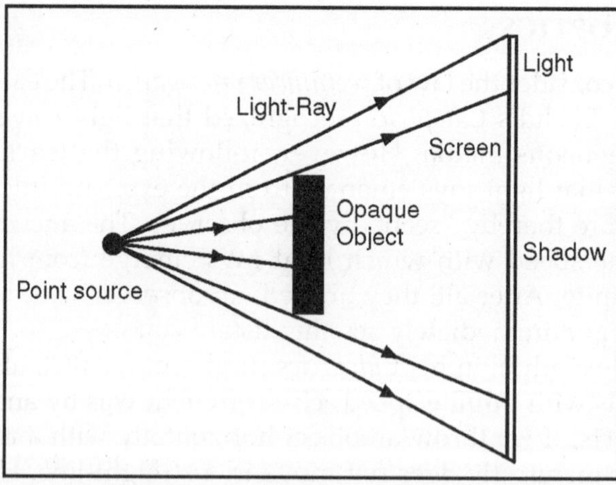

Fig. An Opaque Object Illuminated by a Point Light Source.

This erroneous idea held up progress in optics for over one hundred years, and was not conclusively disproved until the mid-nineteenth century. Incidentally, Fermat's principle of least time can only be justified using wave theory.

PROPERTIES OF LIGHT

In one phenomena, which are bound to distribution of light, its undular properties (interference, diffraction) are shown. In other phenomena its quantum properties (photoeffect, radiation and uptake of a light, process of interaction of a light with substance) are shown. The light has simultaneously undular and quantum properties, which are differently shown in the different phenomena. However, when the wave length of a light is diminished, its undular properties are attenuated and the quantum properties are shown.

COMPTON'S EFFECT

The experience of Arthur Holly Compton on a dispersion of Roentgen rays by substances, which consist of atoms of light devices, have shown, that the dispelled Roentgen rays have a major wave length λ', than impinging λ. Thus the residual $\Delta\lambda = \lambda' - \lambda$ depends on properties of scattering substance and wave length of an impinging light:

$$\Delta\lambda = \lambda' - \lambda = 2\lambda_K \sin^2 \frac{\theta}{2}$$

The quantity λ_K is a stationary value for all substances also is termed as Compton's wave length, $\lambda_K = 2{,}43 \times 10^{12}$ m. This phenomenon, which is termed as Compton's recoil, is impossible to explain by a wave theory of light. In quantum theory the magnification of wave length of radiation at a dispersion is explained by interaction of photons of impinging radiation with electrons of atoms and molecules. The part of energy of an impinging photon, which is transmitted to electrons of scattering substance, depends on a scattering angle of photon. At $\theta = \pi$, that is provided that after a dispersion a photon will fly in the party, which is opposite to a tentative direction, the losses by a photon of energy will be maximal, that will give in maximal magnification of a wave length of radiation.

Light Pressure

Some properties, for example light pressure, can be surveyed simultaneously in the electromagnetic theory and in the quantum theory. According to the electromagnetic theory of light, the pressure p, which renders a luminous flux with power W, impinging normally on a simple surface, is equal:

$$p = \frac{W}{c}(1+R)$$

For an ideal black body ($R = 0$): $p = \frac{W}{c}$

For an absolute reflecting bodyh ($R = 1$): $p = \frac{2W}{c}$

If N of quantums of a luminous flux with frequency ν, which bears (carries) energy W = $N \cdot h\nu$, impinges on a simple surface for 1 second, then the blanket impulse at a reflectivity R will look like:

$$p = \frac{N \bullet h\nu}{c}(1+R) = \frac{W}{c}(1+R)$$

Thus, electromagnetic theory and quantum theory give identical results. However it is essential different representations will be utillized in the electromagnetic and quantum theories of light. The concept of an electromagnetic wave of application to a light guesses existence of a wave in particular field(area) of space, which is capable to contain even some lengths of waves of light.

The light quantums are considered as a particle, which are localized in very small volume of space. The representation about a light wave guesses a continuous distribution of energy in a spatial wave, the representation about quantums guesses localization of energy in a small part of space.

Both these incompatible representations reflect unity of a wave-corpuscle nature of light, which is shown at distribution of light and its interaction with substance.

LAW OF GEOMETRIC PROPAGATION

According to geometric optics, an opaque object illuminated by a point source of light casts a sharp shadow whose dimensions can be calculated using geometry. The method of calculation is very straightforward. The source emits *light-rays* uniformly in all directions. These rays can be represented as straight lines radiating from the source. The light-rays propagate away from the source until they encounter an opaque object, at which point they stop.

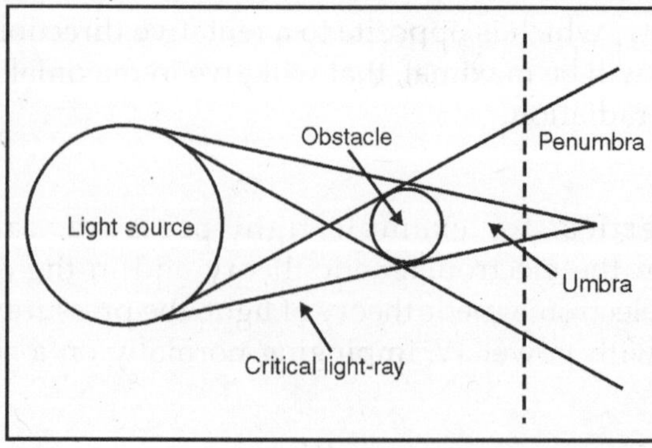

Fig. An Opaque Object Illuminated by an Extended Light Source.

For an extended light source, each element of the source emits light-rays, just like a point source. Rays emanating from different elements of the source are assumed not to interfere with one another. The shadow cast by an opaque sphere illuminated by a spherical light source is calculated using a small number of critical light-rays. The shadow consists of a perfectly black disk called the *umbra,* surrounded by a ring of gradually diminishing darkness called the *penumbra.*

In the umbra, *all* of the light-rays emitted by the source are blocked by the opaque sphere, whereas in the penumbra only *some* of the rays emitted by the source are blocked by the sphere. As was well-known to the ancient Greeks, if the light-source represents the Sun, and the opaque sphere the Moon, then at a point on the Earth's surface which is situated inside the umbra the Sun is totally eclipsed, whereas at a point on the Earth's surface which is situated in the penumbra the Sun is only partially eclipsed.

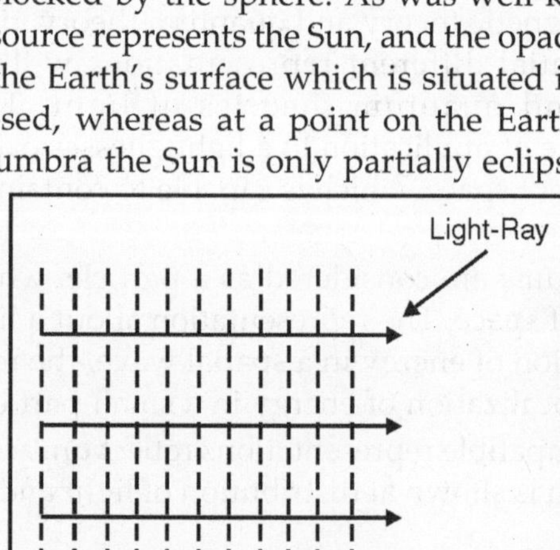

Fig. Relationship between Wave-Fronts and Light-Rays.

In the wave picture of light, a *wave-front* is defined as a surface joining all adjacent points on a wave that have the same phase (*e.g.*, all maxima, or minima, of the electric field). A light-ray is simply a line which runs perpendicular to the wave-fronts at all points along the path of the wave. Thus, the law of rectilinear propagation of light-rays also specifies how wave-fronts propagate through homogeneous media. Of course, this law is only valid in the limit where the wavelength of the wave is much smaller than the dimensions of any obstacles which it encounters.

REFLECTION AND REFRACTION

Most of what we need to know about geometrical optics can be summarized in two rules, the laws of reflection and refraction. These rules may both be inferred by considering what happens when a plane wave segment impinges on a flat surface. If the surface is polished metal, the wave is *reflected*, whereas if the surface is an interface between two transparent media with differing indices of refraction, the wave is partially reflected and partially *refracted*.

Reflection means that the wave is turned back into the half-space from which it came, while refraction means that it passes through the interface, acquiring a different direction of motion from that which it had before reaching the interface.

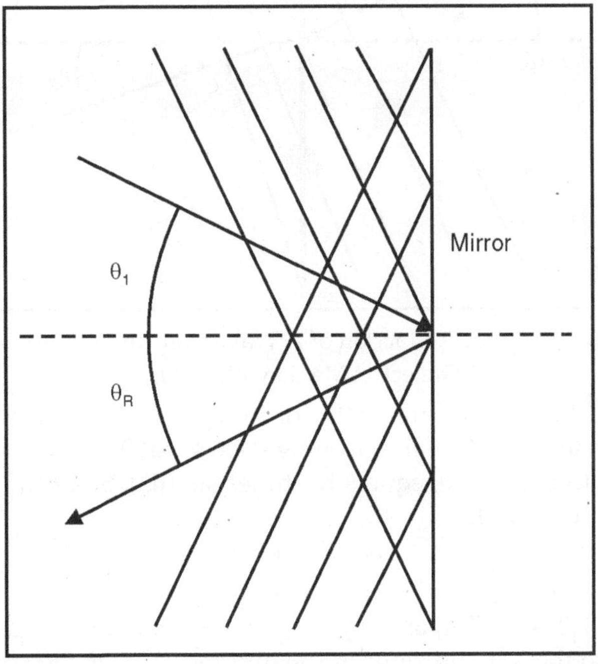

Fig. Sketch Showing the Reflection of a Wave from A Plane Mirror. The Law of Reflection States that $\theta_I = \theta_R$.

The wave vector and wave front of a wave being reflected from a plane mirror. The angles of incidence, θ_I, and reflection, θ_R, are defined to be the angles between the incoming and outgoing wave vectors respectively and the line normal to the mirror. The law of reflection states that $\theta_R = \theta_I$. This is a consequence of the need for the incoming and outgoing wave fronts to be in phase with each other all along the mirror surface. This plus the equality of the incoming and outgoing wavelengths is sufficient to insure the above result. Since $n_R > n_I$, the speed of light in the right-hand medium is less than in the left-hand medium. Recall that the speed of light in a medium with refractive index n is

$$c_{\text{medium}} = c_{vac}/n.$$

The frequency of the wave packet doesn't change as it passes through the interface, so the wavelength of the light on the right side is less than the wavelength on the left side. The side AC is equal to the side BC times $\sin(\theta_I)$.

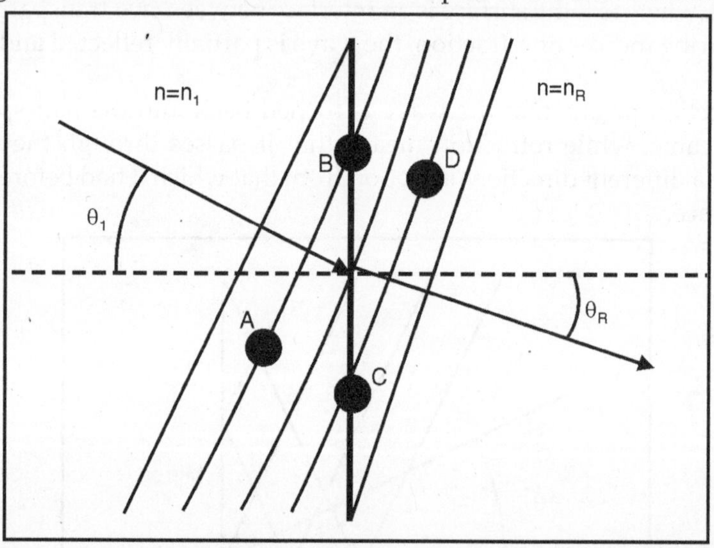

Fig. Sketch Showing the Refraction of A Wave from an Interface between two Dielectric Media with $n_2 > n_1$.

However, AC is also equal to $2\lambda_I$, or twice the wavelength of the wave to the left of the interface. Similar reasoning shows that $2l_R$, twice the wavelength to the right of the interface, equals BC times sin (θ_R). Since the interval BC is common to both triangles

$$\frac{\lambda_I}{\lambda_R} = \frac{\sin(\theta_I)}{\sin(\theta_R)}.$$

Since $\lambda_I = c_I T = c_{vac} T/n_I$ and $\lambda_R = C_R T = c_{vac} T/n_R$ where C_I and C_R are the wave speeds to the left and right of the interface, C_{vac} is the speed of light in a vacuum, and T is the (common) period, we can easily recast the above equation in the form

$$n_I \sin(\theta_I) = n_R \sin(\theta_R).$$

This is called *Snell's law,* and it governs how a ray of light bends as it passes through a discontinuity in the index of refraction. The angle θ_I is called the incident angle and θ_R is called the refracted angle. Notice that these angles are measured from the normal to the surface, not the tangent.

PHENOMENON OF RAINBOWS

The most well-known, naturally occurring phenomenon which involves the dispersion of light is a *rainbow*. A rainbow is an *arc* of light, with an angular radius of 42°, centred on a direction which is *opposite* to that of the Sun in the sky (*i.e.*, it is centred on the direction of propagation of the Sun's rays). Thus, if the Sun is low in the sky (*i.e.*, close to the horizon) we see almost a full semi-circle. If the Sun is higher in the sky we see a smaller arc, and if the Sun is more than 42° above the horizon then there is no rainbow (for viewers on the Earth's surface).

Observers on a hill may see parts of the rainbow below the horizontal: *i.e.*, an arc greater than a semi-circle. Passengers on an airplane can sometimes see a full circle.

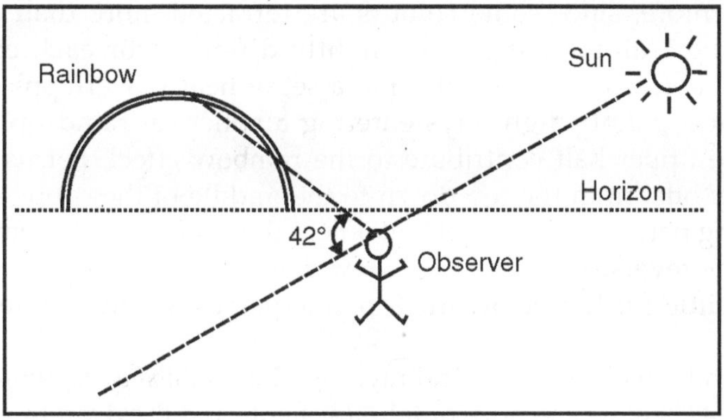

Fig. A Rainbow

The colours of a rainbow vary smoothly from red on the outside of the arc to violet on the inside. A rainbow has a diffuse inner edge, and a sharp outer edge. Sometimes a *secondary arc* is observed. This is fainter and larger (with an angular radius of 50°) than the primary arc, and the order of the colours is reversed (*i.e.*, red is on the inside, and violet on the outside).

The secondary arc has a diffuse outer edge, and a sharp inner edge. The sky between the two arcs sometimes appears to be less bright than the sky elsewhere. This region is called *Alexander's dark band,* in honour of Alexander of Aphrodisias who described it some 1800 years ago. Rainbows have been studied since ancient times. Aristotle wrote extensively on rainbows in his *De Meteorologica,* and even speculated that a rainbow is caused by the reflection of sunlight from the drops of water in a cloud.

The first scientific study of rainbows was performed by Theodoric, professor of theology at Freiburg, in the fourteenth century. He studied the path of a light-ray through a spherical globe of water in his laboratory, and suggested that the globe be thought of as a model of a single falling raindrop. A ray, from the Sun, entering the drop, is refracted at the air-water interface, undergoes internal reflection from the inside surface of the drop, and then leaves the drop in a backward direction, after being again refracted at the surface.

Thus, looking away from the Sun, towards a cloud of raindrops, one sees an enhancement of light due to these rays. Theodoric did not explain why this enhancement is concentrated at a particular angle from the direction of the Sun's rays, or why the light is split into different colours. The first person to give a full explanation of how a rainbow is formed was René Descartes. He showed mathematically that if one traces the path through a spherical raindrop of parallel light-rays entering the drop at different points on its surface, each emerges in a different direction, but there is a concentration of emerging rays at an angle of 42° from the reverse direction to the incident rays, in exact agreement with the observed angular size of rainbows.

Furthermore, since some colours are refracted more than others in a raindrop, the "rainbow angle" is slightly different for each colour, so a raindrop disperses the Sun's light into a set of nearly overlapping coloured arcs. It shows parallel light-rays entering a spherical raindrop. Only rays entering the upper half contribute to the rainbow effect. Let us follow the rays, one by one, from the top down to the middle of the drop. We observe the following pattern. Rays which enter near the top of the drop emerge going in almost the reverse direction, but a few degrees below the horizontal. Rays entering a little further below the top emerge at a greater angle below the horizontal.

Eventually, we reach a critical ray, called the *rainbow ray*, which emerges in an angle 42° below the horizontal. Rays entering the drop lower than the rainbow ray emerge at an angle less than 42°. Thus, the rainbow ray is the one which deviates *most* from the reverse direction to the incident rays. This variation, with 42°being the maximum angle of deviation from the reverse direction, leads to a bunching of rays at that angle, and, hence, to an unusually bright arc of reflected light centred around 42° from the reverse direction. The arc has a sharp outer edge, since reflected light *cannot* deviate by more than 42° from the reverse direction, and a diffuse inner edge, since light *can* deviate by less than 42° from the reverse direction: 42° is just the *most likely* angle of deviation.

Finally, since the rainbow angle varies slightly with wavelength (because the refractive index of water varies slightly with wavelength), the arcs corresponding to each colour appear at slightly different angles relative to the reverse direction to the incident rays. We expect violet light to be refracted more strongly than red light in a raindrop. The red arc deviates slightly more from the reverse direction to the incident rays than the violet arc. In other words, violet is concentrated on the inside of the rainbow, and red is concentrated on the outside.

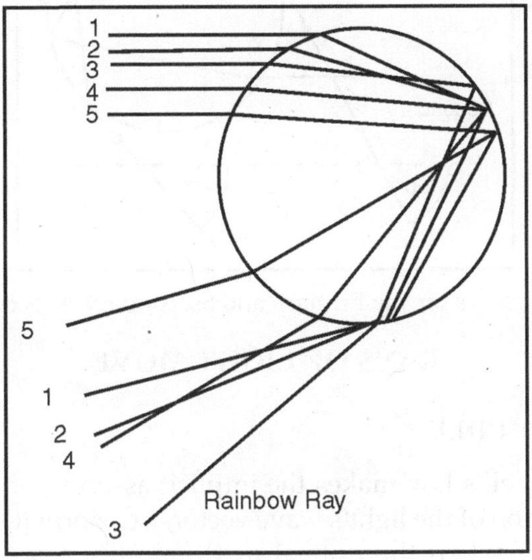

Fig. Descarte's Theory of the Rainbow.

Descartes was also able to show that light-rays which are internally reflected *twice* inside a raindrop emerge concentrated at an angle of 50° from the reverse direction to the incident rays. Of course, this angle corresponds exactly to the angular size of the secondary rainbow sometimes seen outside the first.

This rainbow is naturally less intense than the primary rainbow, since a light-ray loses some of its intensity at each reflection or refraction event. Note that 50°represents the angle of *maximum* deviation of doubly reflected light from the reverse direction (*i.e.*, doubly reflected light can deviate by more than this angle, but not by less). Thus, we expect the secondary rainbow to have a diffuse outer edge, and a sharp inner edge. We also expect doubly reflected violet light to be refracted more strongly in a raindrop than doubly reflected red light.

The red secondary arc deviates slightly less from the reverse direction to the incident rays than the violet secondary arc. In other words, red is concentrated on the inside of the secondary rainbow, and violet on the outside. Since no reflected light emerges between the primary and secondary rainbows (*i.e.*, in the angular range 42° to 50°, relative to the reverse direction), we naturally expect this region of the sky to look slightly less bright than the other surrounding regions of the sky, which explains Alexander's dark band.

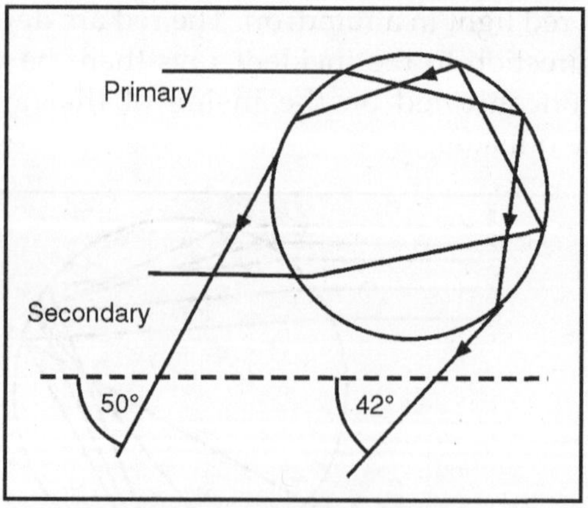

Fig. Rainbow Rays for the Primary and Ssecondary Arcs of a Rainbow.

RAYS OF LIGHT MOVE

ANISOTROPIC MEDIA

Notice that Snell's law makes the implicit assumption that rays of light move in the direction of the light's wave vector, i. e., normal to the wave fronts. This is valid only when the optical medium is isotropic, i. e., the wave frequency depends only on the magnitude of the wave vector, not on its direction.

Certain kinds of crystals, such as those made of calcite, are not isotropic — the speed of light in such crystals, and hence the wave frequency, depends on the orientation of the wave vector. As an example, the angular frequency in an isotropic medium might take the form

$$\omega = \left[\frac{c_1^2 \left(k_x + k_y\right)^2}{2} + \frac{c_2^2 \left(c_x - k_y\right)^2}{2} \right]^{1/2},$$

where c_1 is the speed of light for waves in which $k_y = k_x$, and c_2 is its speed when $k_y = k_x$.

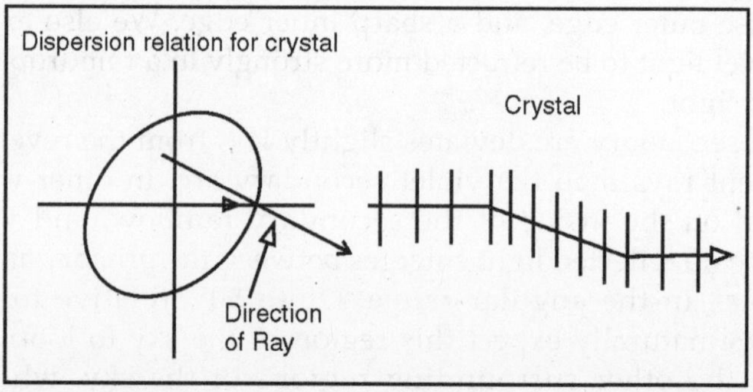

Fig. The Right Panel Shows the Fate of a Light Ray Normally Incident on the Face of a Properly Cut Calcite Crystal.

The Anisotropic Dispersion Relation Which Gives Rise to this Behaviour Is Shown in the Left Panel. An example in which a ray hits a calcite crystal oriented so that constant frequency contours are as specified in equation. The wave vector is oriented normal to the surface of the crystal, so that wave fronts are parallel to this surface. Upon entering the crystal, the wave front orientation must stay the same to preserve phase continuity at the surface. However, due to the anisotropy of the dispersion relation for light in the crystal, the ray direction changes in the right panel. This behaviour is clearly inconsistent with the usual version of Snell's law! It is possible to extend Snell's law to the anisotropic case. However, we will not present this here. The following discussions of optical instruments will always assume that isotropic optical media are used.

THIN LENS FORMULA AND OPTICAL INSTRUMENTS

Given the laws of reflection and refraction, one can see in principle how the passage of light through an optical instrument could be traced. For each of a number of initial rays, the change in the direction of the ray at each mirror surface or refractive index interface can be calculated. Between these points, the ray traces out a straight line.

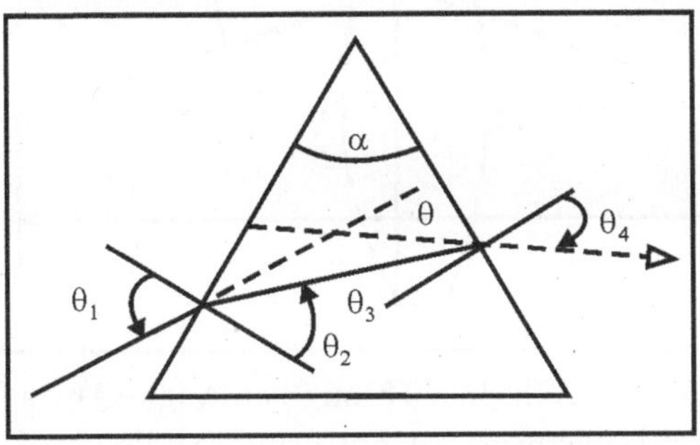

Fig. Bending of a Ray of Light as it Passes Through a Prism.

Though simple in conception, this procedure can be quite complex in practice. However, the procedure simplifies if a number of approximations, collectively called the *thin lens approximation*, are valid. We begin with the calculation of the bending of a ray of light as it passes through a prism. The pieces of information needed to find θ, the angle through which the ray is deflected are as follows: The geometry of the triangle defined by the entry and exit points of the ray and the upper vertex of the prism leads to

$$\alpha + (\pi/2 - \theta_2) + (\pi/2 - \theta_3) = \pi,$$

which simplifies to

$$\alpha = \theta_2 + \theta_3.$$

Snell's law at the entrance and exit points of the ray tell us that

$$n = \frac{\sin(\theta_1)}{\sin(\theta_2)} \quad n = \frac{\sin(\theta_4)}{\sin(\theta_3)},$$

where n is the index of refraction of the prism. One can also infer that

$$\theta = \theta_1 + \theta_4 - \alpha.$$

This comes from the fact that the the sum of the internal angles of the shaded quadrangle is,

$$(\pi/2 - \theta_1) + \alpha + (\pi/2 - \theta_4) + (\pi/2 - \theta_4) + (\pi + \theta) = 2\pi.$$

Combining equations allows the ray deflection θ to be determined in terms of θ_1 and α, but the resulting expression is very messy. However, great simplification occurs if the following conditions are met:

- The angle a << 1.
- The angles $\theta_1 \ll 1$ and $\theta_4 \ll 1$.

With these approximations it is easy to show that

$$\theta = \alpha(n-1) \text{ (small angles).}$$

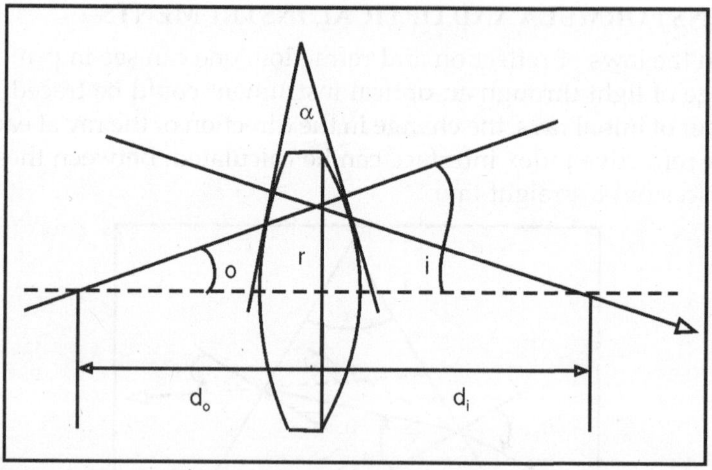

Fig. Light Ray Undergoing Deflection Through an Angle Q by a Lens. The Angle a is the Angle between the Tangents to the Entry and Exit Points of the Ray on the Lens.

Generally speaking, lenses and mirrors in optical instruments have curved rather than flat surfaces. However, we can still use the laws for reflection and refraction by plane surfaces as long as the segment of the surface on which the wave packet impinges is not curved very much on the scale of the wave packet dimensions. This condition is easy to satisfy with light impinging on ordinary optical instruments. In this case, the deflection of a ray of light is given by equation if α is defined as the intersection of the tangent lines to the entry and exit points of the ray.

A *positive lens* is thicker in the center than at the edges. The angle α between the tangent lines to the two surfaces of the lens at a distance τ from the central axis takes the form $\alpha = C_\tau$, where C is a constant. The deflection angle of a beam hitting the lens a distance τ from the center is therefore

$$\theta = C\tau\,(n-1).$$

The angles o and i sum to the deflection angle:

$$o + i = \theta = C\tau\,(n-1).$$

However, to the extent that the small angle approximation holds,

$$o = \tau/d_o \text{ and } i = \tau/d_i$$

where d_o is the distance to the object and d_i is the distance to the image of the object. Putting these equations together and cancelling the τ results in the *thin lens formula*:

$$\frac{1}{d_o} + \frac{1}{d_i} = C(n-1) \equiv \frac{1}{f}.$$

The quantity f is called the *focal length* of the lens. Notice that $f = d_i$ if the object is very far from the lens, i. e., if d_o is extremely large.

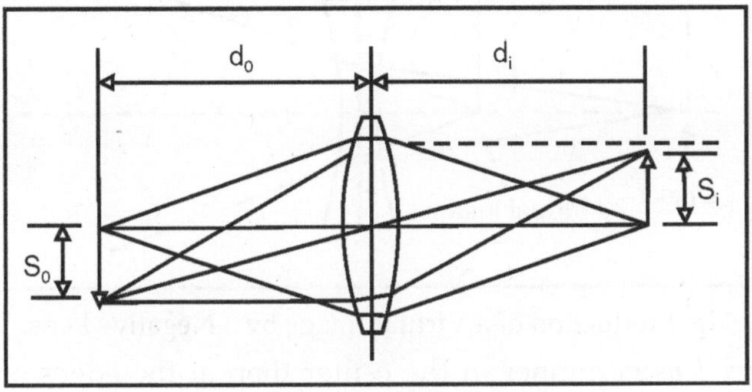

Fig. A Positive Lens Producing an Image on the Right of the Arrow on the Left.

The image is produced by all of the light from each point on the object falling on a corresponding point in the image. If the arrow on the left is an illuminated object, an *image* of the arrow will appear at right if the light coming from the lens is allowed to fall on a piece of paper or a ground glass screen. The size of the object S_o and the size of the image S_i are related by simple geometry to the distances of the object and the image from the lens:

$$\frac{S_i}{S_o} = \frac{d_i}{d_o}.$$

Notice that a positive lens inverts the image.

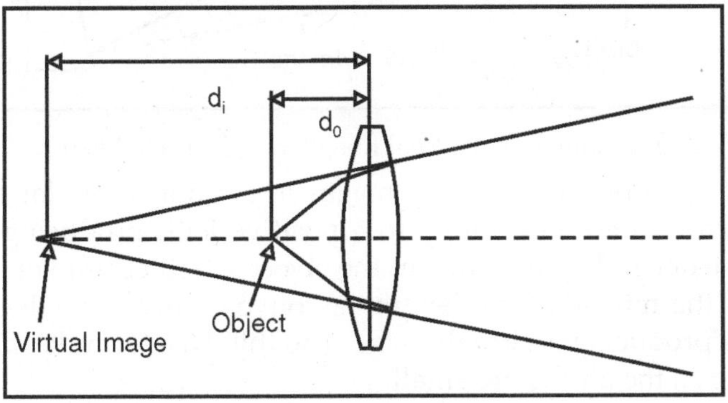

Fig. Production of a Virtual Image by a Positive Lens.

An image will be produced to the right of the lens only if $d_o > f$. If $d_o < f$, the lens is unable to converge the rays from the image to a point. However, in this case the backward extension of the rays converge at a point called a *virtual image,* which in the case of a positive lens is always farther away from the lens than the object. The thin lens formula still applies if the distance from the lens to the image is taken to be negative. The image is called virtual because it does not appear on a ground glass screen placed at this point.

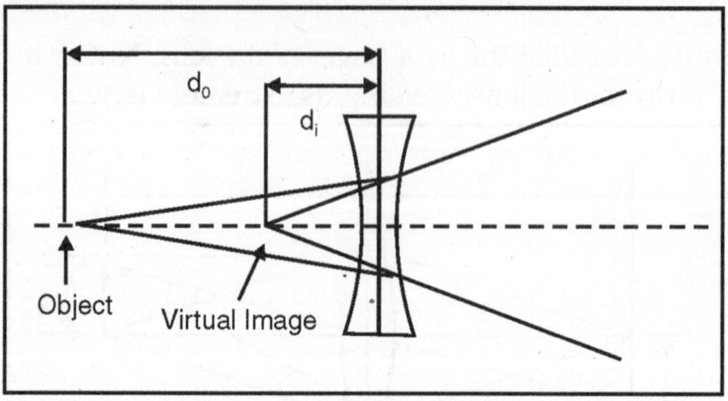

Fig. Production of a Virtual Image by a Negative Lens.

A *negative lens* is thinner in the center than at the edges and produces only virtual images. The virtual image produced by a negative lens is closer to the lens than is the object. Again, the thin lens formula is still valid, but both the distance from the image to the lens and the focal length must be taken as negative. Only the distance to the object remains positive.

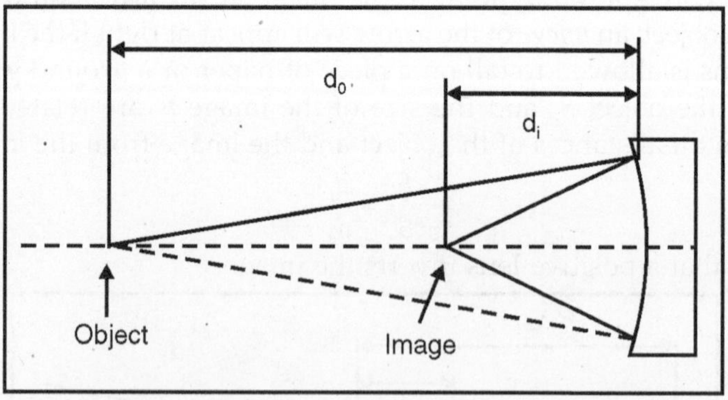

Fig. Production of a Real Image by a Concave Mirror.

Curved mirrors also produce images in a manner similar to a lens. A concave mirror works in analogy to a positive lens, producing a real or a virtual image depending on whether the object is farther from or closer to the mirror than the mirror's focal length. A convex mirror acts like a negative lens, always producing a virtual image. The thin lens formula works in both cases as long as the angles are small.

FERMAT'S PRINCIPLE

An alternate approach to geometrical optics can be developed from Fermat's principle. This principle states that light waves of a given frequency traverse the path between two points which takes the least time. The most obvious example of this is the passage of light through a homogeneous medium in which the speed of light doesn't change with position. In this case shortest time is equivalent to the shortest distance between the points, which, as we all know, is a straight line. Thus, Fermat's principle is consistent with light traveling in a straight line in a homogeneous medium. Fermat's principle can also be used to derive the laws of reflection and refraction.

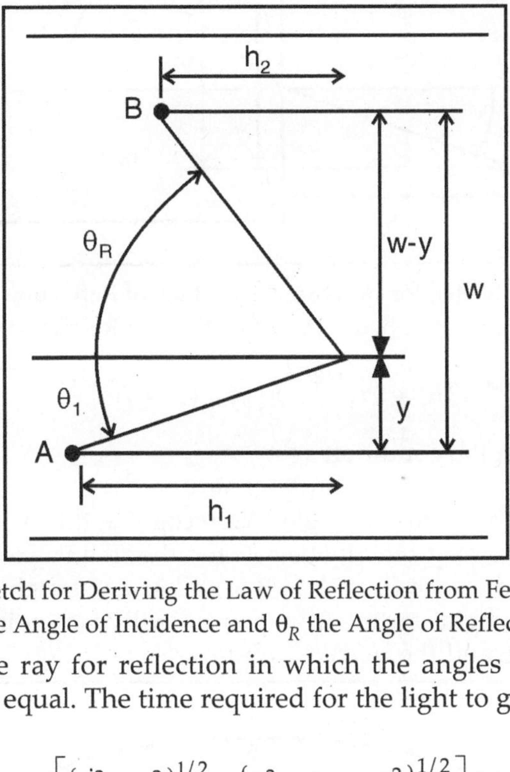

Fig. Definition Sketch for Deriving the Law of Reflection from Fermat's Principle. θ_I is the Angle of Incidence and θ_R the Angle of Reflection.

A candidate ray for reflection in which the angles of incidence and reflection are not equal. The time required for the light to go from point A to point B is

$$t = \left[\left\{h_1^2 + y^2\right\}^{1/2} + \left\{h_2^2 + (w-y)^2\right\}^{1/2}\right]/c$$

where c is the speed of light. We find the minimum time by differentiating t with respect to y and setting the result to zero, with the result that

$$\frac{y}{\left\{h_1^2 + y^2\right\}^{1/2}} = \frac{w-y}{\left\{h_2^2 + (w-y)^2\right\}^{1/2}}.$$

However, we note that the left side of this equation is simply $\sin\theta_I$, while the right side is $\sin\theta_R$, so that the minimum time condition reduces to,

$$\sin\theta_I = \sin\theta_R \text{ or } \theta_I = \theta_R,$$

which is the law of reflection. A similar analysis may be done to derive Snell's law of refraction. The speed of light in a medium with refractive index n is c/n, where is its speed in a vacuum. Thus, the time required for light to go some distance in such a medium is times the time light takes to go the same distance in a vacuum the time required for light to go from A to B becomes

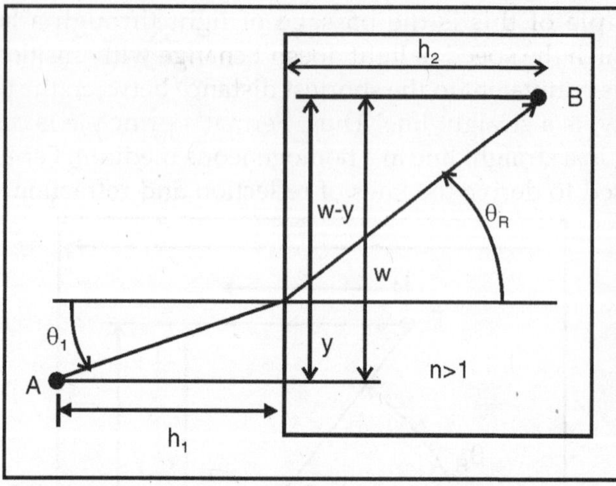

Fig. Definition Sketch for Deriving Snell's Law of Refraction from Fermat's Principle. The Shaded Area has Index of Refraction $n > 1$.

$$t = \left[\left\{ h_1^2 + y^2 \right\}^{1/2} + n \left\{ h_2^2 + (w-y)^2 \right\}^{1/2} \right] / c.$$

This results in the condition

$$\sin \theta_I = n \sin \theta_R$$

where θ_R is now the refracted angle. We recognize this result as Snell's law. The minimum time may actually be a *local* rather than a global minimum the global minimum distance from A to B is still just a straight line between the two points! In fact, light starting from point A will reach point B by *both* routes – the direct route and the reflected route.

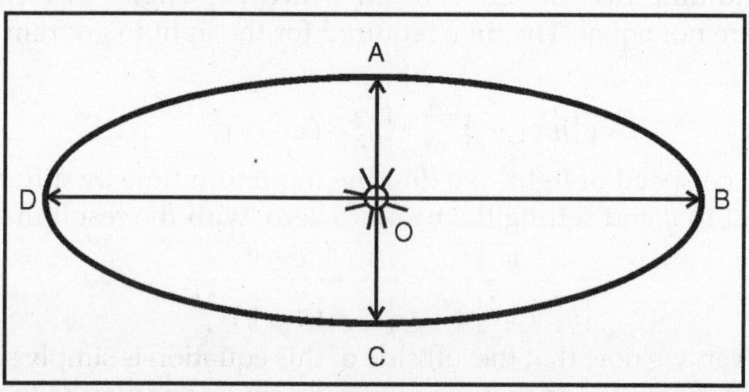

Fig. Ellipsoidal Mirror Showing Minimum and Maximum Time Rays from the Center of the Ellipsoid to the Mirror Surface and Back again.

It turns out that trajectories allowed by Fermat's principle don't strictly have to be minimum time trajectories. They can also be *maximum* time trajectories In this case light emitted at point O can be reflected back to point O from four points on the mirror, A, B, C, and D. The trajectories O-A-O and O-C-O are minimum time trajectories while O-B-O and O-D-O are maximum time trajectories.

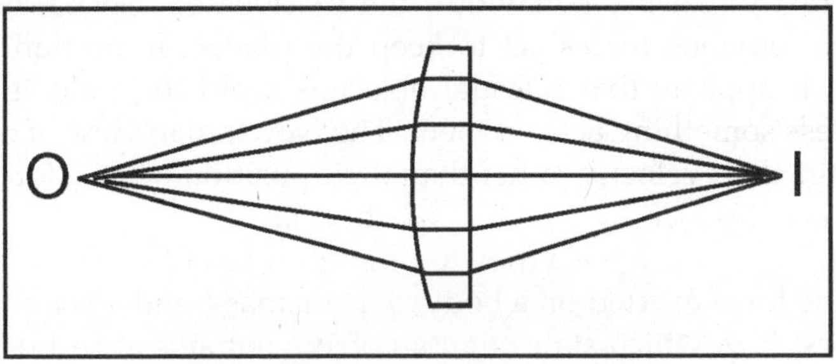

Fig. Ray Trajectories from a Point O being Focused to Another Point I by a Lens.

All the rays from point O which intercept the lens end up at point I. This would seem to contradict Fermat's principle, in that only the minimum (or maximum) time trajectories should occur. However, a calculation trajectories in this particular case take the same time. Thus, the light cannot choose one trajectory over another using Fermat's principle and all of the trajectories are equally favoured. Note that this inference applies not to just any set of trajectories, but only those going from one focal point to another.

NEWTON'S LAWS AND GEOMETRICAL OPTICS

The question that motivates us to study physics is "What makes things go?" The answers we conceive to this question constitute the subject of dynamics. This is in contrast to the question we have primarily addressed so far, namely "How do things go?" As noted earlier, the latter question is about kinematics. Extensive preparation in the kinematics of waves and particles in relativistic spacetime is needed to intelligently address dynamics. This preparation is now complete. Three different dynamical principles based respectively on pre-Newtonian, Newtonian, and quantum mechanical thinking. The last two are still in common use, and we first show that they are consistent with each other in the realm in which Newtonian and quantum mechanics overlap, i. e., in the geometrical optics limit of quantum mechanics.

For simplicity, this relationship is first developed in one dimension in the non-relativistic limit. Higher dimensions require the introduction of partial derivatives, and the relativistic case will be considered later.

Only forces which are "conservative" in Newtonian mechanics, i. e., exhibit a potential energy, have a correspondence in the quantum mechanical world. Forces of this type receive our primary attention. Certain ancillary concepts in Newtonian mechanics such as work and power are introduced at this stage.

FUNDAMENTAL PRINCIPLES OF DYNAMICS

NEWTONIAN DYNAMICS

In contrast to earthly behaviour, the motions of celestial objects seem effortless. No obvious forces act to keep the planets in motion around the sun. In fact, it appears that celestial objects simply coast along at constant velocity unless something acts on them. The Newtonian view of dynamics — objects change their velocity rather than their position when a force is exerted on them — is expressed by Newton's second law:

$$F = ma \text{ (newton's second law)}$$

where F is the force exerted on a body, *m* is its mass, and a is its acceleration. Newton's first law, which states that an object remains at rest or in uniform motion unless a force acts on it, is actually a special case of Newton's second law which applies when F = 0.

It is no wonder that the first successes of Newtonian mechanics were in the celestial realm, namely in the predictions of planetary orbits. It took Newton's genius to realise that the same principles which guided the planets also applied to the earthly realm as well. In the Newtonian view, the tendency of objects to stop when we stop pushing on them is simply a consequence of frictional forces opposing the motion. Friction, which is so important on the earth, is negligible for planetary motions, which is why Newtonian dynamics is more obviously valid for celestial bodies.

Note that the principle of relativity is closely related to Newtonian physics and is incompatible with pre-Newtonian views. After all, two reference frames moving relative to each other cannot be equivalent in the pre-Newtonian view, because objects with nothing pushing on them can only come to rest in one of the two reference frames! Einstein's relativity is often viewed as a repudiation of Newton, but this is far from the truth — Newtonian physics makes the theory of relativity possible through its invention of the principle of relativity. Compared with the differences between pre-Newtonian and Newtonian dynamics, the changes needed to go from Newtonian to Einsteinian physics constitute minor tinkering.

Quantum Dynamics

In quantum mechanics, particles are represented by matter waves, with the absolute square of the wave displacement yielding the probability of finding the particle. The behaviour of particles thus follows from the reflection, refraction, diffraction, and interference of the associated waves.

The connection with Newtonian dynamics comes from tracing the trajectories of matter wave packets. Changes in the speed and direction of motion of these packets correspond to the accelerations of classical mechanics.

When wavelengths are small compared to the natural length scale of the problem at hand, the wave packets can be made small, thus pinpointing the position of the associated particle, without generating excessive uncertainty in the particle's momentum. This is the geometrical optics limit of quantum mechanics.

ONE-DIMENSIONAL NON-RELATIVISTIC CASE

Louis de Broglie[9.1] made an analogy between matter waves and light waves, pointing out that wave packets of light change their velocity as the result of spatial variations in the index of refraction of the medium in which they are travelling. This behaviour comes about because the dispersion relation for light traveling through a medium with index of refraction n is

$$\omega = kc/n,$$

so that the group velocity,

$$u_s = d\omega/dk = c/n.$$

Thus, when n increases, u_s decreases, and vice versa.[9.2]

The problem for matter waves is to determine how analogous modifications can be made to the free particle dispersion relation so as to produce accelerations of wave packets consistent with Newtonian dynamics in the geometrical optics limit.

A simple guess as to how to modify the free particle dispersion relation for matter waves, limiting consideration initially to the one-dimensional, non-relativistic case. As in many situations in physics, the simple guess turns out to be correct! This leads to a connection between wave dynamics and Newtonian mechanics.

The dispersion relation for free matter waves is:

$$\omega = (k^2c^2 + \mu^2)^{1/2}.$$

In the non-relativistic limit $k^2\, c^2 << \mu^2$ and this equation can be approximated as

$$\omega = \mu(1 + k^2c^2/\mu^2)^{1/2} \approx \mu + k^2c^2/(2\mu).$$

Let us modify this equation by adding a term S' which depends on x, and which plays a role analogous to the index of refraction for light:

$$\omega = S'(x) + \mu + k^2c^2/(2\mu) = S(x) + k^2c^2/(2\mu).$$

The rest frequency has been made to disappear on the right side of the above equation by defining $S = S' + \mu$. This is done to simplify the notation.

Let us now imagine that all parts of the wave governed by this dispersion relation oscillate in phase. The only way this can happen is if ω is constant, i. e., it takes on the same value in all parts of the wave. It turns out we can do this if S is not a function of time.

If ω is constant, the only way S can vary with x in equation is if the wavenumber varies in a compensating way. Thus, constant frequency and spatially varying S together imply that $k = k(x)$. Solving equation for k yields

$$k(x) = \left[\frac{2\mu[\omega - S(x)]}{c^2}\right]^{1/2}.$$

Since ω is constant, the wavenumber becomes smaller and the wavelength larger as the wave moves into a region of increased *S*.

In the geometrical optics limit, we assume that *S* doesn't change much over one wavelength so that the wave remains reasonably sinusoidal in shape with approximately constant wavenumber over a few wavelengths.

However, over distances of many wavelengths the wavenumber and amplitude of the wave are allowed to vary considerably.

As yet we have no idea what causes *S* or where it comes from. For now we simply explore the consequences of its presence, especially in the geometric optics limit in which quantum dynamics gives way to Newtonian dynamics. The group velocity calculated from the dispersion relation given by equation is

$$u_g = \frac{d\omega}{dk} = \frac{kc^2}{\mu} = \left(\frac{2c^2(\omega - S)}{\mu}\right)^{1/2}$$

where *k* is eliminated in the last step with the help of equation. The resulting equation tells us how the group velocity varies as a matter wave traverses a region of slowly varying S. Thus, as S increases, u_g decreases and vice versa. We can now calculate the acceleration of a wave packet resulting from the spatial variation in *S*. We assume that $x(t)$ represents the position of the wave packet, so that $u_g = dx/dt$. Using the chain rule $du_g/dt = (du_g/dx)(dx/dt) = (du_g/dx)u_g$, we find

$$a = \frac{du_g}{dt} = \frac{du_g}{dx}u_g = \frac{du_g^2/2}{dx} = -\frac{c^2}{\mu}\frac{dS}{dx} = -\frac{\hbar}{m}\frac{dS}{dx}.$$

The group velocity is eliminated in favour of *S* by squaring equation and substituting the result into equation. Using Newton's second law to infer the force from the acceleration, we find,

$$F = ma = -\frac{dU}{dx} \text{ (force from potential energy)},$$

where we have defined U º $\hbar S$. The quantity *U* is called the potential energy. We have thus established a relationship between quantum mechanical and Newtonian dynamics, in that $U = \hbar S$ dictates the form of the force in Newton's second law, while *S* governs the refraction of matter waves.

A force which equals minus the derivative of some potential energy $U(x)$ is called conservative. Certain forces such as friction are not conservative. At present we will deal mainly with conservative forces.

POTENTIAL ENERGY

Recall that the momentum is related to the group velocity by,

$$\Pi = mu_g \text{ (momentum)}$$

in the nonrelativistic case. Multiplying equation by $\hbar$ and eliminating k and μ in favour of Π and m results in

$$E = U + \Pi^2/(2m) = U + K \text{ (total energy)}$$

where we have used $U = \hbar S$ and $\hbar m = mc^2$, and where

$$K \equiv \Pi^2/(2m) = mu_g^2/2 \text{ (kinetic energy)}$$

is the nonrelativistic kinetic energy.

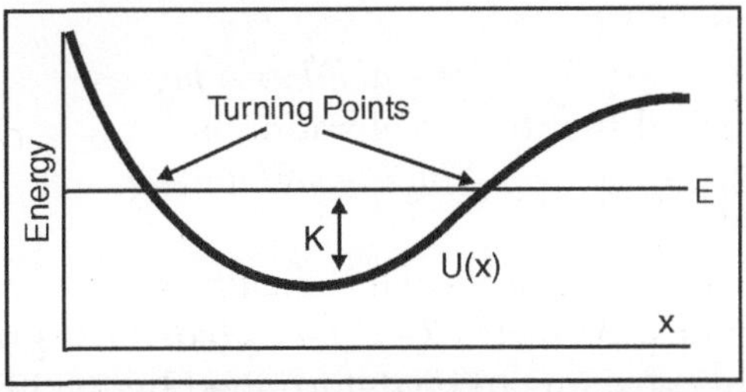

Fig. Example of Spatially Variable Potential Energy $U(x)$ with Fixed Total Energy E.

The Kinetic Energy $k = E - U$ is Zero where the E and U Lines Cross. These points are Called Turning Points.

The energy E, which we now refer to for clarity as the total energy, is constant since it is proportional to the (constant) frequency ω. We say that the total energy is conserved. Thus, increases in the potential energy coincide with decreases in the kinetic energy and vice versa.

In Newtonian dynamics the kinetic energy cannot be negative, since it is the product of half the mass and the square of the velocity, both of which are positive. Thus, a particle with total energy E and potential energy $U$is forbidden to venture into regions in which the kinetic energy $K = E - U$ is negative.

The points at which the kinetic energy is zero are called turning points. This is because a particle decreases in speed as it approaches a turning point, stops there for an instant, and reverses direction. Note also that a particle with a given total energy always has the same speed at some point x regardless of whether it approaches this point from the left or the right:

$$\text{speed} = |u_g| = |\pm[2(E-U)/m]^{1/2}|.$$

Gravity as a Conservative Force

An example of a conservative force is gravity. An object of mass m near the surface of the earth has the gravitational potential energy,

$$U = mgz \text{ (gravity near earth's surface)}$$

where z is the height of the object and g = 9.8 ms^{-2} is the local value of the gravitational field near the earth's surface. Notice that the gravitational potential energy increases upward. The speed of the object in this case is

$$|u_g| \backslash [2(E - mgz)/m]^{1/2}.$$

If $|u_g|$ is known to equal the constant value u_0 at elevation $z = 0$, then equations tell us that $u_0 = (2E/m)^{1/2}$ and $|u_g| = (u_0^2 - 2gz)^{1/2}$.

There are certain types of questions which energy conservation cannot directly answer. For instance if an object is released at elevation h with zero velocity at $t = 0$, at what time will it reach $z = 0$ under the influence of gravity? In such cases it is often easiest to return to Newton's second law. Since the force on the object is

$$F = -\,dU/dz = -\,mg$$

in this case, we find that the acceleration is

$$a = F/m = -\,mg/m = -\,g.$$

However,

$$a = du/dt = d^2z/dt^2, \text{ so}$$

$$u = -gt + C_1 \quad z = -\,gt^2/2 + C_1t + C_2 \text{ (constnat gravity)},$$

where C_1 and C_2 are constants to be determined by the initial conditions. These results can be verified by differentiating to see if the original acceleration is recovered. Since $u = 0$ and $z = h$ at $t = 0$, we have $C_1 = 0$ and $C_2 = h$. With these results it is easy to show that the object reaches $z = 0$ when $t = (2h/g)^{1/2}$.

WORK AND POWER

When a force is exerted on an object, energy is transferred to the object. The amount of energy transferred is called the work done on the object. However, energy is only transferred if the object moves. The work W done is

$$W = F\Delta x$$

where the distance moved by the object is Δx and the force exerted on it is F. Notice that work can either be positive or negative. The work is positive if the object being acted upon moves in the same direction as the force, with negative work occurring if the object moves opposite to the force.

Equation assumes that the force remains constant over the full displacement Δx. If it is not, then it is necessary to break up the displacement into a number of smaller displacements, over each of which the force can be assumed to be constant. The total work is then the sum of the works associated with each small displacement.

If more than one force acts on an object, the works due to the different forces each add or subtract energy, depending on whether they are positive or negative. The total work is the sum of these individual works.

There are two special cases in which the work done on an object is related to other quantities. If F is the total force acting on the object, then

$$W = F\Delta x = ma\Delta x \text{by Newton's second law.}$$

However,

$$a = dv/dt$$

where v is the velocity of the object, and

$$\Delta x = (\Delta x/\Delta t)\Delta t \approx v\Delta t, \text{s}$$

where Δt is the time required by the object to move through distance Δx. The approximation becomes exact when Δx and Δt become very small. Putting all of this together results in

$$W_{total} = m\frac{dv}{dt}v\Delta t = \frac{d}{dt}\left(\frac{mv^2}{2}\right)\Delta t = \Delta K \text{ (total work)},$$

where K is the kinetic energy of the object. Thus, when F is the only force, $W = W_{total}$ is the total work on the object, and this equals the change in kinetic energy of the object. This is called the work-energy theorem, and it demonstrates that work really is a transfer of energy to an object.

The other special case occurs when the force is conservative, but is not necessarily the total force acting on the object. In this case

$$W_{cons} = -\frac{dU}{dx}\Delta x = -\Delta Ua \text{ (conservative force)},$$

where ΔU is the change in the potential energy of the object associated with the force of interest. The power associated with a force is simply the amount of work done by the force divided by the time interval Δt over which it is done. It is therefore the energy per unit time transferred to the object by the force of interest. From equation we see that the power is

$$P = \frac{F\Delta x}{\Delta t} = Fv \text{ (Power)},$$

where v is the velocity at which the object is moving. The total power is just the sum of the powers associated with each force. It equals the time rate of change of kinetic energy of the object:

$$P_{total} = \frac{W_{total}}{\Delta t} = \frac{dK}{dt} \text{ (total power)}.$$

PARTIAL DERIVATIVES

In order to understand the generalization of Newtonian mechanics to two and three dimensions, we first need to understand a new type of derivative called the partial derivative.

To motivate the discussion, let us use the chain rule to take the ordinary derivative of $f(x, y) = x^2 + y^2$ where y is some function of x, say $y = x^2 + x$:

$$\frac{df}{dx} + 2x + 2y\frac{dy}{dx} = 2x + 2y.(3x^2 + 1).$$

(At this point we don't really care what form $y = y\ (x)$ takes.) At times we may wish to take the derivative of f with respect to x while ignoring the possible dependence of y on x. A derivative of this type has special notation — the "d" is replaced by "∂" and the derivative is called a partial derivative. Thus, the partial derivative of f with respect to x is

$$\frac{\partial f}{\partial x} = 2x.$$

Notice that the partial derivative is actually simpler to evaluate than an ordinary derivative, because only the explicit dependence of the function on the differentiating variable need be considered — all other variables are taken to be constant.

TWO AND THREE DIMENSIONS OF MOTION

When a matter wave moves through a region of variable potential energy in one dimension, only the wavenumber changes. In two or three dimensions the wave vector can change in both direction and magnitude. This complicates the calculation of particle movement.

However, we already have an example of how to handle this situation, namely, the refraction of light. In that case Snell's law tells us how the direction of the wave vector changes, while the dispersion relation combined with the constancy of the frequency gives us information about the change in the magnitude of the wave vector.

For matter waves a similar procedure works, though the details are different, because we seek the consequences of a change in potential energy rather than a change in the index of refraction.

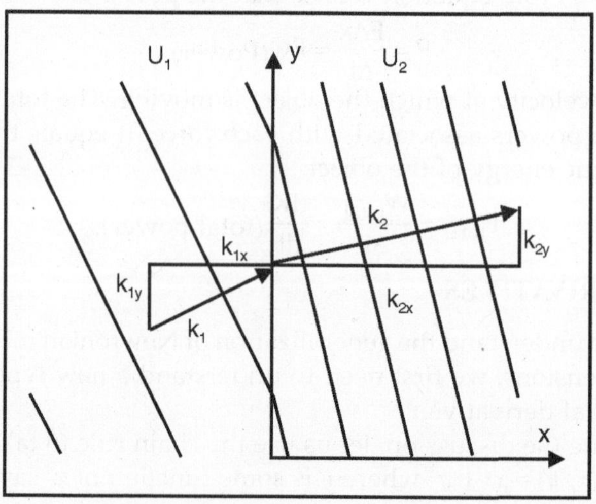

Fig. Refraction of a Matter wave by a Discontinuity in Potential Energy. The Component of the Wave Vector Parallel to the Discontinuity, k_y, Doesn't Change, so $k_{1y} = k_{2y}$.

The refraction of matter waves at a discontinuity in the potential energy. Let us suppose that the discontinuity occurs at $x = 0$. If the matter wave to the left of the discontinuity is

$$\psi_1 = \sin (k_{1x}x + k_{1y}y - \omega_1 t)$$

and to the right is

$$\psi_2 = \sin (k_{2x}x + k_{2y}y - \omega_2 t) ,$$

then the wavefronts of the waves will match across the discontinuity for all time only if $\omega_1 = \omega_2 \equiv \omega$ and $k_{1y} = k_{2y} \equiv k_y$. We are already familiar with the first condition from the one-dimensional problem, so the only new ingredient is the constancy of the y component of the wave vector. In two dimensions the momentum is a vector: $\Pi = mu$ when $|u| << c$, where u is the particle velocity. Furthermore, the kinetic energy is

$$K = m\ |u|^2/2 = |\Pi|^2/(2m)\ (\Pi_x^2 + \Pi_y^2)/(2m).$$

The relationship between kinetic, potential, and total energy is unchanged from the one-dimensional case, so we have

$$E = U + (\Pi_x^2 + \Pi_y^2)/(2m) = \text{constant}.$$

The de Broglie relationship tells us that $\Pi = \hbar k$, so the constancy of k_y across the discontinuity in U tells us that Π_y = constant there.

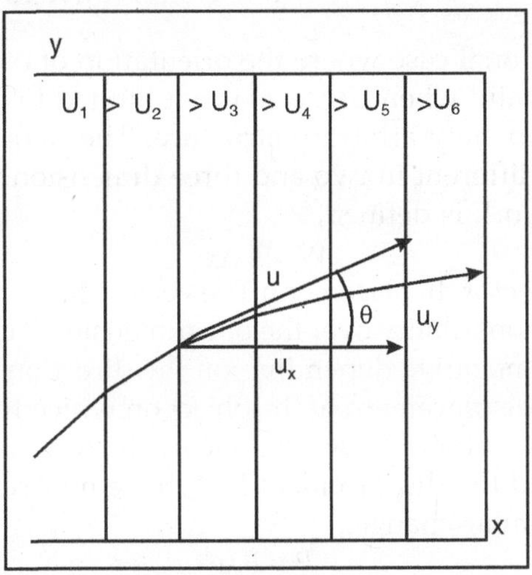

Fig. Trajectory of a Wave Packet Through a Variable Potential Energy, $U(x)$, which Decreases to the Right.

Let us now approximate a continuously variable $U(x)$ by a series of steps of constant U oriented normal to the x axis. The above analysis can be applied at the jumps or discontinuities in U between steps with the result that equations are valid across all discontinuities.

If we now let the step width go to zero, these equations then become valid for Ucontinuously variable in x. An example from classical mechanics of a problem of this type is a ball rolling down an inclined ramp with an initial velocity component across the ramp. The potential energy decreases in the down ramp direction, resulting in a force down the ramp. This accelerates the ball in that direction, but leaves the component of momentum across the ramp unchanged.

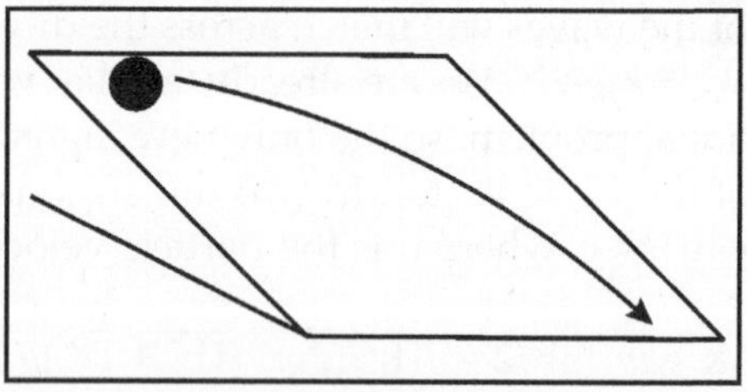

Fig. Classical Mechanics

Using the procedure which we invoked before, we find the force components associated with $U(x)$ in the x and y directions to be $F_x = - dU/dx$ and $F_y = 0$. This generalizes to

$$F = -\left(\frac{\partial U}{\partial x}, \frac{\partial U}{\partial y}, \frac{\partial U}{\partial z}\right) \text{ (3-D conservative force)}$$

in the three-dimensional case where the orientation of constant U surfaces is arbitrary. It is also valid when $U(x, y, z)$ is not limited to a simple ramp form, but takes on a completely arbitrary structure. The definitions of work and power are slightly different in two and three dimensions.

In particular, work is defined,

$$W = F.\Delta x$$

where Δx is now a vector displacement. The vector character of this expression yields an additional possibility over the one dimensional case, where the work is either positive or negative depending on the direction of Δx relative to F. If the force and the displacement of the object on which the force is acting are perpendicular to each other, the work done by the force is actually zero, even though the force and the displacement both have non-zero magnitudes. The power exhibits a similar change:

$$P = \text{F.u}$$

Thus the power is zero if an object's velocity is normal to the force being exerted on it. Energy conservation by itself is somewhat less useful for solving problems in two and three dimensions than it is in one dimension. This is because knowing the kinetic energy at some point tells us only the magnitude of the velocity, not its direction. If conservation of energy fails to give us the information we need, then we must revert to Newton's second law, as we did in the one-dimensional case.

For instance, if an object of mass m has initial velocity $u_0 = (u_0, 0)$ at location $(x, z) = (0, h)$ and has the gravitational potential energy $U = mgz$, then the force on the object is $F = (0, - mg)$. The acceleration is therefore $a = F/m = (0, - g)$. Since $a = du/dt = d^2x/dt^2$ where $x = (x, z)$ is the object's position, we find that

$$u = (C_1, -gt + C_2) \; x = (C_1 t + C_3, -gt^2/2 + C_2 t + C_4),$$

where C_1, C_2, C_3, and C_4 are constants to be evaluated so that the solution reduces to the initial conditions at $t = 0$. The specified initial conditions tell us that $C_1 = u_0$, $C_2 = 0$, $C_3 = 0$, and $C_4 = h$ in this case. From these results we can infer the position and velocity of the object at any time.

KINETIC AND POTENTIAL MOMENTUM

If you have previously taken a physics course then you have probably noticed that a rather odd symbol is used for momentum, namely Π, rather than the more commonly employed P.

The reason for this peculiar usage is that there are actually two kinds of momentum, kinetic momentum and total momentum, just as there are two kinds of energy, kinetic and total. The symbol Π represents total momentum while P represents kinetic momentum. In non-relativistic mechanics we don't need to distinguish between the two quantities, as they are generally equal to each other. However, we will find later in the course that it is crucial to make this distinction in the relativistic case. As a general rule, the total momentum is related to a particle's wave vector via the de Broglie relation, $\Pi = \hbar k$, while the kinetic momentum is related to a particle's velocity, $P = mu/(1 - u^2/c^2)^{1/2}$.

EARLY IDEAS ABOUT LIGHT PROPAGATION

Attempts to measure the speed of light played an important part in the development of the theory of special relativity, and, indeed, the speed of light is central to the theory. The first recorded discussion of the speed of light (I think) is in Aristotle, where he quotes Empedocles as saying the light from the sun must take some time to reach the earth, but Aristotle himself apparently disagrees, and even Descartes thought that light traveled instantaneously. SIMP. Everyday experience shows that the propagation of light is instantaneous; for when we see a piece of artillery fired at great distance, the flash reaches our eyes without lapse of time; but the sound reaches the ear only after a noticeable interval. Of course, Galileo points out that in fact nothing about the speed of light can be deduced from this observation, except that light moves faster than sound. He then goes on to suggest a possible way to measure the speed of light. The idea is to have two people far away from each other, with covered lanterns.

One uncovers his lantern, then the other immediately uncovers his on seeing the light from the first. This routine is to be practiced with the two close together, so they will get used to the reaction times involved, then they are to do it two or three miles apart, or even further using telescopes, to see if the time interval is perceptibly lengthened. Galileo claims he actually tried the experiment at distances less than a mile, and couldn't detect a time lag. From this one can certainly deduce that light travels at least ten times faster than sound.

MEASURING THE SPEED OF LIGHT WITH JUPITER'S MOONS

The first real measurement of the speed of light came about half a century later, in 1676, by a Danish astronomer, Ole Römer, working at the Paris

Observatory. He had made a systematic study of Io, one of the moons of Jupiter, which was eclipsed by Jupiter at regular intervals, as Io went around Jupiter in a circular orbit at a steady rate. Actually, Römer found, for several months the eclipses lagged more and more behind the expected time, until they were running about eight minutes late, then they began to pick up again, and in fact after about six months were running eight minutes early. The cycle then repeated itself. Römer realized the significance of the time involved-just over one year. This time period had nothing to do with Io, but was the time between successive closest approaches of earth in its orbit to Jupiter. The eclipses were furthest behind the predicted times when the earth was furthest from Jupiter. The natural explanation was that the light from Io (actually reflected sunlight, of course) took time to reach the earth, and took the longest time when the earth was furthest away. From his observations, Römer concluded that light took about twenty-two minutes to cross the earth's orbit.

This was something of an overestimate, and a few years later Newton wrote in the *Principia*: "For it is now certain from the phenomena of Jupiter's satellites, confirmed by the observations of different astronomers, that light is propagated in succession and requires about seven or eight minutes to travel from the sun to the earth." This is essentially the correct value. Of course, to find the speed of light it was also necessary to know the distance from the earth to the sun. During the 1670's, attempts were made to measure the parallax of Mars, that is, how far it shifted against the background of distant stars when viewed simultaneously from two different places on earth at the same time. This (very slight) shift could be used to find the distance of Mars from earth, and hence the distance to the sun, since all *relative* distances in the solar system had been established by observation and geometrical analysis. According to Crowe, they concluded that the distance to the sun was between 40 and 90 million miles.

Measurements presumably converged on the correct value of about 93 million miles soon after that, because it appears Römer (or perhaps Huygens, using Römer's data a short time later) used the correct value for the distance, since the speed of light was calculated to be 125,000 miles per second, about three-quarters of the correct value of 186,300 miles per second. This error is fully accounted for by taking the time light needs to cross the earth's orbit to be twenty-two minutes (as Römer did) instead of the correct value of sixteen minutes.

STARLIGHT AND RAIN

The next substantial improvement in measuring the speed of light took place in 1728, in England. An astronomer James Bradley, sailing on the Thames with some friends, noticed that the little pennant on top of the mast changed position each time the boat put about, even though the wind was steady. He thought of the boat as the earth in orbit, the wind as starlight coming from some distant star, and reasoned that the apparent direction the starlight was "blowing" in would depend on the way the earth was moving. Another possible analogy is to imagine the starlight as a steady downpour of rain on a windless day, and to think of yourself as walking around a circular path at a steady pace. The apparent direction of the incoming rain will not be vertically downwards-more will hit your front than your back. In fact, if the rain is falling at, say, 15 mph, and you are walking at 3 mph, to you as observer the rain will be coming down at a slant so that it has a vertical speed of 15 mph, and a horizontal speed towards you of 3 mph. Whether it is slanting down from the north or east or whatever at any given time depends on where you are on the circular path at that moment. Bradley reasoned that the apparent direction of incoming starlight must vary in just this way, but the angular change would be a lot less dramatic. The earth's speed in orbit is about 18 miles per second, he knew from Römer's work that light went at about 10,000 times that speed. That meant that the angular variation in apparent incoming direction of starlight was about the magnitude of the small angle in a right-angled triangle with one side 10,000 times longer than the other, about one two-hundredth of a degree. Notice this would have been just at the limits of Tycho's measurements, but the advent of the telescope, and general improvements in engineering, meant this small angle was quite accurately measurable by Bradley's time, and he found the velocity of light to be 185,000 miles per second, with an accuracy of about one per cent.

ASTRONOMICAL TECHNIQUES

The problem is, all these astronomical techniques do not have the appeal of Galileo's idea of two guys with lanterns. It would be reassuring to measure the speed of a beam of light between two points on the ground, rather than making somewhat indirect deductions based on apparent slight variations in the positions of stars. We can see, though, that if the two lanterns are ten miles apart, the time lag is of order one-ten thousandth of a second, and it is difficult to see how to arrange that. This technical problem was solved in France about 1850 by two rivals, Fizeau and Foucault, using slightly different techniques. In Fizeau's apparatus, a beam of light shone between the teeth of a rapidly rotating toothed wheel, so the "lantern" was constantly being covered and uncovered. Instead of a second lantern far away, Fizeau simply had a mirror, reflecting the beam back, where it passed a second time between the teeth of the wheel. The idea was, the blip of light that went out through one gap between teeth would only make it back through the same gap if the teeth had not had time to move over significantly during the round trip time to the far away mirror. n It was not difficult to make a wheel with a hundred teeth, and to rotate it hundreds of times a second, so the time for a tooth to move over could be arranged to be a fraction of one ten thousandth of a second. The method worked. Foucault's method was based on the same general idea, but instead of a toothed wheel, he shone the beam on to a rotating mirror. At one point in the mirror's rotation, the reflected beam fell on a distant mirror, which reflected it right back to the rotating mirror, which meanwhile had turned through a small angle.

After this second reflection from the rotating mirror, the position of the beam was carefully measured. This made it possible to figure out how far the mirror had turned during the time it took the light to make the round trip to the distant mirror, and since the rate of rotation of the mirror was known, the speed of light could be figured out. These techniques gave the speed of light with an accuracy of about 1,000 miles per second.

ALBERT ABRAHAM MICHELSON

Albert Michelson was born in 1852 in Strzelno, Poland. His father Samuel was a Jewish merchant, not a very safe thing to be at the time. Purges of Jews were frequent in the neighbouring towns and villages. They decided to leave town. Albert's fourth birthday was celebrated in Murphy's Camp, Calaveras County, about fifty miles south east of Sacramento, a place where five million dollars worth of gold dust was taken from one four acre lot. Samuel prospered selling supplies to the miners. When the gold ran out, the Michelsons moved to Virginia City, Nevada, on the Comstock lode, a silver mining town. Albert went to high school in San Francisco. In 1869, his father spotted an announcement in the local paper that Congressman Fitch would be appointing a candidate to the Naval Academy in Annapolis, and inviting applications.

Albert applied but did not get the appointment, which went instead to the son of a civil war veteran. However, Albert knew that President Grant would also be appointing ten candidates himself, so he went east on the just opened continental railroad to try his luck. Unknown to Michelson, Congressman Fitch wrote directly to Grant on his behalf, saying this would really help get the Nevada Jews into the Republican party. This argument proved persuasive. In fact, by the time Michelson met with Grant, all ten scholarships had been awarded, but the President somehow came up with another one. Of the incoming class of ninety-two, four years later twenty-nine graduated. Michelson placed first in optics, but twenty-fifth in seamanship. The Superintendent of the Academy, Rear Admiral Worden, who had commanded the Monitor in its victory over the Merrimac, told Michelson: "If in the future you'd give less attention to those scientific things and more to your naval gunnery, there might come a time when you would know enough to be of some service to your country."

GALILEAN RELATIVITY

Shortly after graduation, Michelson was ordered aboard the USS Monongahela, a sailing ship, for a voyage through the Caribbean and down to Rio. According to the biography of Michelson written by his daughter (The Master of Light, by Dorothy Michelson Livingston, Chicago, 1973) he thought a lot as the ship glided across the quiet Caribbean about whether one could decide in a closed room inside the ship whether or not the vessel was moving. In fact, his daughter quotes a famous passage from Galileo on just this point:

Shut yourself up with some friend in the largest room below decks of some large ship and there procure gnats, flies, and other such small winged creatures. Also get a great tub full of water and within it put certain fishes; let also a certain bottle be hung up, which drop by drop lets forth its water into another narrow-necked bottle placed underneath. Then, the ship lying still, observe how those small winged animals fly with like velocity towards all parts of the room; how the fish swim indifferently towards all sides; and how the distilling drops all fall into the bottle placed underneath.

And casting anything towards your friend, you need not throw it with more force one way than another, provided the distances be equal; and leaping with your legs together, you will reach as far one way as another. Having observed all these particulars, though no man doubts that, so long as the vessel stands still, they ought to take place in this manner, make the ship move with what velocity you please, so long as the motion is uniform and not fluctuating this way and that. You will not be able to discern the least alteration in all the forenamed effects, nor can you gather by any of them whether the ship moves or stands still....in throwing something to your friend you do not need to throw harder if he is towards the front of the ship from you... the drops from the upper bottle still fall into the lower bottle even though the ship may have moved many feet while the drop is in the air...

Of this correspondence of effects the cause is that the ship's motion is common to all the things contained in it and to the air also; I mean if those things be shut up in the room; but in case those things were above the deck in the open air, and not obliged to follow the course of the ship, differences would be observed,... smoke would stay behind.

MICHELSON MEASURES THE SPEED OF LIGHT

On returning to Annapolis from the cruise, Michelson was commissioned Ensign, and in 1875 became an instructor in physics and chemistry at the Naval Academy, under Lieutenant Commander William Sampson. Michelson met Mrs. Sampson's niece, Margaret Heminway, daughter of a very successful Wall Street tycoon, who had built himself a granite castle in New Rochelle, NY. Michelson married Margaret in an Episcopal service in New Rochelle in 1877. At work, lecture demonstrations had just been introduced at Annapolis. Sampson suggested that it would be a good demonstration to measure the speed of light by Foucault's method. Michelson soon realized, on putting together the apparatus, that he could redesign it for much greater accuracy, but that would need money well beyond that available in the teaching demonstration budget. He went and talked with his father in law, who agreed to put up $2,000. Instead of Foucault's 60 feet to the far mirror, Michelson had about 2,000 feet along the bank of the Severn, a distance he measured to one tenth of an inch. He invested in very high quality lenses and mirrors to focus and reflect the beam.

His final result was 186,355 miles per second, with possible error of 30 miles per second or so. This was twenty times more accurate than Foucault, made the New York Times, and Michelson was famous while still in his twenties. In fact, this was accepted as the most accurate measurement of the speed of light for the next forty years, at which point Michelson measured it again.

THE NATURE OF LIGHT

Michelson's efforts in 1879, the speed of light was known to be 186,350 miles per second with a likely error of around 30 miles per second. This measurement, made by timing a flash of light travelling between mirrors in Annapolis, agreed well with less direct measurements based on astronomical observations. Still, this did not really clarify the *nature* of light. Two hundred years earlier, Newton had suggested that light consists of tiny *particles* generated in a hot object, which spray out at very high speed, bounce off other objects, and are detected by our eyes. Newton's arch-enemy Robert Hooke, on the other hand, thought that light must be a kind of *wave motion,* like sound. To appreciate his point of view, let us briefly review the nature of sound.

THE WAVELIKE NATURE OF SOUND

Actually, *sound* was already quite well understood by the ancient Greeks. The essential point they had realized is that sound is generated by a vibrating material object, such as a bell, a string or a drumhead. Their explanation was that the vibrating drumhead, for example, alternately pushes and pulls on the air directly above it, sending out waves of compression and decompression (known as rarefaction), like the expanding circles of ripples from a disturbance on the surface of a pond. On reaching the ear, these waves push and pull on the eardrum with the same frequency (that is to say, the same number of pushes per second) as the original source was vibrating at, and nerves transmit from the ear to the brain both the intensity (loudness) and frequency (pitch) of the sound.

There are a couple of special properties of sound waves (actually any waves) worth mentioning at this point. The first is called *interference*. This is most simply demonstrated with water waves. If you put two fingers in a tub of water, just touching the surface a foot or so apart, and vibrate them at the same rate to get two expanding circles of ripples, you will notice that where the ripples overlap there are quite complicated patterns of waves formed. The essential point is that at those places where the wave-crests from the two sources arrive at the same time, the waves will work together and the water will be very disturbed, but at points where the crest from one source arrives at the same time as the wave trough from the other source, the waves will cancel each other out, and the water will hardly move. You can hear this effect for sound waves by playing a constant note through stereo speakers. As you move around a room, you will hear quite large variations in the intensity of sound. Of course, reflections from walls complicate the pattern.

This large variation in volume is *not* very noticeable when the stereo is playing music, because music is made up of many frequencies, and they change all the time. The different frequencies, or notes, have their quiet spots in the room in different places. The other point that should be mentioned is that high frequency tweeter-like sound is much more *directional* than low frequency woofer-like sound. It really doesn't matter where in the room you put a low-frequency woofer—the sound seems to be all around you anyway. On the other hand, it is quite difficult to get a speaker to spread the high notes in all directions. If you listen to a cheap speaker, the high notes are loudest if the speaker is pointing right at you. A lot of effort has gone into designing tweeters, which are small speakers especially designed to broadcast high notes over a wide angle of directions.

WAVE NATURE OF LIGHT

Bearing in mind the above minireview of the properties of waves, let us now reconsider the question of whether light consists of a stream of particles or is some kind of wave. The strongest argument for a particle picture is that light travels in straight lines. You can hear around a corner, at least to some extent, but you certainly can't see. Furthermore, no wave-like interference effects are very evident for light. Finally, it was long known, as we have mentioned, that sound waves were compressional waves in air. If light is a wave, just what is waving? It clearly isn't just air, because light reaches us from the sun, and indeed from stars, and we know the air doesn't stretch that far, or the planets would long ago have been slowed down by air resistance.

Despite all these objections, it was established around 1800 that light *is* in fact some kind of wave. The reason this fact had gone undetected for so long was that the wavelength is *really* short, about one fifty-thousandth of an inch. In contrast, the shortest wavelength sound detectable by humans has a wavelength of about half an inch. The fact that light travels in straight lines is in accord with observations on sound that the higher the frequency (and shorter the wavelength) the greater the tendency to go in straight lines. Similarly, the interference patterns mentioned above for sound waves or ripples on a pond vary over distances of the same sort of size as the wavelengths involved.

Patterns like that would not normally be noticeable for light because they would be on such a tiny scale. In fact, it turns out, there *are* ways to see interference effects with light. A familiar example is the many colours often visible in a soap bubble. These come about because looking at a soap bubble you see light reflected from both sides of a very thin film of water—a thickness that turns out to be comparable to the wavelength of light. The light reflected from the lower layer has to go a little further to reach your eye, so that light wave must wave an extra time or two before getting to your eye compared with the light reflected from the top layer. What you actually *see* is the *sum* of the light reflected from the top layer and that reflected from the bottom layer. Thinking of this now as the sum of two sets of waves, the light will be bright if the crests of the two waves arrive together, dim if the *crests* of waves reflected from the top layer arrive simultaneously with the *troughs* of waves reflected from the bottom layer. Which of these two possibilities actually occurs for reflection from a particular bit of the soap film depends on just how much further the light reflected from the lower surface has to travel to reach your eye compared with light from the upper surface, and that depends on the angle of reflection and the thickness of the film.

Suppose now we shine white light on the bubble. White light is made up of all the colours of the rainbow, and these different colours have different wavelengths, so we see colours reflected, because for a particular film, at a particular angle, some colours will be reflected brightly (the crests will arrive together), some dimly, and we will see the ones that win. Having established that light is a wave, though, we still haven't answered one of the major objections raised above. Just what is waving? We discussed sound waves as waves of compression in air.

Actually, that is only one case—sound will also travel through liquids, like water, and solids, like a steel bar. It is found experimentally that, other things being equal, sound travels faster through a medium that is harder to compress: the material just springs back faster and the wave moves through more rapidly. For media of equal springiness, the sound goes faster through the less heavy medium, essentially because the same amount of springiness can push things along faster in a lighter material. So when a sound wave passes, the material—air, water or solid—waves as it goes through. Taking this as a hint, it was natural to suppose that light must be just waves in some mysterious material, which was called the *aether*, surrounding and permeating everything.

This aether must also fill all of space, out to the stars, because we can see them, so the medium must be there to carry the light. (We could never *hear* an explosion on the moon, however loud, because there is no air to carry the sound to us.) Let us think a bit about what properties this aether must have. Since light travels so fast, it must be very light, and very hard to compress. Yet, as mentioned above, it must allow solid bodies to pass through it freely, without aether resistance, or the planets would be slowing down. Thus we can picture it as a kind of ghostly wind blowing through the earth. But how can we prove any of this? Can we detect it?

MICHELSON-MORLEY EXPERIMENT

Detecting the aether wind was the next challenge Michelson set himself after his triumph in measuring the speed of light so accurately. Naturally, something that allows solid bodies to pass through it freely is a little hard to get a grip on. But Michelson realized that, just as the speed of sound is relative to the air, so the speed of light must be relative to the aether. This must mean, if you could measure the speed of light accurately enough, you could measure the speed of light travelling upwind, and compare it with the speed of light travelling downwind, and the difference of the two measurements should be twice the windspeed.

Unfortunately, it wasn't that easy. All the recent accurate measurements had used light travelling to a distant mirror and coming back, so if there was an aether wind along the direction between the mirrors, it would have opposite effects on the two parts of the measurement, leaving a very small overall effect. There was no technically feasible way to do a one-way determination of the speed of light. At this point, Michelson had a very clever idea for detecting the aether wind. As he explained to his children, it was based on the following puzzle:

Suppose we have a river of width w (say, 100 feet), and two swimmers who both swim at the same speed v feet per second (say, 5 feet per second). The river is flowing at a steady rate, say 3 feet per second. The swimmers race in the following way: they both start at the same point on one bank. One swims directly across the river to the closest point on the opposite bank, then turns around and swims back. The other stays on one side of the river, swimming upstream a distance (measured along the bank) exactly equal to the width of the river, then swims back to the start. Who wins? Let's consider first the swimmer going upstream and back. Going 100 feet upstream, the speed relative to the bank is only 2 feet per second, so that takes 50 seconds. Coming back, the speed is 8 feet per second, so it takes 12.5 seconds, for a total time of 62.5 seconds.

The swimmer going across the flow is trickier. It won't do simply to aim directly for the opposite bank: the flow will carry the swimmer downstream. To succeed in going directly across, the swimmer must actually aim upstream at the correct angle (of course, a real swimmer would do this automatically). Thus, the swimmer is going at 5 feet per second, at an angle, relative to the river, and being carried downstream at a rate of 3 feet per second. If the angle is correctly chosen so that the net movement is directly across, in one second the swimmer must have moved *four feet* across: the distances covered in one second will form a 3,4,5 triangle. So, at a crossing rate of 4 feet per second, the swimmer gets across in 25 seconds, and back in the same time, for a total time of 50 seconds. The cross-stream swimmer wins. This turns out to true whatever their swimming speed.

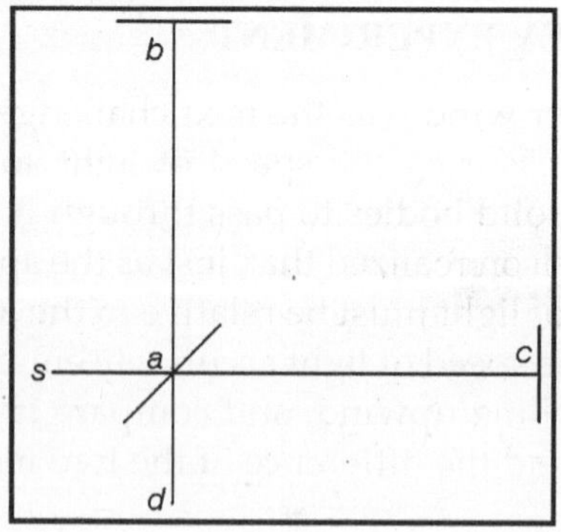

Fig. This Diagram is from the Original paper.

The source of light is at *s*, the 45 degree line is the half-silvered mirror, *b* and *c* are mirrors and d the observer. Michelson's great idea was to construct an exactly similar race for pulses of light, with the aether wind playing the part of the river. The scheme of the experiment is as follows: a pulse of light is directed at an angle of 45 degrees at a half-silvered, half transparent mirror, so that half the pulse goes on through the glass, half is reflected. These two half-pulses are the two swimmers. They both go on to distant mirrors which reflect them back to the half-silvered mirror.

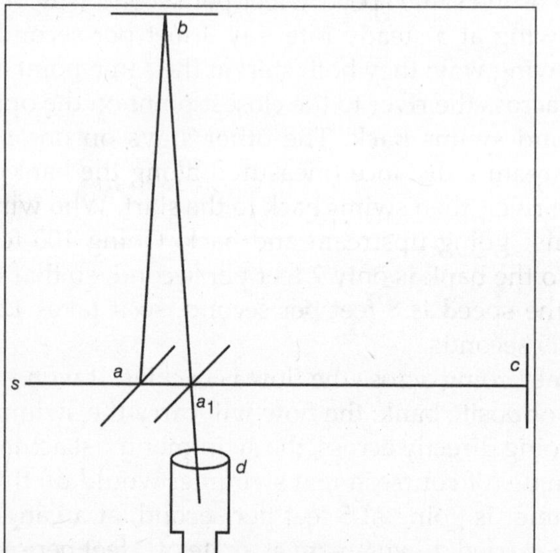

Fig. This is also from the Original paper, and shows the Expected path of light relative to the aether with an aether wind blowing.

At this point, they are again half reflected and half transmitted, but a telescope is placed behind the half-silvered mirror as shown in the figure so that half of each half-pulse will arrive in this telescope. Now, if there is an aether wind blowing, someone looking through the telescope should see the halves of the two half-pulses to arrive at slightly different times, since one would have gone more upstream and back, one more across stream in general. To maximize the effect, the whole apparatus, including the distant mirrors, was placed on a large turntable so it could be swung around.

Let us think about what kind of time delay we expect to find between the arrival of the two half-pulses of light. Taking the speed of light to be c miles per second relative to the aether, and the aether to be flowing at v miles per second through the laboratory, to go a distance w miles upstream will take $w/(c-v)$ seconds, then to come back will take $w/(c+v)$ seconds. The total roundtrip time upstream and downstream is the sum of these, which works out to be $2wc/(c^2-v^2)$, which can also be written $(2w/c)\times 1/(1-v^2/c^2)$. Now, we can safely assume the speed of the aether is much less than the speed of light, otherwise it would have been noticed long ago, for example in timing of eclipses of Jupiter's satellites.

This means v^2/c^2 is a very small number, and we can use some handy mathematical facts to make the algebra a bit easier. First, if x is very small compared to 1, $1/(1-x)$ is very close to $1+x$. (You can check it with your calculator.) Another fact we shall need in a minute is that for small x, the square root of $1+x$ is very close to $1+x/2$. Putting all this together, upstream-downstream roun dtrip time,

$$\cong \frac{2w}{c}\times\left(1+\frac{v^2}{c^2}\right).$$

Now, what about the cross-stream time? The actual cross-stream speed must be figured out as in the example above using a right-angled triangle, with the hypoteneuse equal to the speed c, the shortest side the aether flow speed v, and the other side the cross-stream speed we need to find the time to get across. From Pythagoras' theorem, then, the cross-stream speed is the square root of $(c^2 - v^2)$.

Since this will be the same both ways, the roundtrip cross-stream time will be

$$2w/\sqrt{c^2-v^2}\,.$$

This can be written in the form:

$$\frac{2w}{c}\frac{1}{\sqrt{1-v^2/c^2}}\cong\frac{2w}{c}\frac{1}{1-(v^2/2c^2)}\cong\frac{2w}{c}\left(1+\frac{v^2}{2c^2}\right)$$

where the two successive approximations, valid for $v/c = x << 1$, are

$$\sqrt{1-x}\cong 1-(x/2) \text{ and } 1/(1-x)\cong 1+x.$$

Therefore the, cross-stream roundtrip time

$$\cong \frac{2w}{c}\times\left(1+\frac{v^2}{2c^2}\right).$$

Looking at the two roundtrip times at the ends of the two paragraphs above, we see that they differ by an amount $(2w/c)\times v^2/2c^2$. Now, $2w/c$ is just the time the light would take if there were no aether wind at all, say, a few millionths of a second. If we take the aether windspeed to be equal to the earth's speed in orbit, for example, v/c is about 1/10,000, so v^2/c^2 is about 1/100,000,000. This means the time delay between the pulses reflected from the different mirrors reaching the telescope is about one-hundred-millionth of a few millionths of a second. It seems completely hopeless that such a short time delay could be detected. However, this turns out *not* to be the case, and Michelson was the first to figure out how to do it.

The trick is to use the interference properties of the lightwaves. Instead of sending pulses of light, as we discussed above, Michelson sent in a steady beam of light of a single colour. This can be visualized as a sequence of ingoing waves, with a wavelength one fifty-thousandth of an inch or so. Now this sequence of waves is split into two, and reflected as previously described. One set of waves goes upstream and downstream, the other goes across stream and back. Finally, they come together into the telescope and the eye. If the one that took longer is half a wavelength behind, its troughs will be on top of the crests of the first wave, they will cancel, and nothing will be seen. If the delay is less than that, there will still be some dimming. However, slight errors in the placement of the mirrors would have the same effect. This is one reason why the apparatus is built to be rotated. On turning it through 90 degrees, the upstream-downstream and the cross-stream waves change places.

Now the other one should be behind. Thus, if there is an aether wind, if you watch through the telescope while you rotate the turntable, you should expect to see variations in the brightness of the incoming light. To magnify the time difference between the two paths, in the actual experiment the light was reflected backwards and forwards several times, like a several lap race. Michelson calculated that an aether windspeed of only one or two miles a second would have observable effects in this experiment, so if the aether windspeed was comparable to the earth's speed in orbit around the sun, it would be easy to see. In fact, *nothing* was observed. The light intensity did not vary at all. Some time later, the experiment was redesigned so that an aether wind caused by the earth's daily rotation could be detected. Again, nothing was seen.

Finally, Michelson wondered if the aether was somehow getting stuck to the earth, like the air in a below-decks cabin on a ship, so he redid the experiment on top of a high mountain in California. Again, no aether wind was observed. It was difficult to believe that the aether in the immediate vicinity of the earth was stuck to it and moving with it, because light rays from stars would deflect as they went from the moving faraway aether to the local stuck aether. The only possible conclusion from this series of very difficult experiments was that the whole concept of an all-pervading aether was wrong from the start. Michelson was very reluctant to think along these lines. In fact, new theoretical insight into the nature of light had arisen in the 1860's from the brilliant theoretical work of Maxwell, who had written down a set of equations describing how electric and magnetic fields can give rise to each other. He had discovered that his equations predicted there could be waves made up of electric and magnetic fields, and the speed of these waves, deduced from experiments on how these fields link together, would be 186,300 miles per second.

This is, of course, the speed of light, so it is natural to assume that light is made up of fast-varying electric and magnetic fields. But this leads to a big problem: Maxwell's equations predict a definite speed for light, and it *is* the speed found by measurements. But what is the speed to be measured relative to? The whole point of bringing in the aether was to give a picture for light resembling the one we understand for sound, compressional waves in a medium. The speed of sound through air is measured relative to air. If the wind is blowing towards you from the source of sound, you will hear the sound sooner. If there isn't an aether, though, this analogy doesn't hold up. So what does light travel at 186,300 miles per second relative to? There is another obvious possibility, which is called the emitter theory: the light travels at 186,300 miles per second relative to the source of the light. The analogy here is between light emitted by a source and bullets emitted by a machine gun. The bullets come out at a definite speed (called the muzzle velocity) relative to the barrel of the gun. If the gun is mounted on the front of a tank, which is moving forward, and the gun is pointing forward, then relative to the ground the bullets are moving faster than they would if shot from a tank at rest. The simplest way to test the emitter theory of light, then, is to measure the speed of light emitted in the forward direction by a flashlight moving in the forward direction, and see if it exceeds the known speed of light by an amount equal to the speed of the flashlight.

Actually, this kind of direct test of the emitter theory only became experimentally feasible in the nineteen-sixties. It is now possible to produce particles, called neutral pions, which decay each one in a little explosion, emitting a flash of light. It is also possible to have these pions moving forward at 185,000 miles per second when they self destruct, and to catch the light emitted in the forward direction, and clock its speed. It is found that, despite the expected boost from being emitted by a very fast source, the light from the little explosions is going forward at the usual speed of 186,300 miles per second. In the last century, the emitter theory was rejected because it was thought the appearance of certain astronomical phenomena, such as double stars, where two stars rotate around each other, would be affected. Those arguments have since been criticized, but the pion test is unambiguous.

Index

A

Ability 48, 66
Accuracy 154, 172
Adsorbates 169, 170
Adulterants 84
Algae 80
Alkali-like Spectra 3
Alkalis 3
Ambient 177, 178
Analysis Tools 197
Analytical 150, 153
Architectures 176, 177
Atomic Force 198, 202
Atomic Spectroscopy 15
Atomic States 3

B

Balloons 49
Beryllium 51, 59

C

Calibration 15, 16
Cantilevers 198, 202, 206
Capacitive Techniques 197
Continuum Radiation 20
Convolution 199, 207, 209
Creep 190, 191, 192, 194, 195, 196, 197, 198
Crystallographic 206

D

Depth Profiling 164
Dietary Supplements 81
Distance Curves 174, 175
Doppler Broadening 19
Drugs 80, 81, 82, 83, 84, 85, 96, 97

E

Eigenvector 9, 10, 12
Electrochemical 178
Electromagnetic 63
Emission Intensities 16

F

Feedback Artifacts 199
Food 79, 81, 83, 96
Fourier Transform 85, 86, 89, 92

Frequency, 2
Fungi 80

H

Hardware Correction 196
Herb 78, 79, 80, 81, 82, 83, 84, 85, 97

I

Image Artifacts 206
Image Processing Capabilities 199
Infrared Spectroscopy 78, 85, 96, 97, 98
Instrument 200, 201
Instrumentation 172
Intrinsic Nonlinearity 189, 198
Ionization 150, 159, 161, 166

M

Magnetic flux 12
Metallic Surfaces 152, 169
Mineral 81

N

Neutral Atoms 12
Notations 5, 6, 7, 8

O

Oils 80, 82
Optical Microscope 201, 202, 206
Optical Techniques 197

P

Periodicity 57
Photoelectron 159, 160, 164, 168
Physicochemical 85
Pressure Broadening 19
Probability 66
Probe Handling 202
Programming 154

R

Radiation 63, 64
Regularities 14
Resolution 154, 159, 163, 167, 170, 172, 188, 191, 198, 202, 203, 204, 205

S

Satisfactory 64
Scanner 174, 175, 176, 185, 186, 187, 188, 189, 190, 191, 192, 193, 194, 195, 196, 197
Scanner Linearity 197
Scanning Tunneling 173, 176
Separation 3, 4, 8, 10
Software 178, 194, 195, 196, 199, 200, 202
Species 97
Spectrometric Techniques 91
Spectrophotometers 87, 89, 90, 91
Spectroscopy 150, 151, 152, 154, 158, 159, 161, 163, 164, 168, 169, 170, 173, 174, 176
SPM Environments 176
Strain-Gauge Techniques 197
System Accessibility 202

T

Tip Convolution 207
Tip Shape 202
Traditional Quality 85

U

Ultraviolet 159, 168

V

Vibrational 152, 153, 168, 170, 171

W

Water 83, 84
Wavelength 2, 15, 16, 17, 18, 19

Z

Zeeman Effect 12
Zeeman effect 12